D. BLACKMORE

ALGEBRAIC GEOMETRY

ALGEBRAIC GEOMETRY

Daniel Bump

Department of Mathematics
Stanford University, USA

World Scientific
Singapore • New Jersey • London • Hong Kong

Published by

World Scientific Publishing Co. Pte. Ltd.

P O Box 128, Farrer Road, Singapore 912805

USA office: Suite 1B, 1060 Main Street, River Edge, NJ 07661

UK office: 57 Shelton Street, Covent Garden, London WC2H 9HE

Library of Congress Cataloging-in-Publication Data
Bump, Daniel, 1952–
    Algebraic geometry / by Daniel Bump.
        p. cm.
    Includes bibliographical references and index.
    ISBN 9810235615
    1. Geometry, Algebraic. I. Title.
QA564.B775   1998
516.3'5--dc21   98-42147
                                                    CIP

British Library Cataloguing-in-Publication Data
A catalogue record for this book is available from the British Library.

First published 1998
Reprinted 2001

This book is printed on acid-free paper.

Printed in Singapore by Uto-Print

Preface

In 1989–1990 I taught a course in Algebraic Geometry at Stanford University, writing up lecture notes. These were revised for publication in 1998. In 1989-90 I covered the material in Chapters 1–14 in two quarters, and continued with a quarter on cohomology of coherent sheaves, lecturing out of Hartshorne's book.

The aim is to make this a text that can be used in a one year at the graduate level. I have tried to give complete proofs assuming a background in algebra at the level one expects from a first or second year graduate student. The point of view here is that of Serre [23] or Chapter I of Mumford [21]—a variety is a ringed space locally isomorphic to an affine variety over a field, which is algebraically closed except in Chapter 14. Although I do not treat schemes I trust the reader will not find the transition too difficult.

The first eight sections contain material applicable to varieties of every dimension, the last six contain material which is particular to the theory of curves. We give a portion of the general theory of elliptic curves, the zeta function of a curve and Riemann hypothesis. For most of the book I consider irreducible varieties over an algebraically closed field. In Chapter 14, we work over a finite field.

The most significant omission is intersection theory. Intersection theory is vital to the understanding of surfaces. In the theory of curves, the theory of correspondences on a curve X is just the intersection theory on the surface $X \times X$, which was put on a rigorous foundation by Weil in his *Foundations* [30] in preparation for proving the Riemann hypothesis for curves of higher genus. Although the intersection theory is not covered here, the material in Chapter 5 is very relevant to this topic. I would like to point out Fulton's brief and insightful introduction [9] to the intersection theory, which may be more accessible than his definitive treatise [10].

I will maintain a web page at `http://math.stanford.edu/~bump/ag.html` for this book. In particular, if errors are found, consult this web page for corrections.

I would like to thank Sen Hu and E. H. Chionh of World Scientific for their interest and encouragement, and for efficient editing of the manuscript.

Table of Contents

Information for the Reader

In these notes, k will be an algebraically closed field, which will be fixed throughout the discussion. The term *ring* will always refer to a commutative ring with unit.

Prerequisites for these notes are primarily a good knowledge of field theory, and a certain amount of mathematical maturity. I have made Lang's *Algebra* the primary reference for any facts from the theory of fields or other algebra which are required. However proofs of such facts could be found equally well in other standard references such as Jacobson's *Basic Algebra I and II*.

I have developed most of the commutative algebra required from scratch. I was influenced by Lang [16] in systematically using the Extension Theorem, in the pure commutative algebra as well as in the geometric applications. Because of this decision, some proofs have an unnecessary assumption that a ring be an integral domain. It is not hard to eliminate this assumption, but I find this approach instructive.

The end of proof sign is ■. I use ☐ for the end of the proof of a Lemma inside a larger proof.

The reader is often asked to fill in certain gaps by means of the exercises. When an exercise is required for the development of the material, I have included enough hints that the exercise should be straightforward.

1. Affine Algebraic Sets and Varieties

In these notes, the letter k will be reserved for a fixed algebraically closed field. The term *ring* will always mean a commutative ring with unity.

Let $x_1, \cdots, x_n$ be indeterminates. Then $k[x_1, \cdots, x_n]$ will denote the polynomial ring over k in n variables, and $k(x_1, \cdots, x_n)$ its field of fractions. The elements in $k[x_1, \cdots, x_n]$ may be regarded as functions on affine n-space $\mathbf{A}^n = k^n$. By an *algebraic set* we will mean the set $V(\Sigma)$ of zeros of some subset Σ of $k[x_1, \cdots, x_n]$. It is clear that if $\mathfrak{a}$ is the ideal generated by Σ then $V(\mathfrak{a}) = V(\Sigma)$. Consequently every algebraic set has the form $V(\mathfrak{a})$, where $\mathfrak{a}$ is an ideal. On the other hand, according to the Hilbert Basis Theorem (Exercise 1.4), every ideal in $k[x_1, \cdots, x_n]$ is finitely generated, and so every algebraic set also has the form $V(\Sigma)$, where Σ is a *finite* set.

Do we thus have a bijection between the algebraic sets in $\mathbf{A}^n$ and the ideals of the ring $k[x_1, \cdots, x_n]$? The answer is "no," as we may see by an example. Let f be an irreducible polynomial in $k[x_1, \cdots, x_n]$. Let $\mathfrak{a} = (f^r)$ be the principal ideal generated by f^r, where r is some integer greater than one, and let $\mathfrak{b} = (f)$. Then $V(\mathfrak{a}) = V(\mathfrak{b})$ are the same set, namely, the set of zeros of f. However, $\mathfrak{a}$ and $\mathfrak{b}$ are not the same—$\mathfrak{b}$ is a larger ideal than $\mathfrak{a}$. Clearly $\mathfrak{b}$ should be the ideal associated with this algebraic set, not $\mathfrak{a}$.

In order to obtain an exact bijection between algebraic sets in $\mathbf{A}^n$ and certain ideals of $k[x_1, \cdots, x_n]$, we need the notion of a *radical ideal*. The *radical* $r(\mathfrak{a})$ of any ideal $\mathfrak{a}$ of a commutative ring A is by definition the set of all $f \in A$ such that $f^r \in \mathfrak{a}$ for some sufficiently large r. The properties of $r(\mathfrak{a})$ are the subject of Exercise 1.5 below. We summarize these properties now briefly. The radical $r(\mathfrak{a})$ is an ideal. An ideal which is its own radical is called a *radical ideal*. It is easy to see that $r(r(\mathfrak{a})) = r(\mathfrak{a})$, so the radical of $\mathfrak{a}$ is a radical ideal.

If $A = k[x_1, \cdots, x_n]$, then for ideals $\mathfrak{a}$ of A it is clear that $V(r(\mathfrak{a})) = V(\mathfrak{a})$. Thus in the previous example, the ideal $\mathfrak{b}$ is the radical of $\mathfrak{a}$—it is a radical ideal. The association $\mathfrak{a} \mapsto V(\mathfrak{a})$ is a bijection between radical ideals and algebraic sets. This is the content of a famous Theorem, Hilbert's *Nullstellensatz*. We are not yet ready to prove this. What we will show in this chapter is that the Nullstellensatz actually follows from a special case.

Theorem 1.1 (Weak Nullstellensatz). *If* $\mathfrak{a}$ *is a proper ideal of the polynomial ring* $k[x_1, \cdots, x_n]$, *then* $V(\mathfrak{a})$ *is nonempty.*

We will prove this in Chapter 2. Assuming this, we may prove:

Theorem 1.2 (Strong Nullstellensatz). *If* $\mathfrak{a}$ *is an ideal of* $k[x_1, \cdots, x_n]$, *and if* f *is any element of* $k[x_1, \cdots, x_n]$ *which vanishes on* $V(\mathfrak{a})$, *then* $f \in r(\mathfrak{a})$.

Proof (assuming the weak Nullstellensatz). Let $f_i(x) = f_i(x_1, \cdots, x_n)$ $(i = 1, \cdots, r)$ generate the ideal $\mathfrak{a}$. Let x_0 be an additional indeterminate. Since f vanishes wherever $f_1, \cdots, f_r$ vanish, the polynomials $f_1(x), \cdots, f_r(x), 1 - x_0 f(x)$ in $n + 1$ variables cannot vanish simultaneously, and consequently, by the weak Nullstellensatz, the ideal which they generate is all of $k[x_0, x_1, \cdots, x_n]$. Thus there exists a relation

$$1 = \sum_{i=1}^{r} g_i(x_0, x) \, f_i(x) + h(x_0, x) \left(1 - x_0 \, f(x)\right).$$

Substitute $1/f(x)$ for x_0 in this expression, and multiply by a high power of $f(x)$ to clear the denominator. We obtain a relation of the form

$$f(x)^N = \sum_{i=1}^{r} G_i(x) \, f_i(x).$$

Hence $f \in r(\mathfrak{a})$, as required. ∎

We may thus summarize the situation as follows. Let us define, for any subset S of $\mathbf{A}^n$, the ideal $I(S)$ to be the set of elements of $k[x_1, \cdots, x_n]$ which vanish on S. The operations I and V are *order reversing* in the sense that if $S_1 \subseteq S_2$, then $I(S_1) \supseteq I(S_2)$, and if $\mathfrak{a}_1 \subseteq \mathfrak{a}_2$, then $V(\mathfrak{a}_1) \supseteq V(\mathfrak{a}_2)$. It is clear that $I(S)$ is always a radical ideal. Moreover, it is clear that if X is an algebraic set, then $V(I(X)) = X$. What is deeper—this is the Nullstellensatz—is that if $\mathfrak{a}$ is an ideal, then $I(V(\mathfrak{a})) = r(\mathfrak{a})$. So in particular, if $\mathfrak{a}$ is a radical ideal, then $I(V(\mathfrak{a})) = \mathfrak{a}$.

Thus the content of the Nullstellensatz is that there is an order reversing bijection between radical ideals and algebraic subsets of $\mathbf{A}^n$, where the ideal $I(X)$ corresponds to the algebraic set X, and the algebraic set $V(\mathfrak{a})$ corresponds to the ideal $\mathfrak{a}$.

We come now to the *Zariski Topology*.

Proposition 1.3. *(i) The union of two algebraic sets is an algebraic set.*

(ii) The intersection of an arbitrary family of algebraic sets is an algebraic set.

Proof. Let us prove (i). We will prove that $V(\mathfrak{a}) \cup V(\mathfrak{b}) = V(\mathfrak{ab})$. It is obvious that $V(\mathfrak{a}) \cup V(\mathfrak{b}) \subseteq V(\mathfrak{ab})$. On the other hand, let $x \in V(\mathfrak{ab}) - V(\mathfrak{a})$; then we prove that $x \in V(\mathfrak{b})$. Let $g \in \mathfrak{b}$. Since $x \notin V(\mathfrak{a})$, there exists $f \in \mathfrak{a}$ such that $f(x) \neq 0$. Now $fg \in \mathfrak{ab}$ and $x \in V(\mathfrak{ab})$, so $f(x) g(x) = 0$. Consequently, $g(x) = 0$. This proves that $x \in V(\mathfrak{b})$.

The proof of (ii) is straightforward. If $X_i = V(\mathfrak{a}_i)$ for some family of ideals $\mathfrak{a}_i$, then it is clear that $\bigcap_i X_i = V(\mathfrak{a})$, where $\mathfrak{a}$ is the ideal generated by the $\mathfrak{a}_i$. ∎

Thus the axioms for a topology are satisfied in which the closed sets of $\mathbf{A}^n$ are the algebraic sets. This topology is called the *Zariski Topology*. It is not Hausdorff. If X is an algebraic set, then we topologize it as a subset of $\mathbf{A}^n$, and this topology on X is also called the Zariski Topology.

The algebraic set X is called a *variety* if its ideal $I(X)$ is prime.

We define a topological space X to be *irreducible* if there do not exist two proper closed subsets X_1 and X_2 such that $X = X_1 \cup X_2$. Also, we call a topological space *Noetherian* if every descending chain of closed subsets becomes constant. We will sometimes call this property the *descending chain condition* for closed subspaces of X. If Z is a subset of a topological space X, the *interior* Z° of Z (relative to X) is the largest open subset contained in Z. We will denote by $\overline{Z}$ the closure of Z (in X). A special role is played in the proof of Proposition 1.5 by closed subsets Z such that $Z = \overline{Z^\circ}$.

Lemma 1.4. *Let Z be a Noetherian space. If Z is not irreducible, then $Z = X \cup Y$, where X and Y are proper closed subsets, $X = \overline{X^\circ}$, $Y = \overline{Y^\circ}$, and $X^\circ \cap Y^\circ = \varnothing$.*

Proof. Since Z is reducible, $Z = X_1 \cup Y_1$, where X_1 and Y_1 are proper closed subsets. Let $X_2 = \overline{Z - Y_1}$. Then X_2 is closed, and is contained in X_1, and still $Z = X_2 \cap Y_1$. We repeat the process, replacing Y_1 by $Y_2 = \overline{Z - X_2}$. Eventually the descending chains of closed subsets

$$X_1 \supset X_2 \supset X_3 \supset \cdots , \qquad Y_1 \supset Y_2 \supset Y_3 \supset \cdots$$

will stabilize, so eventually $X_N = X_{N+1} = \ldots$ and $Y_N = Y_{N+1} = \ldots$. We take $X = X_N$ and $Y = Y_N$. Since X_1 and Y_1 are proper closed subsets, so are X and Y. Evidently $X = \overline{Z - Y}$. Since X is the closure of an open set, it is the closure of its interior, i.e. $X = \overline{X^\circ}$, and similarly $Y = \overline{Y^\circ}$. Since $X = \overline{Z - Y} = Z - Y^\circ$, the interiors X° and Y° are disjoint. ∎

Proposition 1.5. *If X is a Noetherian space, then there are a finite number of maximal irreducible subspaces $X_1, \cdots , X_r$ of X. Each X_i is the closure of its interior, and X is the union of the X_i.*

Proof. First let us show that X is a union of irreducible subspaces each of which is the closure of an open set. Indeed, if $P \in X$ is not contained in any such subspace, then we may construct an infinite descending chain of closed subspaces as follows. Since X is not irreducible, it is a union of proper closed subspaces, $X = X_1 \cup Y_1$, where by Lemma 1.4, we may assume that $X_1 = \overline{X_1^\circ}$ and $Y_1 = \overline{X_1^\circ}$. We may also assume that $P \in X_1$. By our assumption, X_1 is not irreducible, so we may again decompose $X_1 = X_2 \cup Y_2$, where $X_2 \subsetneq X_1$ is equals $\overline{X_2^\circ}$, and $P \in X_2$. Continuing this way gives an infinite descending chain $X_1 \supsetneq X_2 \supsetneq X_3 \supsetneq \ldots$ of closed subspaces, contradicting the assumption that X is Noetherian.

Let us call a (nonempty) irreducible subspace Z of X such that $Z = \overline{Z^\circ}$ a *component*. If Z and W are distinct components, then the interior Z° of Z is disjoint from W. Indeed, if $Z^\circ \cap W \neq \varnothing$, then $W = (Z \cap W) \cup (W - Z^\circ)$ is a union of two closed sets, and $W - Z^\circ$ is proper. Since W is irreducible, we have $W = Z \cap W$, i.e. $W \subseteq Z$. Now we see that Z contains the interior of W, so by the same reasoning $Z \subseteq W$. Thus $Z = W$, which is a contradiction.

If there are an infinite number of components, say $Z_1, Z_2, Z_3, \cdots$ are all distinct components, then since we have seen that the interior of each component is disjoint from every other component, $Z_1^\circ \subsetneq Z_1^\circ \cup Z_2^\circ \subsetneq Z_1^\circ \cap Z_2^\circ \cap Z_3^\circ \subsetneq \ldots$ is an infinite ascending chain of open subsets; taking complements gives an infinite descending chain of closed sets, which is a contradiction. Thus X is a finite union of components, $X = \bigcup_{i=1}^r X_i$.

It remains to be shown that the components are the maximal irreducible subsets of X. If V is any irreducible subset, then $V = \bigcup (V \cap X_i)$. Since V is irreducible, one of these closed sets must exhaust V, so V is contained in some X_i. ∎

Proposition 1.6. *Algebraic sets are Noetherian in the Zariski topology.*

Proof. It is clear from the definition that a closed subset of a Noetherian space is Noetherian. Thus it is sufficient to show that $\mathbf{A}^n$ is Noetherian. There is an order-reversing bijection between the closed subsets of $\mathbf{A}^n$ and the radical ideals of $k[x_1, \cdots, x_n]$. Since the ideals of $k[x_1, \cdots, x_n]$ satisfy the ascending chain condition by the Hilbert Basis Theorem (Exercise 1.4) it follows that closed subsets of $\mathbf{A}^n$ satisfy the descending chain condition which is the definition of a Noetherian space. ∎

Any Noetherian space can be decomposed into a union of irreducible subspaces. An algebraic set is irreducible in the Zariski topology if and only if it is irreducible. Thus the term *irreducible algebraic set* is a synonym for variety.

Exercises

The first exercises concern the concept of a *Noetherian ring* and the Hilbert Basis Theorem.

Exercise 1.1. Let R be a ring. Prove that the following three conditions are equivalent:

(i) Any ascending chain $\mathfrak{a}_1 \subseteq \mathfrak{a}_2 \subseteq \mathfrak{a}_3 \subseteq \cdots$ of ideals in R is eventually constant, i.e. there exists an r such that $\mathfrak{a}_r = \mathfrak{a}_{r+1} = \mathfrak{a}_{r+2} = \ldots$;

(ii) Any nonempty set Σ of ideals in R contains a maximal element;

(iii) Every ideal in R is finitely generated.

A ring satisfying the equivalent conditions of Exercise 1.1 is called *Noetherian*. Condition (i) is called the *ascending chain condition*.

Exercise 1.2. If R is a Noetherian ring, then so is $R/\mathfrak{a}$ for any ideal $\mathfrak{a}$ of R.

Exercise 1.3. If R is a Noetherian ring, then any submodule or quotient module of a finitely generated R-module is finitely generated.

Exercise 1.4 (Hilbert Basis Theorem). (i) If R is a Noetherian ring, so is the polynomial ring $R[x]$ in one indeterminate.

Hint: Let $\mathfrak{a}$ be an ideal in $R[x]$. Let $\mathfrak{a}_0$ be the set of leading coefficients of polynomials in $\mathfrak{a}$. Then $\mathfrak{a}_0$ is an ideal of R, hence finitely generated. Let $a_1, \cdots, a_r$ be generators. These occur as leading coefficients of polynomials $f_1, \cdots, f_r$ in $\mathfrak{a}$. Let n be the maximum degree of any f_i. Let $\mathfrak{a}'$ be the ideal of $R[x]$ generated by the f_i. Obviously $\mathfrak{a}'$ is finitely generated. Let M be the R-submodule of $R[x]$ generated by $1, x, x^2, \cdots, x^n$. Let $N = \mathfrak{a} \cap M$. Then N is a finitely generated R-module, and hence $\mathfrak{a}'' = R[x].N$ is a finitely generated ideal of $R[x]$. Prove that $\mathfrak{a} = \mathfrak{a}' + \mathfrak{a}''$.

(ii) If A is a Noetherian ring, then any ring R (for example, the polynomial ring $k[x_1, \cdots, x_n]$ over a field) which is finitely generated as an A-algebra is Noetherian.

Exercise 1.5. If $\mathfrak{a}$ is an ideal of A, then the *radical* $r(\mathfrak{a})$ of $\mathfrak{a}$ is the set of all $f \in A$ such that $f^n \in \mathfrak{a}$ for some $n > 0$.

(i) Prove that $r(\mathfrak{a})$ is an ideal of A.

(ii) Prove that $r(\mathfrak{a})$ is the intersection of all prime ideals of A containing $\mathfrak{a}$.

Hint: The hard part is to show that the intersection of all prime ideals containing $\mathfrak{a}$ is contained in $r(\mathfrak{a})$. If $x \notin r(\mathfrak{a})$, use Zorn's Lemma to construct an ideal $\mathfrak{p}$ which contains $\mathfrak{a}$ and which is maximal with respect to the property that $x^n \notin \mathfrak{p}$ for all $n > 0$. Show that $\mathfrak{p}$ is prime.

(iii) Prove that $r(r(\mathfrak{a})) = r(\mathfrak{a})$. An ideal which is its own radical is called a *radical* ideal. We see that the radical of an ideal is a radical ideal.

(iv) If $\mathfrak{a}$ is a *proper* ideal of A, then so is $r(\mathfrak{a})$.

Exercise 1.6. Prove that an algebraic set $X \subseteq \mathbf{A}^n$ is irreducible if and only if X is a variety, i.e., its ideal $I(X)$ is prime.

Exercise 1.7. Suppose that $f : X \to Y$ is a continuous map of topological spaces, where X is irreducible and $f(X)$ is dense in Y. Prove that Y is irreducible.

2. The Extension Theorem

The purpose of this chapter is to develop some useful commutative algebra. In particular, this will allow us to prove the Nullstellensatz. References for this material are Lang [**18**], Atiyah and Macdonald [**3**], and Zariski and Samuel [**36**] Chapter IV, Section 4.

Let R be a ring. (Recall that the term *ring* always means a commutative ring with unit throughout these notes.) R is called a *local ring* if it has a unique maximal ideal. Let F be a field. A subring $R \subseteq F$ called a *valuation ring* of F if $0 \neq x \in F$ implies that either $x \in R$ or $x^{-1} \in R$. Clearly then F is the field of fractions of R.

Proposition 2.1. *A valuation ring is a local ring.*

Proof. Let R be a valuation ring. Let $\mathfrak{m}$ be the set of nonunits of R. We will prove that $\mathfrak{m}$ is an ideal. It is clear that if $x \in \mathfrak{m}$ and $a \in R$ then $ax \in \mathfrak{m}$. Suppose that x and $y \in \mathfrak{m}$. We must show that $x + y \in \mathfrak{m}$. We may assume that both x and y are nonzero. Since R is a valuation ring, either $xy^{-1} \in R$ or $yx^{-1} \in R$. Without loss of generality, assume that $xy^{-1} \in R$. Then $1 + xy^{-1} \in R$. Since $y \in \mathfrak{m}$, and since we have already observed that $R\mathfrak{m} \subseteq \mathfrak{m}$, we have $x + y = (1 + xy^{-1})y \in \mathfrak{m}$, as required. We have proved that the nonunits of R form an ideal. By Exercise 2.1, R is a local ring. ∎

Theorem 2.2 (the Extension Theorem). *Let F be a field, and let A be a subring. Let k be an algebraically closed field, and let $\phi : A \to k$ be a homomorphism. Then there exists a valuation ring R of F containing A and a homomorphism $\Phi : R \to k$ extending ϕ, such that $\ker(\Phi)$ is the maximal ideal of R.*

Proof. Let Σ be the set of all ordered pairs (R, Φ), where R is a subring of F containing A, and Φ is a homomorphism $R \to k$ extending ϕ. This is a partially ordered set, with the ordering such that $(R_1, \Phi_1) \leq (R_2, \Phi_2)$ if $R_1 \subseteq R_2$, and if Φ_1 and Φ_2 agree on R_1. By Zorn's Lemma, Σ has maximal elements, and it is sufficient to show that a maximal element (R, Σ) is a valuation ring.

7

First we show that R is a local ring. Let $\mathfrak{m} = \ker(\Phi)$. We will show that if $x \in R - \mathfrak{m}$, then x is a unit. It will follow then from Exercise 2.1 that $\mathfrak{m}$ is the unique maximal ideal of R. Let $R_1 = R[x^{-1}]$. We claim that Φ can be extended to R_1. Since $x \notin \mathfrak{m} = \ker(\Phi)$, we have $\Phi(x) \neq 0$. Let $\xi = \Phi(x)^{-1}$. We therefore define on R_1

$$\Phi_1(a_n x^{-n} + a_{n-1} x^{-n+1} + \ldots + a_0) = \Phi(a_n)\,\xi^n + \Phi(a_{n-1})\,\xi^{n-1} + \ldots + \Phi(a_0),$$

where $a_i \in R$. We must check that this is well defined. Suppose that $a_n x^{-n} + a_{n-1} x^{-n+1} + \ldots + a_0 = 0$. Then we have $a_0 x^n + a_1 x^{n-1} + \ldots + a_n = 0$. Applying Φ and multiplying by ξ^n we see that

$$\Phi(a_n)\,\xi^n + \Phi(a_{n-1})\,\xi^{n-1} + \ldots + a_0 = 0.$$

Thus Φ_1 is well defined, and extends Φ to $R[x^{-1}]$. By the maximality of (R, Φ), $R_1 = R$, so $x^{-1} \in R$. This proves that R is a local ring.

Now let $0 \neq x \in F$. We must prove that either $x \in R$ or $x^{-1} \in R$. Let $R_1 = R[x]$, and let $R_2 = R[x^{-1}]$. Consider the ideals $\mathfrak{m}R_1$ and $\mathfrak{m}R_2$ of R_1 and R_2 generated by the maximal ideal $\mathfrak{m}$ of R. We claim that either $1 \notin \mathfrak{m}R_1$ or $1 \notin \mathfrak{m}R_2$. Suppose that this is not true. Then we have relations

$$1 = m_0 + m_1 x + \ldots + m_k x^k = m_0' + m_1' x^{-1} + \ldots + m_h' x^{-h},$$

where the m_i and m_i' are in $\mathfrak{m}$. We assume that the degrees in these relations are minimal, and that $k \geq h$. We have

$$1 - m_0' = m_1' x^{-1} + \ldots + m_h' x^{-h}.$$

Now $1 - m_0'$ is a unit of R—otherwise, it would be contained in $\mathfrak{m}$, and so therefore would 1, which is impossible. We may therefore multiply the last relation by $m_k(1 - m_0')^{-1} x^k$ and subtract it from the first relation, reducing the degree, which was assumed minimal. This is a contradiction, and hence either $1 \notin \mathfrak{m}R_1$ or $1 \notin \mathfrak{m}R_2$.

Without loss of generality, we assume that $1 \notin \mathfrak{m}R_1$, and prove that $x \in R$. First note that x is algebraic over the field of fractions of R. For if not, $R[x]$ is a polynomial ring, and we can extend Φ to $R[x]$ (for example by mapping x to zero), contradicting the maximality of (R, Φ).

Let $\mathfrak{m}_1$ be a maximal ideal of R_1 containing $\mathfrak{m}R_1$. Clearly $\mathfrak{m}_1 \cap R = \mathfrak{m}$. Now we have an injection of fields $R/\mathfrak{m} \to R_1/\mathfrak{m}_1$, and since x is algebraic over the field of fractions of R, this is an algebraic extension. Now Φ induces an isomorphism of $R/\mathfrak{m}$ into k, and since k is algebraically closed, this may be extended to an inclusion of $R_1/\mathfrak{m}_1$. Composing this with the canonical map $R_1 \to R_1/\mathfrak{m}_1$, we obtain an extension of Φ to R_1. It follows from the maximality of (R, Φ) that $R_1 = R$, and hence $x \in R$. This completes the proof that R is a valuation ring of F. ∎

The next statement is a disguised version of the Nullstellensatz.

Proposition 2.3 (Algebraic Nullstellensatz). *Let K be a field, and let E be an extension field which is finitely generated as a K-algebra. Then E/K is a finite algebraic extension.*

Proof. Let $\xi_1, \cdots, \xi_n$ be generators for E as a K-algebra. We may assume that $\xi_1, \cdots, \xi_r$ are transcendental, and that $\xi_{r+1}, \cdots, \xi_n$ are algebraic over $K[\xi_1, \cdots, \xi_r]$, which is isomorphic to a polynomial ring. Thus there exist equations

$$g_{i,m_i}(\xi_1, \cdots, \xi_r)\,\xi_i^{m_i} + g_{i,m_i-1}(\xi_1, \cdots, \xi_r)\,\xi_i^{m_i-1} + \ldots + g_{i,0}(\xi_1, \cdots, \xi_r) = 0,$$

where $g_{i,j} \in K[\xi_1, \cdots, \xi_r]$ and $g_{i,m_i} \neq 0$, for $i = r+1, \cdots, n$. Since the algebraic closure k of K is infinite, there exist values $a_1, \cdots, a_r \in k$ such that $g_{i,m_i}(a_1, \cdots, a_r) \neq 0$ for each $r+1 \leq i \leq n$.

We now apply the Extension Theorem, taking $A = K[\xi_1, \cdots, \xi_r]$, and ϕ to be the homomorphism which is the identity map on K, and which satisfies $\phi(\xi_i) = a_i$. We see that there exists a valuation ring R of E and an extension Φ of ϕ to R. Now let us show that each $\xi_i \in R$. For $1 \leq i \leq r$, we already know this. If $r+1 \leq i \leq n$, then if $\xi_i \notin R$, since R is a valuation ring, ξ_i^{-1} is in the maximal ideal of R, which is the kernel of Φ. Now we have

$$g_{i,m_i}(\xi_1, \cdots, \xi_r) + g_{i,m_i-1}(\xi_1, \cdots, \xi_r)\,\xi_i^{-1} + \ldots + g_{i,0}(\xi_1, \cdots, \xi_r)\xi_i^{-m_i} = 0,$$

and applying Φ, we get a contradiction, since the first term becomes nonzero, while all the others vanish. Consequently, each $\xi_i \in R$, and therefore $E = R$. Since E is a field, the kernel of Φ must be trivial, so Φ is an injection of E into k which is constant on K. Since k is the algebraic closure of K, this proves that E is algebraic over K. ■

We may now prove the weak Nullstellensatz.

Proof of Theorem 1.1. Let $\mathfrak{a}$ be a proper ideal of $k[x_1, \cdots, x_n]$. We wish to show that $V(\mathfrak{a})$ is nonempty. Let $\mathfrak{m}$ be a maximal ideal containing $\mathfrak{a}$. Then $V(\mathfrak{m}) \subseteq V(\mathfrak{a})$, and it is sufficient to show that $V(\mathfrak{m})$ is nonempty. Let $A = k[x_1, \cdots, x_n]/\mathfrak{m}$, and let ξ_i be the image of x_i in A. Clearly the canonical map $k[x_1, \cdots, x_n] \to A$ is injective on k, and so if we identify k with its image, we see that $A = k[\xi_1, \cdots, \xi_n]$ is a finitely generated k-algebra which happens to be a field. By Proposition 2.3, A is algebraic over k. Since k is assumed to be algebraically closed, this means that $A = k$, i.e. each $\xi_i \in k$. Thus the point $\xi = (\xi_1, \cdots, \xi_n)$ is an element of $\mathbf{A}^n$ which by construction is annihilated by every $f \in \mathfrak{m}$, i.e. $\xi \in V(\mathfrak{m})$. ■

Exercises

Exercise 2.1. Prove that if R is a commutative ring, then R is local if and only if the nonunits of R form an ideal.

Exercise 2.2. What are the valuation rings of $\mathbb{Q}$?

Exercise 2.3. Let k be an algebraically closed field, and let $F = k(x)$ be the field of rational functions in one variable over k. What are the valuation rings of F?

Exercise 2.4. Give an example of a local ring which is not a valuation ring.

In the next exercises, we develop some properties of integral extensions of rings. The proofs suggested are based on the Extension Theorem, and so they are only valid for integral domains. The results are valid even for rings with zero divisors, but different proofs would have to be given.

Let R be a subring of a larger ring S. An element x of S is said to be *integral* over R if it satisfies a monic polynomial equation

$$x^n + a_{n-1}x^{n-1} + \ldots + a_0 = 0, \qquad (a_i \in R).$$

Exercise 2.5. (i) Suppose that R is a subring of a field F. Prove that $x \in F$ is integral over R if and only if it is contained in every valuation ring of F which contains R.

Hint: The hard part is to show that if x is not integral then there exists a valuation ring R of F such that $x \notin R$. First prove that there exists a homomorphism $\phi : R[x^{-1}] \to K$, where K is some field, such that $\phi(x^{-1}) = 0$. Then use the Extension Theorem.

(ii) Prove that the set of elements of F which are integral over R form a ring, called the *integral closure* of R in F.

(iii) If $R \subseteq S$ are rings, and every element of S is integral over R, then we say that S is *integral* over R. Suppose that $R \subseteq S \subseteq T$ are integral domains, that S is integral over R, and that T is integral over S. Prove that T is integral over R.

Note: Exercise 2.5 (ii) is true for if F is replaced by a ring S, though a different proof must be supplied. Similarly in (iii), the assumption that R, S and T are integral domains is unnecessary. For proofs in the general case, see Lang [**18**], Section VII.1, or any other standard reference. We followed Lang [**16**] in giving proofs based on the Extension Theorem, even though this method imposes unnecessary assumptions, because we find it instructive.

The ring R is called *integrally closed* if it is its own integral closure in its field of fractions.

Exercise 2.6. (i) Prove that if R is a subring of a field F, then the integral closure of R in F is integrally closed.

(ii) Prove that $\mathbb{Z}$ is integrally closed. (See Proposition 4.9 for a generalization.)

Exercise 2.7. (The "Going Up Theorem") Let $R \subseteq S$ be integral domains such that S is integral over R, and let $\mathfrak{p}$ be a prime ideal of R. Prove that there exists a prime ideal $\mathfrak{q}$ of S such that $R \cap \mathfrak{q} = \mathfrak{p}$.

Hint: Extend the homomorphism $R \to R/\mathfrak{p}$ to a homomorphism of a valuation ring of the field of fractions of S into the algebraic closure of the field of fractions $R/\mathfrak{p}$, by means of the Extension Theorem. Now use Exercise 2.5 (i).

See Theorem 4.5 for a more precise statement. The assumption that R and S are integral domains is not necessary—see Lang [**18**], Proposition VII.1.10 on p. 339 for another proof which does not make this assumption.

3. Maps of Affine Varieties

Let $X \subseteq \mathbf{A}^n$ be an affine variety, and let $\mathfrak{p} = I(X)$, so that $\mathfrak{p}$ is a prime ideal of the polynomial ring $k[x_1, \cdots, x_n]$. In the proof of the weak Nullstellensatz, we have already encountered the quotient ring $A = k[x_1, \cdots, x_n]/\mathfrak{p}$. This is the ring generated by the coordinate functions on $\mathbf{A}^n$, where two functions are identified if their difference is in $\mathfrak{p}$, i.e. if they agree on X. We may think of A as the ring of functions on X generated by the coordinate functions in the ambient affine space, so it is called the *coordinate ring* of X.

Proposition 3.1. *Let X be an affine variety with coordinate ring A. If $\xi \in X$, let $\mathfrak{m}_\xi = \{f \in A | f(\xi) = 0\}$. Then $\mathfrak{m}_\xi$ is a maximal ideal of A, the composition of canonical maps $k \to A \to A/\mathfrak{m}_\xi$ is an isomorphism, and $\xi \to \mathfrak{m}_\xi$ is a bijection between the points of X and the set of maximal ideals of A.*

Proof. We may define a homomorphism $A \to k$ by $f \to f(\xi)$. This map is surjective and has kernel $\mathfrak{m}_\xi$. Hence $A/\mathfrak{m}_\xi \cong k$. Thus $A/\mathfrak{m}_\xi$ is a field and so $\mathfrak{m}_\xi$ is a maximal ideal. It is clear that if ξ and η are distinct points of $\mathbf{A}^n$ then there is a function $f \in k[x_1, \cdots, x_n]$ such that $f(\xi) = 0$, but $f(\eta) \neq 0$. In particular, this is true if ξ and η are distinct points of X. Thus the image of f in A is in $\mathfrak{m}_\xi$ but not $\mathfrak{m}_\eta$. Hence the association $\xi \to \mathfrak{m}_\xi$ is injective.

It remains to be shown that every maximal ideal $\mathfrak{m}$ of X has the form $\mathfrak{m}_\xi$ for some $\xi \in X$. Let $\mathfrak{m}'$ be the preimage of $\mathfrak{m}$ in $k[x_1, \cdots, x_n]$. Then $k[x_1, \cdots, x_n]/\mathfrak{m}' \cong A/\mathfrak{m}$, so $\mathfrak{m}'$ is a maximal ideal of $k[x_1, \cdots, x_n]$. By the weak Nullstellensatz, there exists $\xi \in \mathbf{A}^n$ which is annihilated by all $f \in \mathfrak{m}'$. By construction $\mathfrak{p} \subseteq \mathfrak{m}'$, and so $\xi \in X$. Now if $f \in \mathfrak{m}$ then $f(\xi) = 0$, so $f \in \mathfrak{m}_\xi$. We see that $\mathfrak{m} \subseteq \mathfrak{m}_\xi$, and since $\mathfrak{m}$ is maximal, $\mathfrak{m} = \mathfrak{m}_\xi$. ∎

Proposition 3.2. *Let X be an affine variety with coordinate ring A. Regard A as a ring of functions on X. If $\mathfrak{a} \subseteq A$ is an ideal, let $V(\mathfrak{a}) = \{x \in X \mid f(x) = 0 \text{ for all } f \in \mathfrak{a}\}$. Then $V(\mathfrak{a})$ is an algebraic subset of X, and every algebraic subset arises in this way. Also, if Y is an algebraic subset of X, let $I(Y) = \{f \in A \mid f(Y) = 0\}$. Then $I(Y)$ is a radical ideal of A, and every radical ideal of A arises this way. There is an order reversing bijection between the closed algebraic subsets Y of X and the radical ideals $\mathfrak{a}$ of A, where if Y corresponds to $\mathfrak{a}$, then $Y = V(\mathfrak{a})$, and $\mathfrak{a} = I(Y)$. Y is a closed subvariety if and only if $\mathfrak{a}$ is prime.*

The terminology "closed subvariety" may seem redundant here. The reason for this usage is that the "principal open set" $\{x \in X \mid f(x) \neq 0\}$ for $f \in A$ may be given the structure of an algebraic variety (cf. Exercise 3.1 (iii)). Thus we refer to $V(\mathfrak{a})$ as a *closed* algebraic subset, or $V(\mathfrak{p})$ as a *closed* subvariety to distinguish it from these other subsets of X which are varieties, but not closed in the Zariski topology.

Proof. These statements follow from the case where $X = \mathbf{A}^n$, which was discussed in Chapter 1. Indeed, suppose that $X \subseteq \mathbf{A}^n$, and let $A = k[x_1, \cdots, x_n]/\mathfrak{p}$, where $\mathfrak{p}$ is the ideal of functions vanishing on X. There is, as we pointed out in Chapter 1, an order reversing correspondence between the algebraic subsets of $\mathbf{A}^n$ and the radical ideals of $k[x_1, \cdots, x_n]$. The algebraic sets which are contained in X correspond to ideals $\mathfrak{a}$ which contain $\mathfrak{p}$, which in turn correspond to ideals of $k[x_1, \cdots, x_n]/\mathfrak{p}$. ■

We have so far regarded affine varieties as simply subsets of affine space. We would like them to be a category, and we are now ready to define the morphisms which are allowed between them. Let $X \subseteq \mathbf{A}^n$ and $Y \subseteq \mathbf{A}^m$ be affine varieties. A *morphism* $X \to Y$ is defined to be a map ϕ whose components are polynomial functions. In other words, ϕ has the form

$$(3.1) \qquad \phi(x_1, \cdots, x_n) = \big(f_1(x_1, \cdots, x_n), \cdots, f_m(x_1, \cdots, x_n)\big),$$

where $f_1, \cdots, f_m$ are polynomials. It must have the property that it maps the subset X of $\mathbf{A}^n$ into Y. Note that different choices of $f_1, \cdots, f_n$ may define the same morphism, since if we add an element of $I(X)$ to each f_i, it does not change the value of the function ϕ on X. We also call a morphism a *regular* map, and an isomorphism is called *biregular*.

We now make a crucial observation: the coordinate ring is a *contravariant functor*. This comes about as follows. Let A, B be the coordinate rings of X and Y, respectively, and suppose that a morphism $\phi : X \to Y$ is given. We will define a map $\phi^* : B \to A$. Indeed, if $g \in B$, then we define $\phi^*(g)$ to be the function on X given by $\big(\phi^*(g)\big)(x) = g\big(\phi(x)\big)$. Note that this is a function on X whose components are polynomials, i.e. an element of A. In this way, the coordinate ring is seen to be a contravariant functor from the category of affine varieties to a certain category of k-algebras. More precisely, an *affine algebra* is a finitely generated k-algebra which is an integral domain. The affine algebras form a category, in which the morphisms are k-algebra homomorphisms.

Proposition 3.3. *Let A and B be the coordinate rings of the affine varieties X and Y, let $\phi : X \to Y$ be a morphism, and let $\phi^* : B \to A$ be the induced map of coordinate rings. Let $x \in X$, $y = \phi(x) \in Y$, and let $\mathfrak{m}_x$ and $\mathfrak{m}_y$ be the maximal ideals of A and B respectively, which correspond to the points x and y by the bijection of* Proposition 3.1. *Then $\mathfrak{m}_y = \phi^{*-1}(\mathfrak{m}_x)$.*

Proof is an easy verification which we leave to the reader. ∎

Theorem 3.4. *The coordinate ring functor is a (contravariant) equivalence of categories between affine varieties and affine algebras.*

Consult Exercise 3.6 for the notion of equivalence of categories. Note that since the coordinate ring functor is a contravariant functor, the two categories which are equivalent are actually the categories of affine varieties and the *opposite* category to the category of affine algebras—that is, the category whose objects are affine algebras, but in which the direction of the arrows is reversed from the usual category.

Proof. We must check two things: (1) that every affine algebra is the coordinate ring of an affine variety; and (2) that if A and B are the respective coordinate rings of affine varieties X and Y, then the correspondence $\phi \to \phi^*$ is a bijection $\mathrm{Hom}(X, Y) \to \mathrm{Hom}(B, A)$ (See Exercise 3.5).

First, to see that every affine algebra is a coordinate ring, let A be a finitely generated k-algebra which is an integral domain. Choose generators $\xi_1, \cdots, \xi_n$. There is a k-algebra homomorphism from the polynomial ring $k[x_1, \cdots, x_n]$ to A such that $x_i \to \xi_i$. Since the image is an integral domain, the kernel $\mathfrak{p}$ is a prime ideal, and clearly $A \cong k[x_1, \cdots, x_n]/\mathfrak{p}$.

Now suppose that $X \subseteq \mathbf{A}^n$ and $Y \subseteq \mathbf{A}^m$. Let $\mathfrak{p}$ be the ideal of X in $k[x_1, \cdots, x_n]$, and let $\mathfrak{q}$ be the ideal of Y in $k[y_1, \cdots, y_m]$, so that we may identify $A \cong k[x_1, \cdots, x_n]/\mathfrak{p}$ and $B \cong k[y_1, \cdots, y_m]/\mathfrak{q}$. To see that $\phi \to \phi^*$ is a bijection of the Hom sets, note first that if $\phi_1, \phi_2 \in \mathrm{Hom}(X, Y)$ and $\phi_1^* = \phi_2^*$, then $\phi_1 = \phi_2$. Indeed, this is clear from Proposition 3.1 and Proposition 3.3, which show that the underlying map of sets $\phi : X \to Y$ is determined from the map ϕ^*—the effect on maximal ideals is given by Proposition 3.3, and according to Proposition 3.1, the maximal ideals in A and B are in bijection with the points of the affine varieties X and Y. Thus $\phi \to \phi^*$ is an injection. To see that it is surjective, let $h : B \to A$ be a given k-algebra homomorphism. We will show that there exists a map ϕ such that $h = \phi^*$. Let η_i be the image of y_i in B, and let $f_i = h(\eta_i) \in A$. We may then define ϕ by Eq. (3.1). It is easy to see that $\phi : X \to Y$ is a morphism, and that $h = \phi^*$. ∎

We may generalize the previous setup slightly to include all algebraic sets, whether irreducible or not. An affine algebraic set is a Zariski closed subset of $\mathbf{A}^n$ for some n, which is a union of irreducible components by Proposition 1.5 and Proposition 1.6, and the latter are affine varieties. An affine algebraic set has a coordinate ring $k[x_1, \cdots, x_n]/\mathfrak{a}$, where $\mathfrak{a}$ is the defining ideal. This is a finitely generated k-algebra. If the affine set is not itself irreducible, the coordinate ring is not an integral domain. However, it has no nilpotent elements because, in accordance with the Nullstellensatz, $r(\mathfrak{a}) = \mathfrak{a}$ when $\mathfrak{a}$ is the defining ideal of an affine algebraic set. A ring without nilpotent elements is called *reduced*. It is clear that $k[x_1, \cdots, x_n]/\mathfrak{a}$ is reduced if and only if $r(\mathfrak{a}) = \mathfrak{a}$, and so any reduced finitely generated k-algebra is the coordinate ring of an affine algebraic set. Morphisms of affine sets correspond to k-algebra homomorphisms, just as with varieties. The analog of Theorem 3.4 is true: there is a (contravariant) equivalence of categories between the category of affine algebraic sets, and the category of reduced finitely generated k-algebras.

The terminology which we follow here is somewhat old-fashioned. Nowadays the term "variety" often includes all algebraic sets, whether irreducible or not. In the modern terminology, an *affine algebra* is not necessarily an integral domain, as we have defined it; rather, it is a finitely generated reduced algebra. We will not follow the modern terminology, however; for us, a variety is irreducible, and an affine algebra is an integral domain.

Proposition 3.5. *Let A be an affine ring, and let $\mathfrak{a}$ be an ideal of A. Then there are a finite number of prime ideals $\mathfrak{p}_1, \cdots, \mathfrak{p}_d$ which are minimal among the prime ideals containing $\mathfrak{a}$.*

This result is true for Noetherian rings, and we will prove it again in this greater generality in Proposition 5.17. We are including it here as an application of the equivalence between varieties and rings.

Proof. The ring A is the coordinate ring of a variety X, and $V(\mathfrak{a})$ is a closed subset. By Proposition 1.5, it has a decomposition into a finite number of irreducible components, each of which is $V(\mathfrak{p}_i)$ for some prime ideal $\mathfrak{p}_i$ containing $\mathfrak{a}$. Since these are the maximal irreducible subsets of $V(\mathfrak{a})$, it is clear that the $\mathfrak{p}_i$ are the minimal primes containing $\mathfrak{a}$. ∎

Since the coordinate ring A of an affine variety X is an integral domain, it has a field of fractions F. F is called the *function field* of X. We should think of the elements of A as being "holomorphic functions" on X, and the elements of F as being "meromorphic functions." Support for this claim is given by the following considerations.

A basis for the open sets in the Zariski topology is given by the sets of the form $X_g = \{x \in X | g(x) \neq 0\}$, where $0 \neq g \in A$. An element of F has the form f/g, where f, g are elements of A, and g is not identically zero. Now, as a function, f/g is not defined everywhere, but at least it is defined on the open set X_g. Thus the elements of F are functions which are defined on open sets of the Zariski topology. They are—like the meromorphic functions of complex analysis—*densely defined functions.*

It is possible to be slightly more precise about where an element of F is defined. If $h \in F$, $x \in X$, we will say that h is *regular at* x if there exist f and $g \in A$ such that $h = f/g$ and $g(x) \neq 0$. There may not be a single expression of h as f/g which is valid for all x where h is regular. As the following example shows, it is possible that we will have to choose different expressions for a single given function h corresponding to different values of the point x.

Example 3.6. Consider $X = \{(x_1, x_2, x_3, x_4) \in \mathbf{A}^4 | x_1 x_4 - x_2 x_3 = 0\}$. It may be checked that the principal ideal $(x_1 x_4 - x_2 x_3)$ in $k[x_1, x_2, x_3, x_4]$ is prime, so this is an affine variety. Consider the function $h = x_1/x_2 = x_3/x_4$. At the point $(1, 1, 0, 0)$, the expression $h = x_1/x_2$ shows that this is regular, while the expression x_3/x_4 is useless. At the point $(0, 0, 1, 1)$, the opposite situation pertains.

Nevertheless, it is certainly true that if $f_1/g_1 = f_2/g_2$ in F, where f_1, f_2, g_1 and $g_2 \in A$, and if $g_1(x)$, $g_2(x)$ are both nonzero, then $f_1(x)/g_1(x) = f_2(x)/g_2(x)$. So if h is regular at x, it has a uniquely defined value. The elements of F may be regarded as functions wherever they are defined.

If $x \in X$, let $\mathcal{O}_x$ be the set of all $f \in F$ which are regular at x. This is a ring. It has an abstract interpretation as follows. Let $\mathbf{m} = \mathbf{m}_x$ be the maximal ideal of A consisting of functions vanishing at x, as in Proposition 3.1. Then $\mathcal{O}_x$ is the *localization* of A at the maximal ideal $\mathbf{m}$. This is a very common construction, which we recall.

If A is a commutative ring, a *multiplicative subset* of A is defined to be a set S such that $1 \in S$ but $0 \notin S$, and such that S is closed under multiplication. In this case, there is defined a *ring of fractions* $S^{-1}A$, and if M is an A-module, a *module of fractions* $S^{-1}M$, which is an $S^{-1}A$-module. We define $S^{-1}A$ to be the quotient of the set $S \times A$ by the equivalence relation $(s, a) \sim (t, b)$ if $v(sb - ta) = 0$ for some $v \in S$. We denote the equivalence class of $(s, a) \in S \times A$ by a/s or $s^{-1}a$, and think of this as a fraction. Thus $s/a = t/b$ if and only if $v(sb - ta) = 0$ for some $v \in S$—if A is an integral domain, this is equivalent to $sb = ta$. $S^{-1}A$ becomes a ring with

$$(a/s) + (b/t) = (at + bs)/(st), \qquad (a/s) \cdot (b/t) = ab/st.$$

There is a canonical homomorphism $A \to S^{-1}A$ which sends $a \to a/1$.

As a special case, if A is an integral domain, we may take $S = A - \{0\}$ to obtain the field of fractions F of A. If A is an integral domain and S is any multiplicative set, then we may embed $S^{-1}A$ in the field of fractions A. Moreover in this case the canonical map $A \to S^{-1}A$ is an injection. For much of our concrete work, A will be an integral domain, and we will find it convenient to identify $S^{-1}A$ with a subring of the field of fractions of A, and to identify A as a subring of S.

If M is an A-module, then $S^{-1}M$ is constructed similarly as a quotient of $S \times M$ by the equivalence relation $(s, m) \sim (t, n)$ if and only if $v(sn - tm) = 0$ for some $v \in S$. Then $S^{-1}M$ naturally has the structure of an $S^{-1}A$-module. The module $S^{-1}M$ is isomorphic to the tensor product $S^{-1}A \otimes_A M$ (Exercise 3.4 (vi)).

Two cases arise frequently in algebraic geometry. Firstly, if $\mathfrak{p}$ is a prime ideal of A, then let $S = A - \mathfrak{p}$. This is a multiplicative subset. The localization $S^{-1}A$ is denoted $A_\mathfrak{p}$. It is a local ring—the unique maximal ideal is $\mathfrak{p}A_\mathfrak{p}$. In the case where A is the coordinate ring of an affine variety X and $x \in X$, we see that $\mathcal{O}_x = A_{\mathfrak{m}_x}$ is precisely the localization of A at the maximal ideal $\mathfrak{m}_x$. More generally, if Y is a closed subvariety of X, let $\mathfrak{p}$ be the ideal of Y in A. We may then consider the localization $A_\mathfrak{p}$. This is a local ring, which we denote $\mathcal{O}_{Y,X}$.

The other case which arises in algebraic geometry is when $0 \neq f \in A$ is given, and $S = \{1, f, f^2, f^3, \cdots\}$. Then the ring of fractions is denoted A_f. Because A is an integral domain, A_f may be identified with the subring $A[f^{-1}]$ of the function field F. See Exercise 3.1.

If $U \subseteq X$ is an open set, let $\mathcal{O}(U) = \mathcal{O}_X(U)$ be the set of functions $h \in F$ which are regular at every point of U. This is a ring. By definition, $\mathcal{O}(U) = \bigcap_{x \in U} \mathcal{O}_x$.

Proposition 3.7. $\mathcal{O}(X) = A$.

Proof. It is clear that $A \subseteq \mathcal{O}(X)$. Conversely, let us show that $\mathcal{O}(X) \subseteq A$. Suppose that $h \in F - A$. We must show that $h \notin \mathcal{O}(X)$. That is, we must show that there exists $x \in X$ such that h is not defined at x. Let $\mathfrak{a} = \{a \in A | ah \in A\}$. Then $\mathfrak{a}$ is an ideal of A, and since $h \notin A$, it is proper. Hence it is contained in a maximal ideal $\mathfrak{m}$. By Proposition 3.1, $\mathfrak{m} = \mathfrak{m}_x$ for some $x \in X$. We will show that h is not regular at x. Otherwise, we could write $h = f/g$, where f and $g \in A$ and $g(x) \neq 0$. Then by construction $g \in \mathfrak{a}$, so $g \in \mathfrak{m} = \mathfrak{m}_x$, which is a contradiction since $g(x) \neq 0$. ■

Finally, let us consider a class of mappings which is strictly larger than the class of morphisms. These are called *rational maps*. We should think of a morphism as a map between affine varieties which is *holomorphic* everywhere, and a rational map as a map which is merely *meromorphic*. A rational map is not defined everywhere, but it is densely defined. To specify a rational map $\phi : X \to Y$, we take a nonempty open set U of X, and a map $\phi : U \to Y$ which has the property that if r is any element of the coordinate ring of Y, then $\phi \circ r$ is in $\mathcal{O}(U)$. In this case, we say that ϕ is *regular* on U.

If we are given a different open set U_1 and a function ϕ_1 defined on U_1, and if ϕ and ϕ_1 agree on $U \cap U_1$, then we regard them as defining the same rational map. (Since X is an irreducible topological space, $U \cap U_1$ is nonempty and in fact dense in X.) So if B is the coordinate ring of Y, a rational map induces a map $B \to \mathcal{O}(U)$ for some dense open set U of X.

Affine varieties and rational maps *do not* form a category. We may see this by considering the following example. Let $X = Z = \mathbf{A}^1$, $Y = \mathbf{A}^2$, let $\phi : X \to Y$ be the rational map $\phi(x) = (x, 0)$, and let $\psi : Y \to Z$ be the rational map $\psi(x, y) = x/y$, defined on the complement of the x-axis. The composition $X \to Z$ is not defined, because the image of ϕ is disjoint from the domain of ψ.

However, the rational map ϕ is called *dominant* if there is an open set U of X such that ϕ is regular on U and $\phi(U)$ is dense in Y. There are some things to check about this definition—see Exercise 3.5. The dominant rational maps *do* form a category.

Let F and K be the function fields of X and Y respectively, and let B be the coordinate ring of Y. Since for any open set U of X the field of fractions of $\mathcal{O}(U)$ is the function field F of X, and since (Exercise 3.5) the map $B \to \mathcal{O}(U)$ induced by a dominant rational map ϕ is injective, ϕ gives rise to a map ϕ^* of function fields. This is constant on k.

The dominant rational map ϕ is called a *birational map* or *birational equivalence* if it has an inverse in the category of affine varieties and rational maps. Clearly this will be the case if the induced map of function fields is an isomorphism. If there exists a birational equivalence of X and Y, we say that X and Y are *birationally equivalent*. This is true if and only if their function fields are isomorphic.

We can now state an analog of Theorem 3.4 for the category of affine varieties and rational maps. We define an *abstract function field* over k to be a field containing k, and finitely generated over k. A morphism in this category is a homomorphism of fields which is constant over k.

Theorem 3.8. *The category of affine varieties and dominant rational maps is (contravariantly) equivalent to the category of abstract function fields over k.*

In other words, the category of affine varieties with morphisms the dominant rational maps is equivalent to the opposite category of the category of abstract function fields.

Proof is left to the reader in Exercise 3.7 below. ■

Exercises

Exercise 3.1. Let X be an affine variety, A its coordinate ring, and $f \in A$. The *principal open set* X_f is by definition the set of $x \in X$ such that $f(x) \neq 0$.

(i) Prove that the principal open sets are a basis for the Zariski topology, but there may be open sets which are not principal. (For example, exhibit an open subset of $\mathbf{A}^2$ which is not principal.)

(ii) Prove the following generalization of Proposition 3.7 to principal open sets: $\mathcal{O}(X_f) = A_f$.

(iii) Let X be an affine variety with coordinate ring A, let $f \in A$, and let U be the principal open set X_f. Prove that there exists an algebraic variety Y and an injective morphism $\phi : Y \to X$ such that the image of Y is precisely U, and such that the map ϕ^* of coordinate rings induces an isomorphism of the coordinate ring of Y with $\mathcal{O}(U)$.

Hint: Take Y to be the set of all points $(x, t) \in X \times \mathbf{A}^1$ such that $f(x)t = 1$.

Thus we may regard the principal open set $U = X_f$ as being "isomorphic" to the affine variety Y. (Looking ahead, the principal open set is a variety in the sense defined in Chapter 8, and it is isomorphic to Y in the category of varieties.) We will now see that an arbitrary open set is *not* so isomorphic to an affine variety.

(iv) Show that if $X = \mathbf{A}^2$ and $U = X - \{(0,0)\}$, then U is an open set of X, and that $\mathcal{O}(U) = \mathcal{O}(X)$. Conclude that there *cannot* exist any algebraic variety Y with an injective morphism $\phi : Y \to X$ whose image is precisely U such that ϕ^* induces an isomorphism of the coordinate ring of Y with $\mathcal{O}(U)$.

This example is a variety in the sense of Chapter 8—but it is not affine. See Example 7.6.

Exercise 3.2. Recall that in any category, an *isomorphism* is a morphism which has an inverse. Here are two examples to show that a bijective morphism of affine varieties is not necessarily an isomorphism.

(i) Let $X = \mathbf{A}^1$, and let $Y \subseteq \mathbf{A}^2$ be the singular cubic curve which is the locus of $f(x, y) = 0$, where $f(x, y) = x^3 - y^2$. Let $\phi : X \to Y$ be the map $\phi(t) = (t^2, t^3)$. Prove that this is a bijective morphism of varieties, but that the inverse map is not a morphism. Is it a rational map?

(ii) Let k be an algebraically closed field of characteristic p, and let $V = \mathbf{A}^1$. The *Frobenius map* $\phi : V \to V$ is defined by $\phi(x) = x^p$. Show that this is a bijective morphism of varieties, but that the inverse map is not a morphism.

Exercise 3.3. (i) Prove that a morphism of affine varieties is continuous in the Zariski topology.

(ii) Prove that if $\phi_1, \phi_2 : X \to Y$ are morphisms of affine varieties, and there exists a nonempty open set $U \subseteq X$ such that ϕ_1 and ϕ_2 agree on U, then $\phi_1 = \phi_2$.

We follow the convention in algebra, that if R is a ring and M is a module, and if $\mathfrak{a} \subseteq R$ and $N \subseteq M$ are additive subgroups, then $\mathfrak{a}N$ is the set of all finite sums $\sum a_i n_i$ with $a_i \in \mathfrak{a}$, and $n_i \in N$. If $\mathfrak{a}$ is an ideal of R, or if N is a submodule of M then $\mathfrak{a}N$ is a submodule of M.

Exercise 3.4 (Localization). Let A be an integral domain, S a multiplicative subset.

(i) If $\mathfrak{a} \subseteq A$ is an ideal, consider the ideal $S^{-1}\mathfrak{a} = \mathfrak{a} \cdot S^{-1}A$ of $S^{-1}A$. Prove that this is a proper ideal if and only if $\mathfrak{a} \cap S \neq \emptyset$.

(ii) Prove that every ideal of $S^{-1}A$ has the form $S^{-1}\mathfrak{a}$ for some ideal $\mathfrak{a}$ of A.

(iii) Prove that if $\mathfrak{p}$ is a prime ideal of A such that $S \cap \mathfrak{p} = \emptyset$, then $S^{-1}\mathfrak{p} \cap A = \mathfrak{p}$. Hence there is a bijection between the prime ideals of $S^{-1}A$ and the prime ideals of A which do not meet S.

(iv) If $\mathfrak{p}$ is a prime ideal of A and $S = A - \mathfrak{p}$, then $S^{-1}A$ is denoted $A_{\mathfrak{p}}$. This has a prime ideal $\mathfrak{p}A_{\mathfrak{p}}$ by (iii). Show that this prime ideal is maximal, and that $A_{\mathfrak{p}}/\mathfrak{p}A_{\mathfrak{p}}$ is isomorphic to the field of fractions of the integral domain $A/\mathfrak{p}$. In particular, if $\mathfrak{p}$ is maximal, then the canonical map $A/\mathfrak{p} \to A_{\mathfrak{p}}/\mathfrak{p}A_{\mathfrak{p}}$ is an isomorphism.

(v) If S is a multiplicative subset of R, use Zorn's Lemma to show that there are ideals $\mathfrak{p}$ of R which are maximal with respect to the property that $\mathfrak{p} \cap S = \emptyset$, and that any such ideal is prime.

(vi) If S is a multiplicative subset of R and M is an R-module, show that $S^{-1}M \cong S^{-1}A \otimes_A M$.

Hint: Define inverse maps between these modules by $s^{-1}m \mapsto s^{-1} \otimes m$ and $s^{-1}a \otimes m \mapsto s^{-1}(am)$.

Exercise 3.5. Dominant rational maps. (i) Show that if X and Y are affine varieties with coordinate rings A and B, and if $f : X \to Y$ is a morphism, then $f(X)$ is dense in Y if and only if $f^* : B \to A$ is injective. If this is the case, then the morphism f is called *dominant*.

More generally, if f is a rational map, f is called *dominant* if there exists an open set U of X such that f is regular on U, and $f(U)$ is dense in Y.

(ii) Let X and Y be affine varieties and $\phi : X \to Y$ a rational map. Prove that if $U \subseteq X$ is an open set on which ϕ is regular, and if $\phi(U)$ is dense in Y, then so is $\phi(U')$ for every nonempty open subset U' of X on which ϕ is regular. (Clearly you may assume that $U' \subseteq U$.) Hence the definition of dominant rational map does not depend on the choice of open set.

(iii) Show that if $\phi : X \to Y$ and $\psi : Y \to Z$ are dominant rational maps, then there is a natural way to define a composition $\psi \circ \phi : X \to Z$ which is a dominant rational map.

(iv) If $\phi : X \to Y$ is a dominant rational map, if $U \subseteq X$ is an open set on which ϕ is regular, and B is the coordinate ring of Y, show that the induced map $B \to \mathcal{O}(U)$ is injective.

Exercise 3.6. Equivalence of categories. Let $\mathcal{X}$ and $\mathcal{Y}$ be categories, and let $\mathcal{F} : \mathcal{X} \to \mathcal{Y}$ be a functor. Part of the data specifying a functor is a mapping of Hom sets, so that if X_1 and X_2 are objects of $\mathcal{X}$, then there is specified a mapping $\mathrm{Hom}(X_1, X_2) \to \mathrm{Hom}(\mathcal{F}X_1, \mathcal{F}X_2)$.

$\mathcal{F}$ is called an *equivalence of categories* if (1) given any object Y in $\mathcal{Y}$, there exists an object X in $\mathcal{X}$ such that $\mathcal{F}X$ is isomorphic to Y; and (2) the map of Hom sets $\mathrm{Hom}(X_1, X_2) \to \mathrm{Hom}(\mathcal{F}X_1, \mathcal{F}X_2)$ is always a bijection.

(i) Let $\mathcal{F}$ be an equivalence of categories. Show that we can construct a functor $\mathcal{G}$ from $\mathcal{Y}$ to $\mathcal{X}$ as follows. For each object Y of $\mathcal{Y}$, choose an object $\mathcal{G}Y$ of $\mathcal{X}$ such that Y is isomorphic to $\mathcal{F}\mathcal{G}Y$. We may do this by property (1) in the definition of equivalence of categories. (Here we are invoking the axiom of choice. There is a foundational issue here, because categories tend to be very large, and we are assuming that the objects in $\mathcal{Y}$ form a set. However, assume that we can do this.) Also, for each object Y of $\mathcal{Y}$, let us choose a specific isomorphism $\iota_Y : Y \to \mathcal{F}\mathcal{G}Y$. In defining a functor, we must specify not only the correspondence of objects, but also Hom sets. Thus if $\mathcal{Y}_1$ and $\mathcal{Y}_2$ are given, we must specify a map $\mathrm{Hom}(Y_1, Y_2) \to \mathrm{Hom}(\mathcal{G}Y_1, \mathcal{G}Y_2)$. If $\phi \in \mathrm{Hom}(Y_1, Y_2)$ is given, we define $\mathcal{G}\phi \in \mathrm{Hom}(\mathcal{G}Y_1, \mathcal{G}Y_2)$ as follows. We have $\iota_{Y_2} \circ \phi \circ \iota_{Y_1}^{-1} \in \mathrm{Hom}(\mathcal{F}\mathcal{G}Y_1, \mathcal{F}\mathcal{G}Y_2)$, and by property (2) in the definition of equivalence of categories, there is a unique element $\phi' \in \mathrm{Hom}(\mathcal{G}Y_1, \mathcal{G}Y_2)$ such that $\mathcal{F}\phi' = \iota_{Y_2} \circ \phi \circ \iota_{Y_1}^{-1}$. We define $\mathcal{G}\phi$ to be this ϕ'. Prove that $\mathcal{G}$ defined this way is a functor.

(ii) Prove that if X is an object of $\mathcal{X}$, then $\mathcal{G}\mathcal{F}X$ is isomorphic to X in $\mathcal{X}$. Also, if Y is an object of $\mathcal{Y}$, then $\mathcal{F}\mathcal{G}Y$ is isomorphic to Y in $\mathcal{Y}$. (In fact, one may say more: the functors $\mathcal{G}\mathcal{F}$ and $\mathcal{F}\mathcal{G}$ are *naturally equivalent to the identity functors* in the categories $\mathcal{X}$ and $\mathcal{Y}$, repectively.)

(iii) Prove that $\mathcal{G}$ is an equivalence of categories.

(iv) Prove that equivalence of categories satisfies the usual properties of an equivalence relation: it is symmetric, reflexive and transitive.

Exercise 3.7. Prove Theorem 3.8.

Exercise 3.8. The content of the following exercise is to show that $\mathcal{O}_X$ is a *sheaf*. Let X be an affine variety. Let $\{U_i | i \in I\}$ be some family of open subsets, and let $U = \bigcup_{i \in I} U_i$. Suppose that for every $i \in I$ there is given $f_i \in \mathcal{O}_X(U_i)$, with the property that for $i, j \in I$, the functions f_i and f_j are equal when restricted to $U_i \cap U_j$. Then it is obvious that there exists a unique k-valued function f on U whose restriction to U_i is f_i. Prove that $f \in \mathcal{O}_X(U)$.

Exercise 3.9. Let X be an affine variety, and Y a closed subvariety. Describe the functions in $\mathcal{O}_{Y,X}$ in concrete terms. If $\mathfrak{m}$ is the maximal ideal of $\mathcal{O}_{Y,X}$, describe the residue field $\mathcal{O}_{Y,X}/\mathfrak{m}$.

Exercise 3.10. Let $f : X \to Y$ be a morphism of affine varieties. Prove that the closure of $f(X)$ is irreducible.

4. Dimension and Products

We recall the notion of *transcendence degree*. Let K be a field, F/K an extension which we may assume for our purposes to be finitely generated. If $\xi_1, \cdots, \xi_n$ is any set of generators, we may reorder them so that $\xi_1, \cdots, \xi_d$ are algebraically independent, and $\xi_{d+1}, \cdots, \xi_n$ are algebraic over the subfield $K(\xi_1, \cdots, \xi_d)$. The set $\xi_1, \cdots, \xi_d$ is called a *transcendence basis*. A transcendence basis is simply a maximal set of algebraically independent elements.

It is a basic fact that any two transcendence bases have the same cardinality. (See Lang [18], Theorem VIII.1.1 on p. 356.) The cardinality of a transcendence basis is called the *transcendence degree* of F over K.

Let X be an affine variety over the algebraically closed field k, let A be its coordinate ring, and F its function field. We define the *dimension* $\dim(X)$ of X to be the transcendence degree of F over k. Intuitively, the number of algebraically independent variables needed to parametrize the variety should be the dimension. If Y is a subvariety, the *codimension* of Y in X is $\dim(X) - \dim(Y)$. An affine variety of dimension one is called an *affine curve*, and an affine variety of dimension two is called an *affine surface*.

Proposition 4.1. *Let X be an affine variety, and Y a closed subvariety. Then $\dim(Y) \leq \dim(X)$, with equality if and only if $X = Y$.*

Proof. Let A be the coordinate ring of X, and B the coordinate ring of Y. Let $d = \dim(X)$. Choose generators $\xi_1, \cdots, \xi_n$ for A. We may order these so that $\xi_1, \cdots, \xi_d$ are algebraically independent. Then $\xi_{d+1}, \cdots, \xi_n$ are algebraic over $k(\xi_1, \cdots, \xi_d)$.

The inclusion map $i : Y \to X$ induces a ring homomorphism $i^* : A \to B$, which is surjective. Indeed, to see that i^* is surjective, we may argue as follows. We may regard X and Y as subspaces affine space $\mathbf{A}^n$. Then B is generated by the images of the coordinate functions $x_1, \cdots, x_n$ on $\mathbf{A}^n$. So if $\xi_1, \cdots, \xi_n$ are the images of these in A, B is generated as a ring by the images of $\xi_1, \cdots, \xi_n \in A$ under i^*.

Now let $\eta_j = i^*(\xi_j) \in B$ ($j = 1, \cdots, n$). It is clear, since $\xi_{d+1}, \cdots, \xi_n$ are algebraic over $k(\xi_1, \cdots, \xi_d)$ that $\eta_{d+1}, \cdots, \eta_n$ are algebraic over $k(\eta_1, \cdots, \eta_d)$.

Thus a transcendence basis of the field of fractions of B over k may be chosen from $\eta_1, \cdots, \eta_d$, and thus $\dim(Y) \leq \dim(X)$. Moreover, if $\dim(Y) = \dim(X)$, it must be the case that $\eta_1, \cdots, \eta_d$ are algebraically independent. We must show that this implies that i^* is an isomorphism. Since we know it is surjective, it is enough to show that it is injective. Suppose, then, that ξ is in the kernel. Then since $\xi_1, \cdots, \xi_d$ is a transcendence basis, ξ is algebraic over $k(\xi_1, \cdots, \xi_d)$. It therefore satisfies an algebraic equation

$$a_n(\xi_1, \cdots, \xi_d)\, \xi^n + a_{n-1}(\xi_1, \cdots, \xi_d)\, \xi^{n-1} + \ldots + a_0(\xi_1, \cdots, \xi_d) = 0.$$

We may assume that $a_0(\xi_1, \cdots, \xi_d) \neq 0$. Now after applying i^*, and using the assumption $i^*(\xi) = 0$, we get $a_0(\eta_1, \cdots, \eta_d) = 0$. However, $\eta_1, \cdots, \eta_d$ are supposed to be algebraically independent, so this is a contradiction. ∎

Theorem 4.2 (Noether's Normalization Lemma). *Let A be a finitely generated integral domain over a field K. Suppose that the transcendence degree of the field of fractions F of A over K is d. Then there exists a transcendence basis $\xi_1, \cdots, \xi_d$ of F/K such that the $\xi_i \in A$, and the ring A is integral over $K[\xi_1, \cdots, \xi_d]$. If K is perfect, then we may arrange that F be separable over $K(\xi_1, \cdots, \xi_d)$.*

Proof. Suppose that $A = K[\eta_1, \cdots, \eta_n]$. If $n = d$, then the η_i are a transcendence basis, and we may take $\xi_i = \eta_i$. Assume therefore that $n > d$, and by induction that the result is true for rings admitting strictly fewer than n generators. The η_i are then algebraically dependent over K, and satisfy a relation

$$0 = f(\eta_1, \cdots, \eta_n) = \sum a(k_1, \cdots, k_n)\, \eta_1^{k_1} \cdots, \eta_n^{k_n},$$

where $a(k_1, \cdots, k_n) \in K$. Let N be an integer such that all nonzero coefficients in this relation have all $k_i < N$. Let $\zeta_i = \eta_i - \eta_1^{N^{i-1}}$ for $i = 2, \cdots, n$. Let us order $\mathbb{Z}^n$ by $(k_1, \cdots, k_n) > (k_1', \cdots, k_n')$ if for some $1 \leq i \leq n$, $k_i > k_i'$, and $k_j = k_j'$ for all j such that $i < j \leq n$. Let $(h_1, \cdots, h_n)$ be the largest n-tuple with respect to this ordering such that $a(h_1, \cdots, h_n) \neq 0$. Consider

$$(4.1) \qquad 0 = f(\eta_1, \zeta_2 + \eta_1^N, \zeta_3 + \eta_1^{N^2}, \cdots, \zeta_n + \eta_1^{N^{n-1}}).$$

It is clear that when this relation is expanded and the terms of equal degree in η_1 are collected, the unique term of highest degree will be

$$a(h_1, \cdots, h_n)\, \eta_1^{h_1 + N h_2 + N^2 h_3 + \ldots + N^{n-1} h_n}.$$

Dividing by $a(h_1, \cdots, h_n)$, we get a relation of integral dependence for η_1 over $\zeta_2, \cdots, \zeta_n$. By induction the ring $K[\zeta_2, \cdots, \zeta_n]$ is integral over a transcendence basis of F, and so we may say the same thing about $K[\eta_1, \cdots, \eta_n] = K[\eta_1, \zeta_2, \cdots, \zeta_n]$.

To obtain the statement about separability, we may assume that the characteristic is a prime p. Note that we may assume that $a(k_1, \cdots, k_n) \neq 0$ for some n-tuple with the k_i not all divisible by p. Indeed, if only terms appear where the k_i are all divisible by p then, since K is perfect, we may write $a(pk_1, \cdots, pk_n) = A(k_1, \cdots, k_n)^p$, and we get a relation of lower degree,

$$0 = \sum A(k_1, \cdots, k_n) \eta_1^{k_1} \cdots \eta_n^{k_n}.$$

After rearranging the η_i, we may assume one term $a(f_1, \cdots, f_n) \neq 0$ with $p \nmid f_1$. Assume that $(f_1, \cdots, f_n)$ is maximal with respect to the ordering described above. Then it is clear that the polynomial for η_1 obtained by expanding Eq. (4.1) and collecting the powers of η_1 will have a term of degree $f_1 + N f_2 + \ldots + N^{n-1} f_n$. We choose N so that $p|N$. Then $p \nmid f_1$, which means that the polynomial will be separable. ■

Next, we consider the "Going Up" and "Going Down" theorems of Cohen and Seidenberg. If $A \subseteq B$ are rings, and if $\mathfrak{p}$ and $\mathfrak{q}$ are prime ideals of A and B respectively, we say that $\mathfrak{q}$ *lies above* $\mathfrak{p}$ if $\mathfrak{q} \cap A = \mathfrak{p}$. Note that if $\mathfrak{q}$ is a prime ideal of B, then $\mathfrak{q} \cap A$ is a prime ideal of A.

Proposition 4.3. *Let $A \subseteq B$ be rings, and let $\mathfrak{p}$ be a prime ideal of A. Then there exists a prime ideal $\mathfrak{q}$ of B lying above $\mathfrak{p}$ if and only if $\mathfrak{p}B \cap A = \mathfrak{p}$.*

Proof. First assume that there exists a prime ideal $\mathfrak{q}$ of B lying above $\mathfrak{p}$. Then $\mathfrak{p}B \cap A \subseteq \mathfrak{q} \cap A = \mathfrak{p} \subseteq \mathfrak{p}B \cap A$ as required.

Conversely, assume that $\mathfrak{p}B \cap A = \mathfrak{p}$. Let S be the multiplicative set $A - \mathfrak{p}$, and consider the ring $B' = S^{-1}B$. Since $\mathfrak{p}B$ does not meet S, we have $1 \notin \mathfrak{p}B'$. Consequently, there exists a maximal ideal $\mathfrak{m}$ of B' containing $\mathfrak{p}B'$. Let $\mathfrak{q} = \mathfrak{m} \cap B$. We have $\mathfrak{p} \subseteq \mathfrak{m}$, and so $\mathfrak{p} \subseteq \mathfrak{q} \cap A$. On the other hand, $\mathfrak{m}$ is a proper ideal of B', and so it does not meet S, the elements of which are units in B'. Consequently $\mathfrak{q} \cap A \subseteq \mathfrak{m} \cap A \subseteq \mathfrak{p}$. ■

Proposition 4.4. *Let $A \subseteq B$ be integral domains, and suppose that B is integral over A.*

(i) Let $\mathfrak{p}$ be a prime ideal of A. Then there exists a prime ideal $\mathfrak{q}$ of B lying above A, and $\mathfrak{p}B \cap A = \mathfrak{p}$.

(ii) Let $\mathfrak{p}$ be a prime ideal of A. Suppose that $\mathfrak{q}$ and $\mathfrak{q}'$ are prime ideals of B lying above A. If $\mathfrak{q}' \subseteq \mathfrak{q}$, then $\mathfrak{q}' = \mathfrak{q}$.

The hypothesis that A and B are integral domains is actually unnecessary.

Proof. Part (i) follows from combining Exercise 2.7 and Proposition 4.3.

To prove part (ii), let $\overline{A} = A/\mathfrak{p}$, and $\overline{B} = B/\mathfrak{q}'$. Then $\overline{A}$ is naturally a subring of the integral domain $\overline{B}$, $\overline{B}$ is integral over $\overline{A}$, and if $\overline{\mathfrak{q}}$ is the ideal

$\mathfrak{q}/\mathfrak{q}'$ of $\overline{B}$, then $\overline{\mathfrak{q}}$ is a prime ideal of $\overline{B}$ such that $\overline{\mathfrak{q}} \cap \overline{A} = 0$. We will show that $\overline{\mathfrak{q}} = 0$. For if not, suppose that $0 \neq x \in \overline{\mathfrak{q}}$. Then we have an integral equation $x^n + a_{n-1}x^{n-1} + \ldots + a_0 = 0$ where $a_i \in \overline{A}$, and $a_0 \neq 0$. Now $a_0 = -x(x^{n-1} + a_{n-1}x^{n-2} + \ldots + a_1) \in \overline{\mathfrak{q}}$ since $x \in \overline{\mathfrak{q}}$. Since $\overline{\mathfrak{q}} \cap \overline{A} = 0$, we have $a_0 = 0$, which is a contradiction. Thus $\overline{\mathfrak{q}} = 0$, which implies that $\mathfrak{q}' = \mathfrak{q}$. ∎

Theorem 4.5 (the "Going-Up" Theorem). *Let $A \subseteq B$ be integral domains such that B is integral over A. Let $\mathfrak{p}_1 \subseteq \mathfrak{p}_2$ be prime ideals of A, and let $\mathfrak{q}_1$ be a prime of B lying above $\mathfrak{p}_1$. Then there exists a prime $\mathfrak{q}_2$ of B lying above $\mathfrak{p}_2$ such that $\mathfrak{q}_1 \subseteq \mathfrak{q}_2$.*

The hypothesis that A and B are integral domains is actually unnecessary.

Proof. Let $\overline{A} = A/\mathfrak{p}_1$ and $\overline{B} = B/\mathfrak{q}_1$. Then $\overline{A}$ is canonically a subring of $\overline{B}$. By Proposition 4.4 there exists a prime ideal of $\overline{B}$ lying above the image in $\overline{A}$ of $\mathfrak{p}_2$. The preimage $\mathfrak{q}_2$ of this ideal in B then fits the requirements of the Theorem. ∎

The "Going Down" theorem is more difficult. Let $A \subseteq B$ be rings, and $\mathfrak{a}$ an ideal of A. If $x \in B$, we will say that x is *integral over* $\mathfrak{a}$ if it satisfies a relation

$$x^n + a_{n-1}x^{n-1} + \ldots + a_0 = 0, \qquad a_i \in \mathfrak{a}.$$

Proposition 4.6. *Let A be an integral domain, and let B be an integral domain which is integral over A. Let $\mathfrak{a}$ be an ideal of A. Then the set of elements of B which are integral over $\mathfrak{a}$ is precisely the ideal $r(\mathfrak{a}B)$.*

Proof. First suppose that $x \in B$ is integral over $\mathfrak{a}$. Then we may write $x^n = -a_{n-1}x^{n-1} - \ldots - a_0$, with $a_i \in \mathfrak{a}$, so $x^n \in \mathfrak{a}B$, and therefore $x \in r(\mathfrak{a}B)$.

On the other hand, assume that $x \in r(\mathfrak{a}B)$. Then $x^n \in \mathfrak{a}B$ for some n. Now if we prove that x^n is integral over $\mathfrak{a}$, it will follow that x is integral over $\mathfrak{a}$. Therefore, we may replace x by x^n, and assume that $x \in \mathfrak{a}B$. We may also clearly assume that $x \neq 0$.

Let $\mathfrak{b} = \{y \in A[1/x] \mid xy \in \mathfrak{a}A[1/x]\}$. Thus $\mathfrak{b}$ is an ideal of $A[1/x]$. We will show that $1 \in \mathfrak{b}$. Suppose not. Then we may find a maximal ideal $\mathfrak{m}$ of $A[1/x]$ containing $\mathfrak{b}$, and a homomorphism ϕ from $A[1/x]$ into an algebraically closed field k whose kernel is precisely $\mathfrak{m}$. By the Extension Theorem (Theorem 2.2), we may extend ϕ to a homomorphism of Φ of a valuation ring R of the field F of fractions of B, and whose kernel is precisely the maximal ideal of R. It is clear that $x^{-1}\mathfrak{a} \subseteq \mathfrak{b}$, and so $\Phi(x^{-1}\mathfrak{a}) = 0$. On the other hand, $x \in R$ because x is integral over A and hence contained in every valuation ring of F by Exercise 2.5 (i). This means that x^{-1} cannot be in the maximal ideal of R and hence $\Phi(x^{-1}) \neq 0$. Therefore $\Phi(\mathfrak{a}) = 0$. Now B is contained in

R, again by Exercise 2.5 (i), and so $\Phi(B\mathfrak{a}) = 0$. As $x \in B\mathfrak{a}$, we see that $\Phi(x) = 0$. Therefore x is in the maximal ideal of R, which is a contradiction since $x^{-1} \in R$. This proves that $1 \in \mathfrak{b}$.

Since $1 \in \mathfrak{b}$, we see that $x \in \mathfrak{a}A[1/x]$, and therefore x may be written

$$x = -a_{n-1} - a_{n-2}x^{-1} - \ldots - a_0 x^{1-n},$$

where each $a_i \in \mathfrak{a}$. This implies that x is integral over $\mathfrak{a}$. ∎

We recall that an integral domain A is called *integrally closed* in a field F containing it if the integral closure of A in F is A. We say that the integral domain is *integrally closed* (without qualification) if A is integrally closed in its field of fractions. This does not imply that it is integrally closed in any larger field.

Proposition 4.7. *Let A be an integrally closed integral domain with field of fractions F. Let K be an extension field of F, and let x be an element of K which is integral over A. Then the minimal polynomial of x over F has the form*

$$t^n + a_{n-1}t^{n-1} + \ldots + a_0,$$

where each $a_i \in A$. Moreover, if x is integral over a prime ideal $\mathfrak{p}$, then each $a_i \in \mathfrak{p}$.

Proof. Note that x satisfies a monic polynomial with coefficients in A, and that the conjugates of x all satisfy the same polynomial. Thus the conjugates x_i of x are integral over A. Now the minimal polynomial of x over F has the form

$$t^n + a_{n-1}t^{n-1} + \ldots + a_0 = \prod(t - x_i),$$

where the product is over all conjugates of x. (Factors may be repeated if x is inseparable.) Equating the coefficients, we see that a_i are polynomials in the x_i, and hence are elements of F which are integral over A. Since A is integrally closed, the $a_i \in A$.

Now suppose that x is integral over $\mathfrak{p}$. Then the conjugates of x also satisfy the same monic polynomial with coefficients in $\mathfrak{p}$ which is satisfied by x and so by Proposition 4.6 all lie in $r(\mathfrak{p}B)$, where B is the integral closure of A in K. Thus the a_i, which are polynomials in the x_i, lie in $r(\mathfrak{p}B)$. Thus for some $m > 0$, $a_i^m \in A \cap \mathfrak{p}B$, which equals $\mathfrak{p}$ by Proposition 4.4. Now $\mathfrak{p}$ is prime, and therefore $a_i \in \mathfrak{p}$, as required. ∎

Theorem 4.8 (the "Going Down" Theorem). *Let A be an integrally closed integral domain, and let B be an integral domain which is integral over A. Let $\mathfrak{p}_2 \subseteq \mathfrak{p}_1$ be prime ideals of A, and let $\mathfrak{q}_1$ be a prime ideal of B lying above $\mathfrak{p}_1$. Then there exists a prime ideal $\mathfrak{q}_2 \subseteq \mathfrak{q}_1$ of B lying above $\mathfrak{p}_2$.*

Proof. Our first objective is to establish that $\mathfrak{p}_2 B_{\mathfrak{q}_1} \cap A = \mathfrak{p}_2$. The inclusion $\supseteq$ is obvious. Assume therefore that $x \in \mathfrak{p}_2 B_{\mathfrak{q}_1} \cap A$. We may write $x = y/s$ where $y \in \mathfrak{p}_2 B$ and s is in the multiplicative set $B - \mathfrak{q}_1$. Let

$$t^r + u_1 t^{r-1} + \ldots + u_r$$

be the minimal polynomial of y. Since $y \in \mathfrak{p}_2 B$, it follows from Proposition 4.6 that y is integral over $\mathfrak{p}_2$. Then by Proposition 4.7, the coefficients $u_i \in \mathfrak{p}_2$. Now the minimal polynomial for $s = y/x$ is

$$t^r + v_1 t^{r-1} + \ldots + v_r,$$

where $v_i = u_i/x^i$. Now $s \in B$ is integral over A, so that by Proposition 4.7, the coefficients $v_i \in A$. We will now show that $x \in \mathfrak{p}_2$. For if not, we have $u_i = x^i v_i \in \mathfrak{p}_2$, and since $\mathfrak{p}_2$ is prime, it follows that $v_i \in \mathfrak{p}_2$. Now we see that $s^r = -v_1 s^{r-1} - \ldots - v_r$ is in $\mathfrak{p}_2 B \subseteq \mathfrak{q}_1$, and since $\mathfrak{q}_1$ is prime, it follows that $s \in \mathfrak{q}_1$. This is a contradiction, because $s \in S = B - \mathfrak{q}_1$. This proves that $x \in \mathfrak{p}_2$. We have thus proved that $\mathfrak{p}_2 B_{\mathfrak{q}_1} \cap A = \mathfrak{p}_2$.

It now follows from Proposition 4.3 that there is a prime $\mathfrak{q}$ of $B_{\mathfrak{q}_1}$ lying above $\mathfrak{p}_2$. We now take $\mathfrak{q}_2 = \mathfrak{q} \cap B$. Observe that $\mathfrak{q}_2$ is contained in $\mathfrak{q}_1$ because $\mathfrak{q}$ is a proper ideal of $B_{\mathfrak{q}_1}$, hence does not meet S. ■

Proposition 4.9. *A unique factorization ring is integrally closed. In particular, a principal ideal domain is integrally closed.*

Proof. Recall that a principal ideal domain is a unique factorization domain— see Lang [18], Theorem II.5.2 on p. 112. So the second assertion follows from the first.

Let A be a unique factorization ring. Let $x = f/g$ be an element of its field of fractions which is assumed to be integral over A. We may assume that f and g are elements of A which have no common irreducible factor. Let

$$x^n + a_{n-1} x^{n-1} + \ldots + a_0 = 0$$

be the monic irreducible polynomial satisfied by x. Then

$$f^n = g\left(-a_{n-1}f^{n-1} - a_{n-2}gf^{n-2} - \ldots - a_0 g^{n-1}\right).$$

Consequently $g | f^n$. If g is not a unit, this implies that g and f have a common irreducible factor, which is a contradiction. Hence $x \in A$. ■

We may now return attention to the matter of dimension.

Proposition 4.10. *Let X be an affine variety, let $\mathfrak{p}$ be a minimal nonzero prime ideal in its coordinate ring A, and let Y be the closed subvariety corresponding to $\mathfrak{p}$. Then $\dim(Y) = \dim(X) - 1$.*

Proof. By Noether's Normalization Lemma (Theorem 4.2), we may find algebraically independent elements $\xi_1, \cdots, \xi_d$ in A such that A is integral over $B = k[\xi_1, \cdots, \xi_d]$. Thus $d = \dim(X)$. Note that B is integrally closed by Proposition 4.9, and so we may apply the "Going Down" Theorem (Theorem 4.8) to conclude that $\mathfrak{p} \cap B$ is a minimal nonzero prime ideal of B. Indeed, $\mathfrak{p} \cap B$ is not zero because the norm of any nonzero element of $\mathfrak{p}$ will be a nonzero element. Furthermore, there can be no nonzero prime ideal of B properly contained in $\mathfrak{p}$, because by the Going Down Theorem (Theorem 4.8), such a prime ideal could be lifted to A, contradicting the assumed minimality of $\mathfrak{p}$. Now, in the unique factorization ring B, the minimal nonzero prime ideal $\mathfrak{p} \cap B$ must have the form (f) where f is an irreducible polynomial in $\xi_1, \cdots, \xi_d$. We may assume that f involves ξ_d nontrivially. In that case, the images of $\xi_1, \cdots, \xi_{d-1}$ in $B/(\mathfrak{p} \cap B)$ are algebraically independent, while the image of ξ_d becomes dependent. Now $A/\mathfrak{p}$, which is the coordinate ring of Y, is algebraic over $B/(\mathfrak{p} \cap B)$, and so the dimension of Y, which by definition is the transcendence degree of the field of fractions of $A/\mathfrak{p}$, equals the transcendence degree of the field of fractions of $B/(\mathfrak{p} \cap B)$, which we have seen to be equal to $d - 1$. ∎

By a *chain* of irreducible subspaces, we mean a nested sequence $X_0 \subsetneq X_1 \subsetneq X_2 \subsetneq \cdots \subsetneq X_d$. The irreducible spaces X_i are called the *members* of the chain, and d is called its *length*. A chain is called a *refinement* of another chain if every member of the first chain is a member of the second. We call the chain *saturated* if it does not admit any refinement but itself. The *combinatorial dimension* of X is the largest length of any chain of irreducible subspaces.

If R is a ring, we make a similar definition. By a *chain* of prime ideals, we mean a nested sequence $\mathfrak{p}_0 \subsetneq \mathfrak{p}_1 \subsetneq \cdots \subsetneq \mathfrak{p}_d$ of distinct primes. The primes $\mathfrak{p}_i$ are called the *members* of the chain, and d is called its *length*. A chain is called a *refinement* of another chain if every member of the first chain is a member of the second. We call the chain *saturated* if it does not admit any refinement but itself. The *Krull dimension* of R is defined to be the largest length of any chain of prime ideals.

Theorem 4.11. *(i) Let X be an affine variety of dimension d. Then d is equal to the combinatorial dimension of X as a Noetherian topological space. Moreover, every saturated chain of irreducible subspaces has length d. It is equal to the Krull dimension of the coordinate ring of X.*

(ii) Let A be an affine algebra. Then every saturated chain of prime ideals has the same length, equal to the Krull dimension of A.

Proof. For (i), it follows from Proposition 4.1 that if $X_0 \subsetneq X_1 \subsetneq X_2 \subsetneq \cdots \subsetneq X_r$ is a chain of irreducible subspaces, then $\dim(X_0) < \dim(X_1) < \cdots < \dim(X_r) \leq d$. Both assertions will follow if we show that for a saturated chain we have $\dim(X_i) = i$, and $r = d$. Firstly, if the chain is saturated, then X_0 is reduced to a point. For if not, we could take any point which is contained in X_0, and add it to the left of the chain. Similarly, it is obvious that for a saturated chain, $X_r = X$. Let us show that if X_i and X_{i+1} are adjacent members of the chain, then $\dim(X_{i+1}) = \dim(X_i) + 1$. Indeed, let A be the coordinate ring of X_{i+1}. It follows from Proposition 3.2 that the ideal in A corresponding to X_i is a minimal nonzero prime ideal. It follows by Proposition 4.10 that $\dim(X_i) = \dim(X_{i+1}) - 1$. It is now clear that for a saturated chain of irreducible subspaces, $r = d$ and $\dim(X_i) = i$.

Since there is an order-reversing bijection between the irreducible subsets of X and the primes of its coordinate ring A, it is clear that the combinatorial dimension of X equals the Krull dimension of A.

As for (ii), we may assume that A is the coordinate ring of X. the fact that every saturated chain of prime ideals of A has the same length follows from (i) in view of the order-reversing bijection between prime ideals and irreducible subsets. ∎

Thus if A is an affine algebra over k, the Krull dimension of A is equal to the transcendence degree over k of the field of fractions of A. For local rings, other concepts of dimension will be introduced in Chapter 5.

Theorem 4.12 (Krull's Principal Ideal Theorem). *Let A be an affine algebra, and let $0 \neq f \in A$. If $\mathfrak{p}$ is a prime ideal of A which is minimal among those containing f, then $\mathfrak{p}$ is a minimal nonzero prime ideal of A.*

Another proof of this will be given in Exercise 5.12.

Proof. Firstly, let us reduce to the case where $\mathfrak{p}$ is the only prime minimal among those containing f. There are a finite number of such primes by Proposition 3.5; call them $\mathfrak{p} = \mathfrak{p}_1, \mathfrak{p}_2, \cdots, \mathfrak{p}_r$. There are no inclusion relations between these, so for $i = 2, \cdots, r$ we may find $g_i \in \mathfrak{p}_i - \mathfrak{p}$. Now let $g = \prod g_i$. Then $g \in \mathfrak{p}_2, \cdots, \mathfrak{p}_r$ but $g \notin \mathfrak{p}$. We now pass to the ring $A_g = A[g^{-1}]$. By Exercise 3.4, there is a one-to-one correspondence between the proper ideals of A_g and the ideals of A which do not contain any power of g. Thus $\mathfrak{p}A_g$ is the unique smallest prime of A_g which contains (f). Moreover, if $\mathfrak{p}A_g$ is a minimal nonzero prime of A_g, then it will follow from Exercise 3.4 that $\mathfrak{p}$ is a minimal nonzero prime of A. Thus replacing A by A_g, we may assume that $\mathfrak{p}$ is the unique smallest prime of A containing f. Thus $\mathfrak{p} = r\big((f)\big)$ by Exercise 1.5 (ii).

We invoke Noether's Normalization Lemma to find algebraically independent elements $\xi_1, \cdots, \xi_d$ in A such that A is integral over $B = k[\xi_1, \cdots, \xi_d]$.

Now B is integrally closed because it is a unique factorization domain. Let L and K be the respective fields of fractions of A and B, and let $f_0 = N_{L/K}(f)$ be the norm of f. Note that this is a power of the constant coefficient in the minimal polynomial of f over K, and so by Proposition 4.7, $f_0 \in B$.

Let $\mathfrak{q} = \mathfrak{p} \cap B$. Let us show that $f_0 \in \mathfrak{q}$. Let A' be the integral closure of A in its field of fractions L. Then A' is also the integral closure of B in L. All of the conjugates of f are integral over B, hence lie in A', and one of them, namely f itself, lies in $\mathfrak{p}$. Consequently f_0, which is a product of these conjugates (with possible repetitions) lies in $\mathfrak{p}A'$. Thus $f_0 \in \mathfrak{p}A' \cap B \subseteq \mathfrak{p}A' \cap A$, and by Proposition 4.4, $\mathfrak{p}A' \cap A = \mathfrak{p}$. So $f_0 \in \mathfrak{p} \cap B = \mathfrak{q}$.

Now we will show that $\mathfrak{q}$ is the radical of (f_0). Since $f_0 \in \mathfrak{q}$, $r((f_0)) \subseteq \mathfrak{q}$. On the other hand if $g \in \mathfrak{q}$, then $g \in \mathfrak{p}$, which is the radical of (f) in A, so some power $g^k = fh$ where $h \in A$. Taking norms, we have $g^{mk} = f_0 . N_{L/K}(h)$, where $m = [L : K]$. Thus $g \in r((f_0))$. We have thus proved that $\mathfrak{q} = r((f_0))$.

The polynomial ring B is a unique factorization domain, and so if we factor $f_0 = c \prod_{i=1}^{r} f_i^{k_i}$, where $0 \neq c \in k$ is a constant and the f_i are irreducible polynomials, then the radical of $((f))$ is clearly the principal ideal generated by $\prod_{i=1}^{r} f_i$. Since this ideal is prime, $r = 1$ and so we see that $f_0 = c f_1^{k_1}$ is a unit times a power of an irreducible polynomial. Thus $\mathfrak{q} = r((f_0)) = (f_1)$ is the principal ideal generated by an irreducible polynomial. Consequently it is a minimal nonzero prime ideal of B.

The minimality of $\mathfrak{q}$ implies the minimality of $\mathfrak{p}$ by Proposition 4.4 (ii). ∎

If X is an affine variety with coordinate ring A, by a *hypersurface* in X, we mean $V((f))$, where (f) is a principal ideal in the coordinate ring.

Proposition 4.13. *(i) Let X be an affine variety, and let Y be a closed subvariety of dimension d. If H is a hypersurface in X, then either $Y \subseteq H$, or else every irreducible component of $H \cap Y$ has dimension $d - 1$.*

(ii) If $H_1, \cdots, H_r$ are hypersurfaces in X, and Y is a subvariety, then every irreducible component of $Y \cap H_1 \cap \cdots \cap H_r$ has codimension $\leq r$ in Y.

Proof. We prove (i). Let A and B be the coordinate rings of X and Y respectively, and let $i : Y \to X$ be the inclusion map. Let $H = V((f))$ for some $f \in A$. If Y is not a subset of H, then f does not vanish identically on Y, so $g = i^*(f) \neq 0$ in B. Then $Y \cap H$ is the hypersurface of Y corresponding to g. Let Z be an irreducible component. Then the ideal of Z in B is a prime ideal, minimal among those containing g, and so by Theorem 4.12 it is a minimal nonzero prime ideal of B. Hence by Proposition 4.10, the dimension of Z is $d - 1$.

We prove (ii). Let Z_i be the irreducible components of $Y \cap H_1 \cap \ldots \cap H_{r-1}$; by induction Z_i has codimension $\leq r-1$ in Y. Since $Y \cap H_1 \cap \cdots \cap H_r = \bigcup Z_i \cap H_r$, every component of this intersection is contained in some $Z_i \cap H_r$, and the irreducible components of $Z_i \cap H_r$ have codimension ≤ 1 in Z_i by (i).   ■

Proposition 4.14. *Let A be a ring, $\mathfrak{p}_1, \cdots, \mathfrak{p}_n$ prime ideals of A, and $\mathfrak{a}$ an ideal which is not contained in any $\mathfrak{p}_i$. Then $\mathfrak{a}$ is not contained in $\bigcup \mathfrak{p}_i$.*

Proof. This is obvious if $n = 1$. Assume that $n > 1$, and by induction that it is true for $n - 1$. If $\mathfrak{a}$ is not contained in any of $\mathfrak{p}_1, \cdots, \mathfrak{p}_n$, then the induction hypothesis implies that for each $1 \leq j \leq n$, there exists an $x_j \in \mathfrak{a} - \bigcup_{i \neq j} \mathfrak{p}_i$. Now suppose that $\mathfrak{a} \subseteq \bigcup \mathfrak{p}_i$. Then each $x_j \in \mathfrak{p}_j$. Now consider

$$x = \sum_{j=1}^{n} \prod_{i \neq j} x_i.$$

This is in $\mathfrak{a}$, so it is in some $\mathfrak{p}_k$. Now each of the terms $\prod_{i \neq j} x_i$ for $j \neq k$ is in $\mathfrak{p}_k$, and so the remaining term $\prod_{i \neq k} x_i \in \mathfrak{p}_k$. However each of the factors x_i in this product is not in $\mathfrak{p}_k$, and since $\mathfrak{p}_k$ is prime, this is a contradiction.   ■

By the *height* of a prime ideal $\mathfrak{p}$ of a ring A, we mean the length r of the longest possible chain $\mathfrak{p}_0 \subsetneq \mathfrak{p}_1 \subsetneq \mathfrak{p}_2 \subsetneq \cdots \subsetneq \mathfrak{p}_r = \mathfrak{p}$ of prime ideals ending with $\mathfrak{p}$. Clearly such a chain is saturated. If A is a domain, the ideal 0 is prime, so $\mathfrak{p}_0 = 0$ in such a chain. Thus the Krull dimension of the ring is the maximal height of any prime ideal.

Proposition 4.15. *(i) Let A be an affine algebra, and let $\mathfrak{p}$ be a prime ideal of height r. Then there exist $f_1, \cdots, f_r$ such that $\mathfrak{p}$ is minimal among the prime ideals of A containing $f_1, \cdots, f_r$.*

(ii) Let X be an algebraic variety, and let Y be a variety of codimension r. Then there exist hypersurfaces $H_1, \cdots, H_r \subset X$ such that Y is an irreducible component of $H_1 \cap \cdots \cap H_r$.

Proof. Let $0 = \mathfrak{p}_0 \subsetneq \mathfrak{p}_1 \subsetneq \cdots \mathfrak{p}_r = \mathfrak{p}$ be a chain of prime ideals of maximal length. Thus $\mathfrak{p}_i$ has height i. We will choose the f_i so that $f_i \in \mathfrak{p}_i - \mathfrak{p}_{i-1}$. That alone is not sufficient, however. We also require that each of the minimal primes containing $f_1, \cdots, f_i$ has height i, and that one of these is $\mathfrak{p}_i$. Let us convince ourselves that the f_i may be chosen in this way.

We may choose f_1 to be any nonzero element of $\mathfrak{p}_1$. Every minimal prime containing f_1 has height one by Theorem 4.12. Suppose now that $f_1, \cdots, f_i$ are chosen. Let $\mathfrak{q}_1, \cdots, \mathfrak{q}_d$ be the minimal primes of A containing $f_1, \cdots, f_i$; by assumption these have height i. If $i < r$, we show how to choose f_{i+1}. Since the height of $\mathfrak{p}_{i+1}$ is greater than the height of each $\mathfrak{q}_i$, the ideal $\mathfrak{p}_{i+1}$ is not contained in any $\mathfrak{q}_i$. By Proposition 4.14, there exists an element $f_{i+1} \in \mathfrak{p}_{i+1}$

which is not in any $\mathfrak{q}_i$. We must show that any minimal prime $\mathfrak{p}'$ containing $f_1, \cdots, f_{i+1}$ has height $i+1$. Since $f_1, \cdots, f_i \in \mathfrak{p}'$, we see that $\mathfrak{p}'$ contains one of the minimal primes $\mathfrak{q}_j$ containing $f_1, \cdots, f_i$, each of which has height i, yet $\mathfrak{p}'$ also contains an element f_{i+1} not in $\mathfrak{q}_j$, so its height is greater than or equal to $i+1$. Let $\overline{f}_{i+1}$ be the image of f_{i+1} in $\overline{A} = A/\mathfrak{q}_j$. Then $\overline{\mathfrak{p}}' = \mathfrak{p}'/\mathfrak{q}_j$ is a minimal prime of $\overline{A}$ containing $\overline{f}_{i+i}$, hence it is a minimal nonzero prime of this ring by Theorem 4.12. This means that the chain $\mathfrak{q}_j \subsetneq \mathfrak{p}'$ is saturated, and since $\mathfrak{q}_j$ has height i, this means that $\mathfrak{p}'$ has height $i+1$, as required. We still have to show that $\mathfrak{p}_{i+1}$ is a minimal prime of A containing $f_1, \cdots, f_{i+1}$. Since it does contain these elements, it will at least contain such a minimal prime, and its height is the same as the height $i+1$ of that prime, so they are equal.

Part (ii) follows from part (i). Indeed, let A be the coordinate ring of X, and let $\mathfrak{p}$ be the prime ideal of Y. Choosing f_i as in (i), let H_i be the hypersurface $\{f_i = 0\}$. Since $\mathfrak{p}$ is a minimal prime containing the f_i, Y is an irreducible component of $V\big((f_1, \cdots, f_r)\big) = \bigcap H_i$. ∎

Our next topic is the product of two affine varieties. Under the equivalence of categories between affine varieties and affine algebras, this corresponds to the tensor product of algebras. Let A and B be two k-algebras. The tensor product $A \otimes B$ may be given the structure of an algebra so that $(a \otimes b) \cdot (a' \otimes b') = aa' \otimes bb'$ (Exercise 4.1). In the particular case where $A = k[x_1, \cdots, x_n]$ and $B = k[y_1, \cdots, y_m]$ are polynomial rings, so is $A \otimes B \cong k[x_1, \cdots, x_n, y_1, \cdots, y_m]$ (Exercise 4.2). Our objective is to generalize this to affine algebras: if A and B are affine algebras, we want to show that $A \otimes B$ is also. The difficult part is to show that it is an integral domain, and this is our next objective.

Proposition 4.16. *Let R be a valuation ring of a field F, and let $\mathfrak{m}$ be the maximal ideal of R. Let $y_1, \cdots, y_n$ be nonzero elements of F. Then there exists $\lambda \in F$ such that $\lambda y_1, \cdots, \lambda y_n \in R$, but not all of $\lambda y_1, \cdots, \lambda y_n$ are in $\mathfrak{m}$.*

Proof. Let $i < n$, and suppose that we have found λ_i such that $\lambda_i y_1, \cdots, \lambda_i y_i \in R$, but not all of $\lambda_i y_1, \cdots, \lambda_i y_i$, are in $\mathfrak{m}$. We will construct λ_{i+1} such that $\lambda_{i+1} y_1, \cdots, \lambda_{i+1} y_{i+1} \in R$, but not all of $\lambda_{i+1} y_1, \cdots, \lambda_{i+1} y_{i+1}$, are in $\mathfrak{m}$. If $\lambda_i y_{i+1} \in R$, we take $\lambda_{i+1} = \lambda_i$. Otherwise, $\lambda_i^{-1} y_{i+1}^{-1} \in R$, and we take $\lambda_{i+1} = y_{i+1}^{-1}$. Then for $j \leq i$, $\lambda_{i+1} y_j = y_j / y_{i+1} = (\lambda_i y_j)(\lambda_i^{-1} y_{i+1}^{-1}) \in R$, while $\lambda_{i+1} y_{i+1} = 1$. We continue this construction inductively, and take $\lambda = \lambda_n$. ∎

Proposition 4.17. *Let A and B be affine algebras over the algebraically closed field k. Then $A \otimes B$ is also an affine algebra. If d and e are the dimensions of A and B respectively, then $d + e$ is the dimension of $A \otimes B$.*

Proof. If a_i are a set of generators of A and b_j are a set of generators of B, then it is easy to see that $A \otimes B$ is generated by the $a_i \otimes 1$ together with the $1 \otimes b_j$. Thus $A \otimes B$ is finitely generated. We must show that it is an integral domain.

Let $A = k[\xi_1, \cdots, \xi_n]$ and $B = k[\eta_1, \cdots, \eta_m]$. We assume that $\xi_1, \cdots, \xi_d$ are algebraically independent, and that $\xi_{d+1}, \cdots, \xi_n$ are algebraic over the field of fractions of $A_0 = k[\xi_1, \cdots, \xi_d]$. Similarly, we assume that $\eta_i, \cdots, \eta_e$ are algebraically independent, and that $\eta_{e+1}, \cdots, \eta_m$ are algebraic over the field of fractions of $B_0 = k[\eta_{e+1}, \cdots, \eta_m]$.

By Exercise 4.2, $A_0 \otimes B_0$ is a polynomial ring. Let F be its field of fractions, so that F has transcendence degree $d + e$ over k. We embed the k-algebras A and B into the algebraic closure $\overline{F}$ of F in an arbitrary way. We consider the subring $AB = \{\sum_i a_i b_i | a_i \in A, b_i \in B\}$ of $\overline{F}$. We will prove that AB is isomorphic to $A \otimes B$. It will follow that $A \otimes B$ is an integral domain, since it is isomorphic to a subring of a field.

Multiplication $A \times B \to AB$ is a k-bilinear map, and hence induces a homomorphism of k-algebras $A \otimes B \to AB$ such that $a \otimes b \to ab$. We will show that this is injective. Suppose not. Then we can find a nonzero element $\sum_i a_i \otimes b_i$ of the kernel, so that $\sum_i a_i b_i = 0$. We may assume that the a_i are linearly independent. (If not, we may express one of them, say a_n, in terms of the others, distribute over b_n, and regroup the terms, shortening the length of the expression.)

Let ϕ be any homomorphism of B_0 into k which is the identity map on k. By the Extension Theorem (Theorem 2.2), we may find a valuation ring R of the field of fractions L of B, and a homomorphism $\phi : R \to k$ whose kernel is the maximal ideal of R, and which extends the original homomorphism ϕ. Now $\xi_1, \cdots, \xi_d$ are algebraically independent over L, for if they were not, the transcendence degree of the field of fractions of AB would be strictly less than $d + e$, which is impossible since the subring $A_0 B_0$ already contains $d + e$ algebraically independent elements. Therefore, we may extend ϕ to a homomorphism of $R[\xi_1, \cdots, \xi_d] \to \overline{F}$ in such a way that $\phi(\xi_i) = \xi_i$ for $i = 1, \cdots, d$. By the Extension Theorem, we may further extend ϕ to a valuation ring R_1 of the field of fractions of AB. Now ϕ is injective on A_0—in fact, it is the identity map—and it follows that the whole field of fractions K_0 of A_0 is contained in R_1. (Indeed, if $x \in K_0 - R_1$ we would have $x^{-1} \in \ker(\phi)$, and if $x = y/z$, where $y, z \in A_0$, then $0 \neq z \in \ker(\phi)$, which is a contradiction.) Since ϕ is the identity map on A_0, it is the identity map on K_0. Let K be the field of fractions of A. Then $R_1 \cap K$ is a valuation ring of K containing a subfield K_0 over which K is algebraic. Since K_0 is a field, this is the same as saying that K is integral over the subring K_0, and it follows from Exercise 2.5 (i) that $R_1 \cap K = K$. Thus K does not meet the maximal ideal consisting of the nonunits of R_1, and hence ϕ is injective on K. Because the a_i are linearly independent over k, it follows that the $\phi(a_i)$ are also linearly independent over k. Now applying Proposition 4.16 to the valuation ring $R \cap L$ of L, we may find $\lambda \in L$ such that $\lambda b_i \in R$, and not all of them are in the maximal ideal. Then $\sum_i a_i \lambda b_i = 0$, and applying ϕ we obtain $\sum_i \phi(\lambda b_i)\,\phi(a_i) = 0$. The coefficients $\phi(\lambda b_i)$ are in k, and not all of them are zero, so this contradicts the linear independence of the $\phi(a_i)$.

Since AB is algebraic over $A_0 B_0$, the transcendence degree of the field of fractions of AB is also $d + e$. ∎

Lemma 4.18. *Suppose we have a commutative diagram with exact rows and columns:*

$$
\begin{array}{ccccccc}
A' & \xrightarrow{\phi'} & B' & \xrightarrow{\psi'} & C' & \longrightarrow & 0 \\
\alpha \downarrow & & \beta \downarrow & & \gamma \downarrow & & \\
A & \xrightarrow{\phi} & B & \xrightarrow{\psi} & C & \longrightarrow & 0 \\
\alpha' \downarrow & & \beta' \downarrow & & \gamma' \downarrow & & \\
A'' & \xrightarrow{\phi''} & B'' & \xrightarrow{\psi''} & C'' & \longrightarrow & 0 \\
\downarrow & & \downarrow & & \downarrow & & \\
0 & & 0 & & 0 & &
\end{array}
$$

Then the kernel of $\gamma'\psi = \psi''\beta' : B \to C''$ is $\operatorname{im}(\beta) + \operatorname{im}(\phi)$.

Proof. It is evident that $\operatorname{im}(\beta) + \operatorname{im}(\phi) \subset \ker(\gamma'\psi)$. Let $b \in \ker(\gamma'\psi)$. Then $\psi(b) \in \ker(\gamma') = \operatorname{im}(\gamma)$, so let $c' \in C'$ such that $\gamma(c') = \psi(b)$. Since ψ' is surjective, we may find $b' \in B'$ such that $\psi'(b') = c'$. Then $b = (b - \beta(b')) + \beta(b')$, where $b - \beta(b') \in \ker(\psi) = \operatorname{im}(\phi)$, and $\beta(b') \in \operatorname{im}(\beta)$. Thus $b \in \operatorname{im}(\phi) + \operatorname{im}(\beta)$. ∎

Theorem 4.19. *Let $X \subseteq \mathbf{A}^n$ and $Y \subseteq \mathbf{A}^m$ be affine varieties. Let A_0 and B_0 be the polynomial rings $k[x_1, \cdots, x_n]$ and $B_0 = k[y_1, \cdots, y_m]$, which are the coordinate rings of $\mathbf{A}^n$ and $\mathbf{A}^m$. We identify these rings with their images in $C_0 = k[x_1, \cdots, x_n, y_1, \cdots, y_m] \cong A_0 \otimes B_0$, which is the coordinate ring of $\mathbf{A}^{n+m}$. Let $\mathfrak{p} \subseteq A_0$ and $\mathfrak{q} \subseteq B_0$ be the ideals of X and Y, respectively, and let $A = A_0/\mathfrak{p}$ and $B = B_0/\mathfrak{q}$ be their coordinate rings. Then $X \times Y \subset \mathbf{A}^{n+m}$ is an affine variety. Its ideal in C_0 is $\mathfrak{p}B_0 + \mathfrak{q}A_0$, which is prime. The coordinate ring $C_0/(\mathfrak{p}B_0 + \mathfrak{q}A_0)$ of $X \times Y$ is isomorphic to $A \otimes B$. $X \times Y$ is a product in the category of affine varieties and morphisms. We have $\dim(X \times Y) = \dim(X) + \dim(Y)$.*

Proof. It is clear that $X \times Y$ is an algebraic set, and $X \times Y = V(\mathfrak{p}B_0 + \mathfrak{q}A_0)$. To show that it is a variety, and that $\mathfrak{p}B_0 + \mathfrak{q}A_0 = I(X \times Y)$, it is sufficient

to check that $\mathfrak{p}B_0 + \mathfrak{q}A_0$ is prime. $C = k[x_1, \cdots, x_n, y_1, \cdots, y_m]/(\mathfrak{p}B_0 + \mathfrak{q}A_0)$.
We have canonical k-algebra homomorphisms $A_0 \to C_0$ and $B_0 \to C_0$, and by
Exercise 4.1 (ii), there is induced a k-algebra homomorphism $A_0 \otimes B_0 \to C_0$.
It is obvious that it is surjective. We show that it is injective. Since the tensor
product functor is right exact (Lang [18], Proposition XVI.2.6 on p. 610),
tensoring the two exact sequences $0 \to \mathfrak{p} \to A_0 \to A \to 0$ and $0 \to \mathfrak{q} \to B_0 \to$
$B \to 0$ gives a commutative diagram with exact rows and columns:

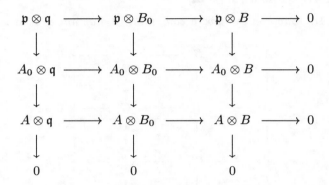

Now by Lemma 4.18, the kernel of the natural map $A_0 \times B_0 = C_0 \to A \otimes B$
is $\mathfrak{p}B_0 + \mathfrak{q}A_0$. Since $A \otimes B$ is an integral domain by Proposition 4.17, it follows
that this ideal is prime. Hence this ideal is $I(X \times Y)$. It is also immediate
that the coordinate ring of $X \times Y$ is isomorphic to $A \otimes B$.

It now follows from Exercise 4.1 (ii) and Theorem 3.4 that $X \times Y$ is a
product in the category of affine varieties and morphisms. The assertion about
dimensions follows from the final statement of Proposition 4.17. ∎

Proposition 4.20. *Let X and Y be affine varieties, and let $V \subseteq X$, $W \subseteq Y$
be closed subvarieties. Let A and B be the coordinate rings of X and Y be
respectively. According to* Theorem 4.19, *the coordinate ring of $X \times Y$ may be
identified with $A \otimes B$. Let $\mathfrak{p} \subseteq A$ and $\mathfrak{q} \subseteq B$ be the prime ideals corresponding
to V and W.*

(i) $V \times W$ is a closed subvariety of $X \times Y$, and $\mathfrak{p} \otimes B + A \otimes \mathfrak{q}$ is its prime ideal.

*(ii) $(V \times Y) \cup (X \times W)$ is a closed (reducible) subset of $X \times Y$. Its ideal is
$\mathfrak{p} \otimes \mathfrak{q}$.*

Proof. The proof of (i) is very similar to the proof of Theorem 4.19, and is
left to the reader.

As for (ii), it is clear that the zero set of $\mathfrak{p} \otimes \mathfrak{q}$ is precisely $(V \times Y) \cup (X \times W)$.
We will show that if $\sum_{i=1}^{r} f_i \otimes g_i \in A \otimes B$ vanishes on this set, then it is in
$\mathfrak{p} \otimes \mathfrak{q}$. We choose a representation of this element minimal length, which

implies that the f_i are linearly independent over k, as are the g_i. Since the f_i are linearly independent over k, there exist $x_1, \cdots, x_r \in X$ such that the $r \times r$ matrix $f_i(x_j)$ is nonsingular. (If this were not the case, there would be constants $\lambda_1, \cdots, \lambda_r \in k$ such that $\sum \lambda_i f_i$ vanished identically on X, contrary to the assumption that the f_i are linearly independent over k.) By assumption $\sum f_i(x_j) g_i$ vanish on W. It follows that $\sum f_i(x_j) g_i \in \mathfrak{q}$. Since the matrix $f_i(x_j)$ is nonsingular, it follows that each $g_i \in \mathfrak{q}$. Similarly, the f_i are in $\mathfrak{p}$, and so $\sum f_i \otimes g_i \in \mathfrak{p} \otimes \mathfrak{q}$. ∎

Theorem 4.21. *Let X and Y be closed subvarieties of $\mathbf{A}^n$ of dimensions r and s, respectively. Then the dimension of every irreducible component of $X \cap Y$ is $\geq r + s - n$.*

Proof. We embed $X \times Y$ in $\mathbf{A}^n \times \mathbf{A}^n$. Let Δ be the diagonal in $A^n(k) \times A^n(k)$. It is the intersection of n hyperplanes, and we may identify $X \cap Y$ with $(X \times Y) \cap \Delta$. Now each time we intersect a hyperplane with an algebraic set, the dimension of the smallest irreducible subset can drop by at most one, by Proposition 4.13. Since by Theorem 4.19 $\dim(X \times Y) = r + s$, every irreducible component of $X \cap Y$ must have dimension at least $r + s - n$. ∎

Exercises

In the following exercises, the unadorned symbol $\otimes$ always means $\otimes_k$.

Exercise 4.1. Let A and B be two k-algebras.

(i) Prove that there exists a bilinear mapping $m : (A \otimes B) \times (A \otimes B) \to A \otimes B$ such that $m(a \otimes b, a' \otimes b') = aa' \otimes bb'$. (Use the universal property of the tensor product.) Prove that this operation, interpreted as multiplication, makes $A \otimes B$ into a k-algebra, so that $(a \otimes b) \cdot (a' \otimes b') = aa' \otimes bb'$. This algebra contains copies of A and B under the isomorphisms $a \to a \otimes 1$ and $b \to 1 \otimes b$.

(ii) Prove that $A \otimes B$ is a *coproduct* in the category of k-algebras. This means that you must verify the following *universal property*. Let $i : A \to A \otimes B$ and $j : B \to A \otimes B$ be the canonical injections $i(a) = a \otimes 1$ and $j(b) = 1 \otimes b$. If $\phi : A \to C$ and $\psi : B \to C$ are any k-algebra homomorphisms, prove that there exists a unique k-algebra homomorphism $r : A \otimes B \to C$ such that $\phi = r \circ i$ and $\psi = r \circ j$.

Exercise 4.2. If $A = k[x_1, \cdots, x_n]$ and $B = k[y_1, \cdots, y_m]$ are polynomial rings, then so is

$$A \otimes B \cong k[x_1, \cdots, x_n, y_1, \cdots, y_m].$$

Exercise 4.3. Let X and Y be affine varieties, and let A and B be their respective coordinate rings. Let $\phi : X \to Y$ be a morphism. ϕ is called *finite* if the induced map $\phi^* : B \to A$ is injective, and if A is integral over the subring B. Assume that this is the case.

(i) Prove that ϕ is surjective.

(ii) If Z is a closed subset of X, prove that $\phi(Z)$ is a closed subset of Y. Thus a finite map is *closed*.

Hint: use the Going Up Theorem (Theorem 4.5) for parts (i) and (ii).

(iii) Prove that if $y \in Y$, then $\phi^{-1}(y)$ is a finite set.

Exercise 4.4. Let X be an affine variety. Let $\Delta = \{(x,x) | x \in X\} \subseteq X \times X$ be the diagonal. Prove that Δ is a closed subset of $X \times X$. It is well known that a topological space X is Hausdorff if and only if the diagonal is closed in $X \times X$ with the product topology. Yet if $\dim(X) > 0$ the Zariski topology on X is not Hausdorff. Why is this not a contradiction?

Exercise 4.5. Let k be an algebraically closed field, and let A and B be integral domains containing k. Prove that $A \otimes B$ is an integral domain.

Hint: If A and B are finitely generated over k, then this is Proposition 4.17. Deduce the general case by showing that if $\sum a_i \otimes b_i \in A \otimes B$ is a zero divisor, then there are finitely generated subalgebras A_0 and B_0 of A and B respectively, such that $a_i \in A_0$ and $b_i \in B_0$, and such that $\sum a_i \otimes b_i$ is a zero divisor in $A_0 \otimes B_0$.

5. Local Algebra

Let us begin with an observation. Let X be an affine variety, and $x \in X$. We consider the local ring $\mathcal{O}_x$. This ring contains all elements of the function field F of X which are regular in some neighborhood of x. Thus, it is particularly adapted to the study of the variety near x. On the other hand, one should not think that the ring $\mathcal{O}_x$ contains only local data. For example, F is the field of fractions of $\mathcal{O}_x$, and so the birational equivalence class of the variety is determined by the local ring $\mathcal{O}_x$. If X and Y are affine varieties, and if $x \in X$ and $y \in Y$, we expect local rings $\mathcal{O}_x$ and $\mathcal{O}_y$ to be isomorphic if and only if X and Y are birationally equivalent, and some neighborhood U of x is biregularly isomorphic to some neighborhood V of y. That is, there is a rational map ϕ from X to Y such that ϕ is regular on U and ϕ^{-1} is regular on V. Thus we may say that the local ring $\mathcal{O}_x$ contains a mixture of local and global data, but the global data is birational in nature, while the local data is more precise. Of course we could make a similar observation about the local ring $\mathcal{O}_{Z,X}$ along a closed subvariety Z of X.

Let us begin by considering the definition of the tangent space to a differentiable manifold M of dimension d at a point x. Let $\mathcal{O}_x$ be the ring of germs of complex valued differentiable functions at x. Thus, $\mathcal{O}_x$ is a local ring, whose maximal ideal $\mathfrak{m}_x$ consists of the germs of differentiable functions which vanish at x. We have $\mathcal{O}_x/\mathfrak{m}_x \cong \mathbb{C}$. This may be seen as follows. We have a ring homomorphism $\mathbb{C} \to \mathcal{O}_x \to \mathcal{O}_x/\mathfrak{m}_x$, where the first arrow is the inclusion of the constant functions. This is an isomorphism, and the inverse map $\mathcal{O}_x/\mathfrak{m}_x \to \mathbb{C}$ is induced by evaluation of functions (or germs) at x. If f is the germ of a differentiable function at x, we will denote its value at x by $f(x)$.

To construct a tangent vector, we start with a parametrized curve $\rho : (-\epsilon, \epsilon) \to M$ through x, so that $\rho(0) = x$. We wish to associate with this an element $d_x\rho$ of the tangent space $T_x(M)$, which is to be a d-dimensional vector space. We can *differentiate* a germ along the tangent vector $d_x\rho$. The derivative is

$$d_x\rho(f) := \frac{d}{dt}f\big(\rho(t)\big)|_{t=0}.$$

With this definition, $d_x\rho$ is a mapping from $\mathcal{O}_x \to \mathbb{C}$ which is a derivation in the sense that $d_x\rho(fg) = f(x)\, d_x\rho(g) + g(x)\, d_x\rho(f)$. This motivates us to define the *tangent space* $T_x(M)$ to be the vector space of all such derivations $\mathcal{O}_x \to \mathbb{C}$.

This definition works perfectly well in algebraic geometry. Let X be an affine variety, $x \in X$, let $\mathcal{O}_x$ be the local ring at x, and let $\mathfrak{m}_x$ be its maximal ideal. Evaluation at x is an isomorphism $\mathcal{O}_x/\mathfrak{m}_x \to k$ which is the inverse of the composite map $k \to \mathcal{O}_x \to \mathcal{O}_x/\mathfrak{m}_x$, where the first arrow is inclusion of the constant functions. By a *k-derivation of $\mathcal{O}_x$*, we mean a function $D : \mathcal{O}_x \to k$ which is k-linear, satisfies $D(a) = 0$ if $a \in k$ is a constant function, and which satisfies the identity

$$(5.1) \qquad\qquad D(fg) = f(x)\, D(g) + g(x)\, D(f).$$

We define $T_x(X)$ to be the vector space of all such derivations.

Proposition 5.1. *If $D : \mathcal{O}_x \to k$ is a k-derivation, then D is determined by its values on a set of generators on the maximal ideal $\mathfrak{m}_x$ of x. A k-derivation D is zero on $\mathfrak{m}_x^2$, and consequently gives rise to a k-linear functional on $\mathfrak{m}_x/\mathfrak{m}_x^2$. Conversely, any such linear functional gives rise to a k-derivation of $\mathcal{O}_x$. Consequently $T_x(X)$ may naturally be identified with the dual space of the k-vector space $\mathfrak{m}_x/\mathfrak{m}_x^2$.*

Proof. Suppose that $f_1, \cdots, f_d$ generate $\mathfrak{m}_x$. If $f \in \mathcal{O}_x$, and if $a = f(x)$, then we may write $f = a + \sum g_i f_i$ with $g_i \in \mathcal{O}_x$. Now $Df = \sum g_i(x)\, Df_i$ since $D(a) = 0$ and $f_i(x) = 0$. Thus D is determined by its values on $\mathfrak{m}_x$.

If $f, g \in \mathfrak{m}_x$ then since $f(x) = g(x) = 0$, the condition Eq. (5.1) implies that $D(fg) = 0$. Therefore D is trivial on $\mathfrak{m}_x^2$, and induces a linear form on $\mathfrak{m}_x/\mathfrak{m}_x^2$.

To see that any linear form arises this way, let D be any linear functional on $\mathfrak{m}_x$ which is trivial on $\mathfrak{m}_x^2$. We extend D to a linear form on $\mathcal{O}_x$ as follows. If $f \in \mathcal{O}_x$, write $f = f_0 + f_1$ where $f_0 \in k$ is the constant function taking value $f(x)$, and f_1 vanishes at x, so $f_1 \in \mathfrak{m}_x$. Now define $D(f) = D(f_1)$. To see that this is a derivation, let $g \in \mathcal{O}_x$ be similarly decomposed as $g_0 + g_1$. Then $fg = f_0 g_0 + f_0 g_1 + g_0 f_1 + f_1 g_1$. $D(f_0 g_0) = 0$ since $f_0 g_0$ is constant, while $D(f_1 g_1) = 0$ since $f_1 g_1 \in \mathfrak{m}_x^2$. Thus $D(fg) = f_0\, D(g_1) + g_0\, D(f_1) = f(x)\, D(g) + g(x)\, D(f)$. ∎

If $\phi : X \to Y$ is a morphism, and if $x \in X$, $y = \phi(x) \in Y$, then there is induced a map $\phi_* : T_x(X) \to T_y(Y)$. Indeed, recall from Chapter 3 that ϕ induces a map $\phi^* : B \to A$, where A and B are the respective coordinate rings of X and Y. If $\mathfrak{m}_x$ and $\mathfrak{m}_y$ are the maximal ideals of the local rings at x and y, then there is induced a map, also denoted $\phi^* : B_{\mathfrak{m}_y} = \mathcal{O}_y(Y) \to A_{\mathfrak{m}_x} = \mathcal{O}_x(X)$.

So if D is a k-derivation of $\mathcal{O}_x(X)$, we may compose it with ϕ^* to obtain a k-derivation of $\mathcal{O}_y(Y)$. Alternatively, we have $\phi^*(\mathfrak{m}_y) \subseteq \mathfrak{m}_x$, so ϕ^* induces a map $\mathfrak{m}_y/\mathfrak{m}_y^2 \to \mathfrak{m}_x/\mathfrak{m}_x^2$, hence a map of the dual spaces $T_x(X) \to T_y(Y)$.

We would like to know that $T_x(X)$ is finite-dimensional. By Proposition 5.1, it is sufficient to show that $\mathfrak{m}_x/\mathfrak{m}_x^2$ is finite dimensional. As it turns out, the k-vector spaces $\mathfrak{m}_x^n/\mathfrak{m}_x^{n+1}$ are important for every n, so we prove this for them, too. The first step is the following result:

Proposition 5.2. *Let A be a local ring with maximal ideal $\mathfrak{m}$. Let K be the field $A/\mathfrak{m}$. Then if $n \geq 0$, $\mathfrak{m}^n/\mathfrak{m}^{n+1}$ is naturally a K-vector space. If $\mathfrak{a}$ is an additive abelian subgroup of $\mathfrak{m}^n$ containing $\mathfrak{m}^{n+1}$ such that $\mathfrak{a}/\mathfrak{m}^{n+1}$ is a K-vector subspace of $\mathfrak{m}^n/\mathfrak{m}^{n+1}$, then $\mathfrak{a}$ is an ideal of A.*

Proof. Since the A-module $\mathfrak{m}^n/\mathfrak{m}^{n+1}$ is annihilated by $\mathfrak{m}$, it is canonically an $A/\mathfrak{m}$-module, which is to say a K-vector space. The assumption that $\mathfrak{a}/\mathfrak{m}^{n+1}$ is an $A/\mathfrak{m}$-vector subspace of $\mathfrak{m}^n/\mathfrak{m}^{n+1}$ amounts to $\mathfrak{a}/\mathfrak{m}^{n+1}$ being closed under multiplication by A, which means that $A\mathfrak{a} \subseteq \mathfrak{a} + \mathfrak{m}^{n+1}$. Since we are assuming that $\mathfrak{m}^{n+1} \subseteq \mathfrak{a}$, this implies that $A\mathfrak{a} \subseteq \mathfrak{a}$, and so $\mathfrak{a}$ is an ideal. ■

Proposition 5.3. *Let A be a Noetherian local ring with maximal ideal $\mathfrak{m}$, and let $K = A/\mathfrak{m}$. If $n \geq 0$, the K-vector space $\mathfrak{m}^n/\mathfrak{m}^{n+1}$ is finite-dimensional.*

Proof. If $\mathfrak{m}^n/\mathfrak{m}^{n+1}$ were not finite-dimensional, we could construct an infinite increasing chain of K-vector subspaces

$$\mathfrak{a}_1/\mathfrak{m}^{n+1} \subsetneq \mathfrak{a}_2/\mathfrak{m}^{n+1} \subsetneq \mathfrak{a}_3/\mathfrak{m}^{n+1} \subsetneq \cdots$$

between, and by Proposition 5.2, $\mathfrak{a}_1 \subsetneq \mathfrak{a}_2 \subsetneq \cdots$ would be an infinite ascending chain of ideals. This is a contradiction since A is a Noetherian ring. ■

Proposition 5.4. *Let X be an affine variety, and let $x \in X$. Let $\mathfrak{m}_x$ be the maximal ideal in the local ring $\mathcal{O}_x$. Then $\mathfrak{m}_x^n/\mathfrak{m}_x^{n+1}$ is finite-dimensional, and any k-vector subspace of $\mathfrak{m}_x^n$ which contains $\mathfrak{m}_x^{n+1}$ is an ideal. Furthermore the tangent space $T_x(X)$ is finite-dimensional.*

Proof. We apply Proposition 5.3 with $A = \mathcal{O}_x$, and $K = k$. We see that $\mathfrak{m}_x^n/\mathfrak{m}_x^{n+1}$ is finite-dimensional. The assertion that any k-vector subspace of $\mathfrak{m}_x^n$ which contains $\mathfrak{m}_x^{n+1}$ is an ideal follows from Proposition 5.2. Since by Proposition 5.1 $T_x(X) \cong (\mathfrak{m}_x/\mathfrak{m}_x^2)^*$, it too is finite-dimensional. ■

We would like to know that a k-basis for $\mathfrak{m}_x/\mathfrak{m}_x^2$ actually generates $\mathfrak{m}_x$ as an ideal. This is a consequence of

Proposition 5.5 (Nakayama's Lemma). *Let A be a local ring, and let $\mathfrak{m}$ be its maximal ideal. Let $K = A/\mathfrak{m}$.*

(i) Let M be a finitely generated A-module, and let N be a submodule such that $\mathfrak{m}M + N = M$. Then $M = N$.

(ii) Let M be a finitely generated A-module. Then $V = M/\mathfrak{m}M$ is naturally a vector space over the field K. If $x_1, \cdots, x_n \in M$ are such that the residue classes mod $\mathfrak{m}M$ are a vector space basis of V, then $x_1, \cdots, x_n$ generate M as an A-module.

Proof. Note that (ii) follows from (i) with $N = Ax_1 + \ldots + Ax_n$. We will prove part (i).

First we prove this with $N = 0$. Thus $\mathfrak{m}M = M$, and we wish to prove that $M = 0$. Let $u_1, \cdots, u_r$ be a minimal set of generators of M. Since $u_r \in M = \mathfrak{m}M$, we may write $u_r = \sum_{i=1}^{r} m_i u_i$ where $m_i \in \mathfrak{m}$. Now $(1 - m_r)u_r = \sum_{i=1}^{r-1} m_i u_i$. Here $1 - m_r$ is a unit of A, for if it were not, it would be in $\mathfrak{m}$, and hence so would $1 = (1 - m_r) + m_r$, which is a contradiction. Hence we may express u_r in terms of $u_1, \cdots, u_{r-1}$, contradicting the assumed minimality of the chosen set of generators.

The general case follows from the case $N = 0$, since $\mathfrak{m}(M/N) = (\mathfrak{m}M + N)/N$. If $\mathfrak{m}M + N = M$ we have $\mathfrak{m}(M/N) = M/N$, and so by the case just established $M/N = 0$, whence $M = N$. ∎

Before starting in earnest to study local rings, we would like to have some examples in mind.

Example 5.6. Let $X = \mathbf{A}^d$, and let $x = (0, \cdots, 0)$ be the origin. This is an affine variety of dimension d. The coordinate ring in this case is the polynomial ring $A = k[x_1, \cdots, x_d]$. The local ring $\mathcal{O}_x$ consists of all functions f/g with $f, g \in A$ such that $g(0) \neq 0$. Since A is a unique factorization domain, we may assume that the polynomials f and g are relatively prime. Then f/g is in the maximal ideal $\mathfrak{m}_x$ if and only if $f(0) = 0$. As a k-vector space, $\mathfrak{m}_x/\mathfrak{m}_x^2$ is generated by the coordinate functions $x_1, \cdots, x_d$, and it follows from Nakayama's Lemma that these also generate $\mathfrak{m}_x$ as an ideal. More generally, $\mathfrak{m}_x^n/\mathfrak{m}_x^{n+1}$ is generated as a vector space by the $\binom{d+n-1}{n}$ monomials of degree n, and so

$$\dim(\mathfrak{m}_x^n/\mathfrak{m}_x^{n+1}) = \binom{d+n-1}{n}.$$

These same $\binom{d+n-1}{n}$ monomials generate the ideal $\mathfrak{m}^n$. We have equivalently,

$$\dim(\mathcal{O}_x/\mathfrak{m}^n) = \sum_{i=0}^{n-1} \binom{d+i-1}{i} = \binom{d+n-1}{n-1}.$$

Note that $\dim(\mathcal{O}_x/\mathfrak{m}^n)$ is a polynomial of degree d in n. The tangent space $T_x(X) \cong (\mathfrak{m}_x/\mathfrak{m}_x^2)^*$ is d-dimensional.

The previous example is one in which we localized at a *smooth point*. A smooth point x on a variety X of dimension d is one where the tangent space $T_x(X)$ has dimension d. In Chapter 6, we will prove that the smooth points are dense. For varieties over the complex numbers, x is smooth if the variety looks like a manifold in a neighborhood of x. A *singular point* is one which is not smooth. Thus *smooth* and *nonsingular* are synonyms. We would also like to have in mind an example of a local ring at a singular point.

Example 5.7. Let X be the curve $\eta^2 = \xi^2(\xi + 1)$ in $\mathbf{A}^2$. This is an affine variety of dimension $d = 1$. The coordinate ring of X is $k[\xi, \eta]$, where the generators ξ and η satisfy one relation $\eta^2 = \xi^2(\xi + 1)$. It may be shown that this algebraic set in $\mathbf{A}^2$ is irreducible, i.e. an affine variety. The origin $x = (0,0)$ is a singular point. The maximal ideal $\mathfrak{m}_x$ of $\mathcal{O}_x$ has two generators ξ and η. These are also a basis for $\mathfrak{m}_x/\mathfrak{m}_x^2$ as a k-vector space. $\mathfrak{m}_x^2$ may be generated as an $\mathcal{O}_x$-module by ξ^2, $\xi\eta$ and η^2. However, the images of these in the k-vector space $\mathfrak{m}_x^2/\mathfrak{m}_x^3$ are not all distinct, because $\eta^2 \equiv \xi^2 \mod \mathfrak{m}_x^3$. Thus $\dim(\mathfrak{m}_x^2/\mathfrak{m}_x^3) = 2$. Similarly if $n \geq 1$, ξ^n and $\xi^{n-1}\eta$ form a basis of $\mathfrak{m}_x^n/\mathfrak{m}_x^{n+1}$. In this example,

$$\dim(\mathfrak{m}_x^n/\mathfrak{m}_x^{n+1}) = \begin{cases} 1 & \text{if } n = 0; \\ 2 & \text{if } n > 0. \end{cases}$$

Also,

$$\dim(\mathcal{O}_x/\mathfrak{m}^n) = \sum_{i=0}^{n-1} \dim(\mathfrak{m}^i/\mathfrak{m}^{i+1}) = 2n - 1.$$

Thus for n sufficiently large $\dim(\mathcal{O}_x/\mathfrak{m}^n)$ is a polynomial of degree $d = 1$ in n. Note that in this example, the number of generators required for $\mathfrak{m}_x$ is strictly greater than d. Thus in contradistinction to the previous example, the tangent space $T_x(X)$ has dimension $2 > d = 1$. This example shows that the dimension of the tangent space may be strictly greater than the dimension of the variety, and indeed this may be taken to be the definition of a singular point. Even though the number of generators needed for the ideal $\mathfrak{m}$ is strictly greater than d, nevertheless there exists an ideal $\mathfrak{q}$ having $\mathfrak{m}_x$ as its radical, and which has precisely $d = 1$ generators. For example, we may take $\mathfrak{q} = (\xi)$. We have then $\mathfrak{m}_x \supset \mathfrak{q} \supset \mathfrak{m}_x^2$, and $r(\mathfrak{q}) = \mathfrak{m}$ because $\mathfrak{m}_x = r(\mathfrak{m}_x) \supseteq r(\mathfrak{q}) \supseteq r(\mathfrak{m}_x^2) = \mathfrak{m}_x$.

Intuitively, the curve $y^2 = x^2(x+1)$ may be thought of having *two tangent lines with distinct tangents* at the singular point $(x,y) = (0,0)$. This can be seen by embedding the curve in $\mathbf{A}^2$, as in the figure at right. The tangent space, being a vector space (two-dimensional in this case), does not carry this type of detailed information about the tangent lines. But see Exercise 6.2, where it is shown that this variety admits a desingularization, that is, a proper map from a nonsingular variety, and that in the desingularization the singular point at $(0,0)$ splits into two points. This type of singularity is called a *node*.

Finally, we have an example in which we localize not at a point, but along a closed subvariety.

Example 5.8. Let $X = \mathbf{A}^2$, and let $A = k[x,y]$ be the coordinate ring, which is a polynomial ring in two variables. This time, let us not localize at a point, but along a closed subvariety. Let $p \in A$ be an irreducible polynomial, and let $\mathfrak{p} = (p)$ be the corresponding prime ideal. The curve $Y = V(\mathfrak{p})$ is thus a closed subvariety of X. The local ring $\mathcal{O}_{Y,X} = A_{\mathfrak{p}} = \{f/g \mid f, g \in A \text{ and } p \nmid g\}$. Its maximal ideal $\mathfrak{m} = \{f/g \mid f, g \in A, p \mid f \text{ and } p \nmid g\}$. The quotient $K = A_{\mathfrak{p}}/\mathfrak{m}$ is isomorphic to the function field of Y, i.e. the field of fractions of its coordinate ring $B = A/\mathfrak{p}$. We have now $\dim_K(\mathfrak{m}^n/\mathfrak{m}^{n+1}) = 1$, a K-basis being just the polynomial p^n. However, it no longer makes any sense to ask for $\dim_K(\mathcal{O}_{Y,X}/\mathfrak{m}^n)$, for $\mathcal{O}_{Y,X}/\mathfrak{m}^n$ is no longer a vector space over K. In the two previous examples, in order to regard $\mathcal{O}_x/\mathfrak{m}_x^n$ as a vector space over $\mathcal{O}_x/\mathfrak{m}_x \cong k$, we took advantage of the fact that a copy of k was embedded in $\mathcal{O}_x$, namely, the subfield of constant functions. In the example at hand, it is no longer the case that $\mathcal{O}_{Y,X}$ contains an isomorphic copy of K.

Because of this difficulty, we must sometimes replace the notion of dimension with the notion of *length*, which we will now recall. Let A be a ring, and M an A-module. Recall that a *simple* module is one which is nonzero, and has no proper nonzero submodules. A *composition series* for M is a filtration by submodules

$$0 = M_0 \subsetneq M_1 \subsetneq M_2 \subsetneq \cdots \subsetneq M_l = M$$

in which each quotient module M_i/M_{i-1} is simple. The *length* of the composition series is l. According to the well-known Jordan-Hölder Theorem, if a module has a composition series, then every composition series has the same length, and the isomorphism classes of the *composition factors* M_i/M_{i-1} are independent of the composition series. The Jordan-Hölder Theorem is proved in Lang [18] for groups in Section I.3, and the proof for modules is identical.

If M has a composition series, then its length l is called the *length* of the module M, and is denoted $l(M)$. If the module has no composition series, we take the length to be ∞. The length function is *additive* in the sense that if we have a short exact sequence $0 \to M' \to M \to M'' \to 0$, then $l(M) = l(M') + l(M'')$. To see this, note that splicing the image in M of a composition series of M' with the preimage in M of a composition series of M'' gives a composition series of M, proving this additivity property.

Proposition 5.9. *Let A be a local ring with maximal ideal $\mathfrak{m}$ and quotient field $K = A/\mathfrak{m}$, and let M be an A-module.*

(i) If $\mathfrak{m}M = 0$, then M is naturally a K-vector space, and $l(M) = \dim_K(M)$.

(ii) If $\mathfrak{m}^n M = 0$ for some M, then let $M_i = \{x \in M | \mathfrak{m}^i x = 0\}$. Then $0 = M_0 \subseteq M_i \subseteq \cdots \subseteq M_n = M$, and for $i = 1, \cdots, n$, the quotient $M_i' = M_i/M_{i-1}$ has the property that $\mathfrak{m}M_i' = 0$. Thus M_i' is a K-vector space, and $l(M_i') = \dim_K(M_i')$. So $l(M) = \sum_i \dim_K(M_i')$.

Proof. It is clear that if $\mathfrak{m}M = 0$ then M is naturally a K-vector space. Furthermore, M is simple if and only if $\dim_K(M) = 1$, because the submodules of M are simply the K-vector subspaces. Thus it is clear that $l(M) = \dim_K(M)$. Part (ii) follows from the additivity of the length function. ∎

We may now remedy the unsatisfying incompleteness of the last example.

Example 5.8 (continued). *To finish off Example 5.8, we see that even though $\mathcal{O}_{Y,X}/\mathfrak{m}^n$ is not a K-vector space, it is nevertheless an $\mathcal{O}_{Y,X}$-module of finite length, and*

$$l(\mathcal{O}_{Y,X}/\mathfrak{m}^n) = \sum_{i=0}^{n-1} \dim_K(\mathfrak{m}^i/\mathfrak{m}^{i+1}) = n.$$

This is a polynomial, whose degree is equal to $1 = \dim(X) - \dim(Y)$.

A *graded ring* is a ring G together with an infinite direct sum decomposition $G = \bigoplus_{n=0}^{\infty} G_n$ such that $G_n G_m \subseteq G_{n+m}$. The additive subgroup G_n is called the *homogeneous part of degree n* of G, and its elements are said to be *homogeneous of degree n*. We will assume at first that $G_0 = K$ is a field, and that as a K-algebra, G is finitely generated. We will call such a graded ring a *Noetherian graded K-algebra*. In this case each G_n is a finite dimensional K-vector space.

We will also consider *graded modules* over G. By a graded module, we mean a G-module M together with a direct sum decomposition $M = \bigoplus_{n=0}^{\infty} M_n$, where $G_n M_m \subseteq M_{n+m}$. We will assume that M is finitely generated as a G-module. This implies that M_n is a finite-dimensional vector space over $K = G_0$. The additive subgroup M_n is called the *homogeneous part of degree* n of M, and its elements are said to be *homogeneous of degree* n. For example, G is a graded module over itself.

Let A be a Noetherian local ring with maximal ideal $\mathfrak{m}$, and let $K = A/\mathfrak{m}$. It was observed by Krull that one may construct from this data a *graded ring* to be denoted $G_{\mathfrak{m}}(A)$. In the case at hand, we take the homogeneous part of degree n in $G_{\mathfrak{m}}(A)$ to be the finite-dimensional K-vector space $\mathfrak{m}^n/\mathfrak{m}^{n+1}$. Note that multiplication in A induces a bilinear mapping $\mathfrak{m}^n/\mathfrak{m}^{n+1} \times \mathfrak{m}^m/\mathfrak{m}^{m+1} \to \mathfrak{m}^{n+m}/\mathfrak{m}^{n+m+1}$, which we take to be the multiplication in $G_{\mathfrak{m}}(A)$.

We will prove by studying this graded ring that for sufficiently large n, $\dim(\mathfrak{m}^n/\mathfrak{m}^{n+1})$ is a polynomial function of n, a phenomenon which we saw in the preceding examples. We will work in somewhat more generality than we need for the particular problem at hand, because the methods we will consider have applications to other problems besides just this one.

Let G be a Noetherian graded K-algebra, and let M be a finitely generated graded module over G. We form the formal power series

$$P(M, t) = \sum_{n=0}^{\infty} \dim(M_n)\, t^n,$$

which is called the *Hilbert series* of M.

We recall that if $0 \to M_1 \to M_2 \to \dots M_n \to 0$ is an exact sequence of vector spaces over some field, then the alternating sum of the dimensions,

$$\sum_{i=1}^{n} (-1)^n \dim(M_n) = 0.$$

To prove this, let M_i' be the kernel of the map $M_i \to M_{i+1}$. (We let $M_{n+1}' = 0$.) Then by exactness, M_i' is also the image of the map $M_{i-1} \to M_i$. Thus we have a series of short exact sequences

$$0 \to M_i' \to M_i \to M_{i+1}' \to 0.$$

Thus $\dim(M_i) = \dim(M_i') + \dim(M_{i+1}')$. Since $M_1' = M_{n+1}' = 0$, summing from $i = 1$ to $i = n$ with alternating sign gives zero, as required. More generally, if the M's are modules of finite length over a ring A, then we have

$$\sum_{i=1}^{n} (-1)^n l(M_n) = 0,$$

and the proof is the same.

Proposition 5.10. *Let M be a finitely generated graded module over a finitely-generated Noetherian graded K-algebra G. Suppose that G is generated by elements $x_1, \cdots, x_r$, where x_i is homogeneous of degree k_i. Then the series $\prod_{i=1}^{r}(1 - t^{k_i}) P(M, t)$ is a polynomial in t.*

Proof is by induction on r. Multiplication by x_r is a linear transformation of vector spaces $M_n \to M_{n+k_r}$. Let M_n' be the kernel, and M_{n+k_r}'' be the cokernel. Thus we have an exact sequence

$$0 \to M_n' \to M_n \xrightarrow{x_r} M_{n+k_r} \to M_{n+k_r}'' \to 0.$$

Consequently,

$$\dim(M_n') - \dim(M_n) + \dim(M_{n+k_r}) - \dim(M_{n+k_r}'') = 0.$$

We may express this in terms of the Hilbert series by the formula

$$t^{k_r} P(M', t) - t^{k_r} P(M, t) + P(M, t) - P(M'', t) = 0.$$

Thus

$$\prod_{i=1}^{r}(1 - t^{k_i}) P(M, t) = \prod_{i=1}^{r-1}(1 - t^{k_i}) [P(M'', t) - t^{k_r} P(M', t)].$$

Now, M' and M'' are graded sub- and quotient modules respectively of the finitely generated module M, and hence are finitely generated as modules over $G = K[x_1, \cdots, x_r]$ by Exercise 1.3. But x_r annihilates both graded modules M' and M'', and so these are finitely generated as modules over the graded subalgebra of G generated by $x_1, \cdots, x_{r-1}$. By induction on r, each factor on the right of the last equation is a polynomial. ■

Proposition 5.11. *Let M be a finitely generated graded module over a Noetherian graded K-algebra G. Suppose that G is generated by the homogeneous elements of degree one. Then there exists a polynomial $\lambda(n)$ of degree $d \le \dim(G_1) - 1$ such that for sufficiently large n, $\dim(M_n) = \lambda(n)$. The leading coefficient of $\lambda(n)$, i.e. the coefficient of n^d, equals $1/d!$ times an integer.*

Proof. We apply Proposition 5.10. Since G is generated by the homogeneous elements of degree one, $r = \dim(G_1)$, and each $k_i = 1$. We see that

$$P(M, t) = f(t) (1 - t)^{-r} = f(t) \sum_{i=0}^{\infty} \binom{r + i - 1}{i} t^i,$$

where $f(t) = \sum_{i=0}^{m} a_i t^i$ is a polynomial. Now if $n \ge m$, we have

$$\dim(M_n) = \sum_{i=0}^{m} a_i \binom{r + n - i - 1}{n - i}.$$

Each term is a polynomial of degree $\le r - 1$.

The fact that the leading coefficient equals $1/d!$ times an integer follows from Exercise 5.6. ■

Proposition 5.12. *Let A be a Noetherian local ring with maximal ideal $\mathfrak{m}$, and let $K = A/\mathfrak{m}$. Suppose that the dimension of the K-vector space $\mathfrak{m}/\mathfrak{m}^2$ is r. Then there exists a polynomial $\chi_{\mathfrak{m}}(n)$ of degree $d \leq r$ such that for sufficiently large n, $l(A/\mathfrak{m}^n) = \chi_{\mathfrak{m}}(n)$. The leading coefficient of the χ_m (i.e. the coefficient of n^d) is $1/d!$ times an integer.*

Proof. Denote $l(A/\mathfrak{m}^n) = \mu(n)$. We may apply Proposition 5.11 to the graded ring $G = G_{\mathfrak{m}}(A)$, which is a module over itself, and is evidently generated by the homogeneous elements of degree one. We see that $\mu(n + 1) - \mu(n) = \dim_K(\mathfrak{m}^n/\mathfrak{m}^{n+1})$ is a polynomial of degree $\leq r - 1$ for sufficiently large n. From this it follows that $\mu(n)$ is a polynomial of degree $\leq r$ for sufficiently large n. Since it is integer valued, the leading coefficient is $1/d!$ times an integer, by Exercise 5.6. ∎

We recall the notion of a *primary* ideal. A *primary ideal* of a ring A is an ideal $\mathfrak{q}$ with the property that if $xy \in \mathfrak{q}$ then either $x \in \mathfrak{q}$ or $y^n \in \mathfrak{q}$ for some n. In this case, the radical $\mathfrak{p} = r(\mathfrak{q})$ is prime (Exercise 5.3), and indeed $\mathfrak{p}$ is the unique smallest prime ideal containing $\mathfrak{q}$. Thus we also say that $\mathfrak{q}$ is $\mathfrak{p}$-*primary*. There are examples of ideals $\mathfrak{q}$ such that $r(\mathfrak{q})$ is prime, but $\mathfrak{q}$ is not primary (Exercise 5.5). However, if $r(\mathfrak{q})$ is maximal, then $\mathfrak{q}$ is always primary (Exercise 5.4). If A is a Noetherian local ring with maximal ideal $\mathfrak{m}$, then $\mathfrak{q}$ is $\mathfrak{m}$-primary if and only if $\mathfrak{m} \supseteq \mathfrak{q} \supseteq \mathfrak{m}^n$ for some n.

Example 5.7 shows however that the degree of $\chi_{\mathfrak{m}}(n)$ need not exactly equal the dimension of $\mathfrak{m}/\mathfrak{m}^2$ in this Proposition. In that example, the degree of $\chi_{\mathfrak{m}}(n)$ was one, while $\mathfrak{m}$ required two generators. Nevertheless, we found in that example that there was another $\mathfrak{m}$-primary ideal $\mathfrak{q}$, which only required one generator. We wish therefore to investigate more generally the dimensions of $A/\mathfrak{q}^n$, where $\mathfrak{q}$ is any $\mathfrak{m}$-primary ideal. We may prove the following generalization of Proposition 5.12.

Proposition 5.13. *Let A be a Noetherian local ring with maximal ideal $\mathfrak{m}$, and let $\mathfrak{q}$ be an $\mathfrak{m}$-primary ideal. Suppose that $\mathfrak{q}$ may be generated as an ideal by r elements. Then there exists a polynomial $\chi_{\mathfrak{q}}(n)$ of degree $d \leq r$ such that for sufficiently large n, $l(A/\mathfrak{q}^n) = \chi_{\mathfrak{q}}(n)$. The leading coefficient, i.e. the coefficient of n^d, equals $1/d!$ times an integer.*

Proof. We will explain what modifications must be made to the proof of Proposition 5.12 to obtain this result. As before, we may form a graded ring $G_{\mathfrak{q}}(A)$ whose homogeneous part of degree n is $\mathfrak{q}^n/\mathfrak{q}^{n+1}$. Now, however, G_0 is not a field, but $A/\mathfrak{q}$. We may no longer use the dimension of $\mathfrak{q}^n/\mathfrak{q}^{n+1}$, which is indeed no longer a vector space over $A/\mathfrak{m}$, but consider instead the length $l(\mathfrak{q}^n/\mathfrak{q}^{n+1})$. This may be interpreted as the length either as A-modules or as $A/\mathfrak{q}$-modules.

If G is a graded algebra, but G_0 is no longer a field, we can still associate a Hilbert series with a graded G-module M, provided each homogeneous part M_n is of finite length. We define

$$P(M,t) = \sum_{n=0}^{\infty} l(M_n)\, t^n,$$

and Proposition 5.10 is valid. Applying this as in Proposition 5.11, if G is generated by the homogeneous elements of degree one, then $l(M_n)$ is a polynomial in n for sufficiently large n, and the degree of this polynomial is no greater than $r-1$, where r is the cardinality of a set of generators of G in G_1. Now as in Proposition 5.12, We may apply this with $G = M = G_{\mathfrak{q}}(A)$, and see that $l(\mathfrak{q}^n/\mathfrak{q}^{n+1})$ is a polynomial in n, and that the degree of this polynomial is $\leq r-1$, where r is the number of elements required to generate $\mathfrak{q}$. It follows from this that $l(A/\mathfrak{q}^n)$ is a polynomial in n when n is sufficiently large, of degree no greater than r. ∎

Proposition 5.14. *Let A be a Noetherian local ring with maximal ideal $\mathfrak{m}$, and let $\mathfrak{q}$ be an $\mathfrak{m}$-primary ideal. Then the degrees of the polynomials $\chi_{\mathfrak{m}}$ and $\chi_{\mathfrak{q}}$ are equal.*

Proof. Since $\mathfrak{q}$ is finitely-generated and $\mathfrak{m}$-primary, there exists an m such that $\mathfrak{m} \supseteq \mathfrak{q} \supseteq \mathfrak{m}^m$. Then for all n, we have $\mathfrak{m}^n \supseteq \mathfrak{q}^n \supseteq \mathfrak{m}^{nm}$. Thus for sufficiently large n, we have $\chi_{\mathfrak{m}}(n) \geq \chi_{\mathfrak{q}}(n) \geq \chi_{\mathfrak{m}}(nm)$. This implies that the degrees of the polynomials χ_m and $\chi_{\mathfrak{q}}$ are equal. ∎

The most important theorem in dimension theory is the theorem of Krull asserting the equality of three notions of dimension for Noetherian local rings. We will prove this below in Theorem 5.18.

The first of the three notions of dimension is the Krull dimension $\dim(A)$ which is defined for any ring to be the maximum possible length d for a chain of prime ideals $\mathfrak{p}_0 \subsetneq \mathfrak{p}_1 \subsetneq \mathfrak{p}_2 \subsetneq \cdots \subsetneq \mathfrak{p}_d$. If there is no maximum possible length, we let $\dim(A) = \infty$. This can happen, even for Noetherian rings. However, we will prove that for a Noetherian local ring, $\dim(A)$ is finite. In this definition, if A is an integral domain, then the maximality of such a chain obviously implies that $\mathfrak{p}_0 = 0$, and if A is a local ring, then $\mathfrak{p}_d$ must be the unique maximal ideal of A.

. For example, if A is the coordinate ring of an affine variety X, by Theorem 4.11, $\dim(A) = \dim(X)$. Also by Theorem 4.11, if $x \in X$, then $\dim(\mathcal{O}_x) = \dim(X)$, and more generally if $Y \subseteq X$ is any closed subvariety, then $\dim(\mathcal{O}_{Y,X}) = \dim(X) - \dim(Y)$.

The second notion of dimension for a Noetherian local ring A with maximal ideal $\mathfrak{m}$ defines $\delta(A)$ to be the minimal number of generators required for an $\mathfrak{m}$-primary ideal.

The third notion of dimension for a Noetherian local ring A defines $d(A)$ to be the degree of $\chi_{\mathfrak{m}}$. By Proposition 5.14, this is equal to the degree of $\chi_{\mathfrak{q}}$ for any $\mathfrak{m}$-primary ideal $\mathfrak{q}$.

Proposition 5.15 (the Artin-Rees Lemma). *Let A be a Noetherian ring, and let $\mathfrak{a}$ be an ideal of A. Let M be a finitely generated A-module, and let N be a submodule of M. Then there exists an r such that if $n \geq r$, then $\mathfrak{a}^n M \cap N = \mathfrak{a}^{n-r}(\mathfrak{a}^r M \cap N)$.*

Proof. It is clear that $\mathfrak{a}^{n-r}(\mathfrak{a}^r M \cap N) \subseteq \mathfrak{a}^n M \cap N$. We must prove that $\mathfrak{a}^n M \cap N \subseteq \mathfrak{a}^{n-r}(\mathfrak{a}^r M \cap N)$, for r appropriately chosen.

Let R be the polynomial ring $A[x]$. Let R' be the subring consisting of polynomials $\sum_{i=0}^{n} a_i x^i$ where $a_i \in \mathfrak{a}^i$. We will show that R' is Noetherian. Since A is Noetherian, the ideal $\mathfrak{a}$ of A is finitely generated, and if $a_1, \cdots, a_d$ is a set of generators, it is clear that R' is the A-subalgebra of R generated by $a_1 x, \cdots, a_d x$. Thus R' is a finitely generated algebra over a Noetherian ring A, and so by Exercise 1.4, R' is Noetherian.

Now we may construct an $A[x]$-module $M[x]$ from M as follows. The elements of $M[x]$ will be polynomials $\sum m_i x^i$ with the $m_i \in M$, and the multiplication will be the obvious one. Let M' be the submodule of $M[x]$ consisting of elements $\sum m_i x^i$ where $m_i \in \mathfrak{a}^i M$. Then M' is an R'-submodule of $M[x]$. It is finitely generated as an R'-module, because $M' = R'M$, so that any finite set of elements which generate M as an A-module will clearly also generate M' as an R'-module.

Let N' be the R'-submodule consisting of elements of the form $\sum m_i x^i$, where $m_i \in \mathfrak{a}^i M \cap N$. Since R' is Noetherian, this submodule of a finitely generated module is also finitely generated, by Exercise 1.3. Suppose that $\mu_1 x^{r_1}, \cdots, \mu_k x^{r_k}$ is a set of generators, where $\mu_i \in \mathfrak{a}^{r_i} M \cap N$. Let $r = \max(r_i)$.

We may now prove that $\mathfrak{a}^n M \cap N \subseteq \mathfrak{a}^{n-r}(\mathfrak{a}^r M \cap N)$. Indeed, let $\mu \in \mathfrak{a}^n M \cap N$. Then $\mu x^n \in N'$, so we may write $\mu x^n = \sum \xi_i \mu_i x^{r_i}$ where $\xi_i \in R'$. Comparing degrees, we see that $\xi_i = a_i x^{n-r_i}$, where $a_i \in \mathfrak{a}^{n-r_i}$. Then $a_i \mu_i \in \mathfrak{a}^{n-r_i}(\mathfrak{a}^{r_i} M \cap N) \subseteq \mathfrak{a}^{n-r}(\mathfrak{a}^r M \cap N)$, and so $\mu = \sum_i a_i \mu_i \in \mathfrak{a}^{n-r}(\mathfrak{a}^r M \cap N)$ also. ∎

Proposition 5.16. *Let A be a Noetherian local ring with maximal ideal $\mathfrak{m}$. Then $d(A) \geq \dim(A)$. In particular, $\dim(A)$ is finite.*

Proof. We prove this by induction on $d(A)$. If $d(A) = 0$, then $l(A/\mathfrak{m}^n)$ is eventually constant. This means that $\mathfrak{m}^n = \mathfrak{m}^{n+1}$ for sufficiently large n. By Nakayama's Lemma, this implies that $\mathfrak{m}^n = 0$. Hence $\mathfrak{m} = r((0))$ is the set of nilpotent elements in A, i.e. the radical of the zero ideal. It follows from Exercise 1.5 (ii) and the fact that $\mathfrak{m}$ is maximal that $\mathfrak{m}$ is the unique prime ideal of A, so $\dim(A) = 0$.

Assume by induction that $d(A) > 0$, and that the theorem is true for all Noetherian local rings B with $d(B) < d(A)$. Let

$$\mathfrak{p}_0 \subsetneq \mathfrak{p}_1 \subsetneq \cdots \subsetneq \mathfrak{p}_d$$

be a chain of prime ideals in A. We will prove that $d(A) \geq d$. This will prove that $d(A) \geq \dim(A)$, as required.

Note that we may replace A by $A/\mathfrak{p}_0$. This certainly does not increase $d(A)$, so if $d(A/\mathfrak{p}_0) \geq d$ then $d(A) \geq d$ also. We may therefore assume that $\mathfrak{p}_0 = 0$, and that A is an integral domain.

Let $0 \neq a \in \mathfrak{p}_1$, and let $\mathfrak{a} = (a)$. Let $\overline{A} = A/\mathfrak{a}$, let $\overline{\mathfrak{m}} = \mathfrak{m}/\mathfrak{a}$, and let $\overline{\mathfrak{p}}_i = \mathfrak{p}_i/\mathfrak{a}$. Then $\overline{\mathfrak{p}}_1 \subsetneq \cdots \subsetneq \overline{\mathfrak{p}}_d = \overline{\mathfrak{m}}$ is a chain of prime ideals in $\overline{A}$ of length $d - 1$. We will prove that $d(A) > d(\overline{A})$. Note that since by induction $d(\overline{A}) \geq \dim(\overline{A})$, this implies that $d(A) - 1 \geq d(A') \geq d - 1$, so $d(A) > d$, as required.

We have $\overline{A}/\overline{\mathfrak{m}}^n \cong A/(\mathfrak{a} + \mathfrak{m}^n)$. Indeed, it is clear that the kernel of the canonical surjection $\overline{A} \to A/(\mathfrak{a} + \mathfrak{m}^n)$ has kernel $\overline{\mathfrak{m}}^n$, which implies the stated isomorphism. On the other hand, we have a short exact sequence $0 \to (\mathfrak{a} + \mathfrak{m}^n)/\mathfrak{m}^n \to A/\mathfrak{m}^n \to A/(\mathfrak{a} + \mathfrak{m}^n) \to 0$, and a canonical isomorphism $(\mathfrak{a} + \mathfrak{m}^n)/\mathfrak{m}^n \cong \mathfrak{a}/(\mathfrak{a} \cap \mathfrak{m}^n)$. Thus

$$l(\overline{A}/\overline{\mathfrak{m}}^n) = l(A/\mathfrak{m}^n) - l(\mathfrak{a}/(\mathfrak{a} \cap \mathfrak{m}^n)).$$

By the Artin-Rees Lemma, there exists an r such that if $n \geq r$, we have $\mathfrak{a} \cap \mathfrak{m}^n = \mathfrak{m}^{n-r}(\mathfrak{a} \cap \mathfrak{m}^r)$.

The length of $\overline{A}/\overline{\mathfrak{m}}^n$ as an A-module is clearly equal to its length as an $\overline{A}$-module, and so we have, for sufficiently large n, that

$$\chi_{\overline{\mathfrak{m}}}(n) = \chi_{\mathfrak{m}}(n) - l(\mathfrak{a}/\mathfrak{m}^{n-r}(\mathfrak{a} \cap \mathfrak{m}^r)).$$

This proves that for sufficiently large n, $f(n) = l(\mathfrak{a}/\mathfrak{m}^{n-r}(\mathfrak{a} \cap \mathfrak{m}^r))$ is a polynomial function of n. What we must show is that its degree, and its leading coefficient, are the same as those of $\chi_{\mathfrak{m}}(n)$. For if this is the case, the leading terms in the last expression will cancel, so that the degree $d(\overline{A})$ of $\chi_{\overline{\mathfrak{m}}}(n)$ is strictly less than the degree $d(A)$ of $\chi_{\mathfrak{m}}(n)$.

Note that $\mathfrak{a}\mathfrak{m}^r \subseteq \mathfrak{a} \cap \mathfrak{m}^r \subseteq \mathfrak{a}$, and consequently,

$$l(\mathfrak{a}/\mathfrak{m}^{n-r}\mathfrak{a}) \leq l(\mathfrak{a}/\mathfrak{m}^{n-r}(\mathfrak{a} \cap \mathfrak{m}^r)) \leq l(\mathfrak{a}/\mathfrak{a}\mathfrak{m}^n).$$

Recall that $\mathfrak{a} = (a)$ is principal. Thus multiplication by a induces an isomorphism $A/\mathfrak{m}^n \to \mathfrak{a}/\mathfrak{a}\mathfrak{m}^n$. We may thus write

$$\chi_{\mathfrak{m}}(n - r) \leq f(n) \leq \chi_{\mathfrak{m}}(n).$$

This implies that the polynomial which is equal to $f(n)$ for sufficiently large n has the same degree and leading coefficient as $\chi_{\mathfrak{m}}(n)$, as required. ■

Proposition 5.17 (Primary Decomposition). *Let A be a Noetherian ring, and let $\mathfrak{a}$ be any ideal.*

(i) $\mathfrak{a}$ is a finite intersection $\bigcap \mathfrak{q}_i$ of primary ideals.

(ii) If $\mathfrak{a} = \bigcap \mathfrak{q}_i$ expresses $\mathfrak{a}$ as a finite intersection of primary ideals, then any prime ideal which is minimal among those which contain $\mathfrak{a}$ equals $r(\mathfrak{q}_i)$ for some i.

Proof. We define an *irreducible ideal* to be an ideal $\mathfrak{q}$ with the property that if $\mathfrak{b}$ and $\mathfrak{c}$ are ideals of A such that $\mathfrak{q} = \mathfrak{b} \cap \mathfrak{c}$, then either $\mathfrak{q} = \mathfrak{b}$ or $\mathfrak{q} = \mathfrak{c}$. Let us show that every ideal in a Noetherian ring is a finite intersection of irreducible ideals. For if this were not the case, since A is Noetherian, by Exercise 1.1 there would be an ideal $\mathfrak{a}$ which is maximal among the set of counterexamples. Since $\mathfrak{a}$ is not irreducible, $\mathfrak{a} = \mathfrak{b} \cap \mathfrak{c}$, where $\mathfrak{b}$ and $\mathfrak{c}$ are strictly larger. Then by the maximality of $\mathfrak{a}$, $\mathfrak{b}$ and $\mathfrak{c}$ both can be expressed as finite intersections of irreducible ideals, hence so can $\mathfrak{a}$, which is a contradiction.

Now let us show that if $\mathfrak{q}$ is irreducible and A is Noetherian, then $\mathfrak{q}$ is primary. Let $\overline{A}$ be the quotient ring $A/\mathfrak{q}$. Then $\mathfrak{q}$ is irreducible in A if and only if (0) is irreducible in $\overline{A}$, and $\mathfrak{q}$ is primary if and only if (0) is primary in $\overline{A}$. Thus we may assume that $\mathfrak{q}$ is the zero ideal. Suppose that $xy = 0$, and that $y \neq 0$. We must show that $x^n = 0$ for some n. Let $\mathfrak{a}_n = \{z \in A | x^n z = 0\}$. Then $\mathfrak{a}_1 \subseteq \mathfrak{a}_2 \subseteq \cdots$, and since A is Noetherian, for some n we have $\mathfrak{a}_n = \mathfrak{a}_{n+1}$. Now we claim that $x^n = 0$. Since (0) is irreducible and $(y) \neq 0$, it is sufficient to show that $(y) \cap (x^n) = 0$. Indeed, suppose that $a \in (y) \cap (x^n)$. Then $a = yz = wx^n$ for some z, w. Now $wx^{n+1} = xyz = 0$, so $w \in \mathfrak{a}_{n+1}$. Since $\mathfrak{a}_{n+1} = \mathfrak{a}_n$, this implies that $wx^n = 0$. Hence $a = 0$. This proves that $(y) \cap (x^n) = 0$.

We now have proved (i), since we see that $\mathfrak{a}$ is a finite intersection of irreducible ideals, and that the irreducible ideals are primary. To prove (ii), suppose that $\mathfrak{a} = \bigcap_i \mathfrak{q}_i$ expresses $\mathfrak{a}$ as a finite intersection of primary ideals. Let $\mathfrak{p}_i = r(\mathfrak{q}_i)$. We will show that every prime ideal $\mathfrak{p}$ which is minimal among those containing $\mathfrak{a}$ is one of the $\mathfrak{p}_i$. First let us show that $\mathfrak{p} \supseteq \mathfrak{q}_i$ for some $\mathfrak{q}_i$. For if not, let $x_i \in \mathfrak{q}_i - \mathfrak{p}$, and consider $x = \prod x_i$. This is an element of each $\mathfrak{q}_i$, hence is in $\mathfrak{q}$. But since $\mathfrak{p}$ is prime, and since none of the factors x_i is in $\mathfrak{p}$, $x \notin \mathfrak{p}$, which is a contradiction.

Now $\mathfrak{p}_i$ is the unique smallest prime containing $\mathfrak{q}_i$. So if $\mathfrak{p} \supseteq \mathfrak{q}_i$, then $\mathfrak{p} \supseteq \mathfrak{p}_i \supseteq \mathfrak{a}$. Since $\mathfrak{p}$ is minimal among the primes of A containing $\mathfrak{a}$, this proves that $\mathfrak{p} = \mathfrak{p}_i$. ∎

We may now prove the dimension theorem of Krull.

Theorem 5.18 (the Krull Dimension Theorem). *Let A be a Noetherian local ring. Then $d(A) = \delta(A) = \dim(A)$.*

Proof. We have established that $\delta(A) \geq d(A)$ by Proposition 5.13 and Proposition 5.14. Moreover $d(A) \geq \dim(A)$, by Proposition 5.16. We must therefore prove that $\dim(A) \geq \delta(A)$. Let $\dim(A) = d$.

We will construct $x_1, \cdots, x_r \in \mathfrak{m}$ such that the only prime ideal containing $x_1, \cdots, x_r$ is $\mathfrak{m}$ itself. Let us assume that $x_1, \cdots, x_i$ have been found with the property that *every prime ideal containing* $(x_1, \cdots, x_i)$ *has height* $\geq i$. If $\mathfrak{m}$ is the only prime ideal containing $(x_1, \cdots, x_i)$, then we take $r = i$ and terminate the process. Note that since $\mathfrak{m}$ is the unique prime ideal of height d, this must occur before $i > d$. Therefore, $r \leq d$.

On the other hand, if $\mathfrak{m}$ is not the only prime ideal of A containing $(x_1, \cdots, x_i)$, we construct x_{i+1} as follows. Let $\mathfrak{p}_1, \cdots, \mathfrak{p}_n$ be the minimal prime ideals containing $(x_1, \cdots, x_i)$. There are only finitely many such prime ideals by Proposition 5.17 (ii). Clearly $\mathfrak{m}$ is not among the $\mathfrak{p}_i$. By Proposition 4.14, there exists an element x_{i+1} which is in $\mathfrak{m}$ but not in any of the $\mathfrak{p}_k$. Now let $\mathfrak{p}$ be any prime ideal containing $(x_1, \cdots, x_{i+1})$. We will show that the height of $\mathfrak{p}$ is at least $i + 1$. Indeed, since the $\mathfrak{p}_i$ are minimal prime ideals containing $(x_1, \cdots, x_i)$, $\mathfrak{p}$ must contain one of the $\mathfrak{p}_i$. By construction, $\mathfrak{p} \neq \mathfrak{p}_i$. Now by induction, the height of $\mathfrak{p}_i$ is $\geq i$, and so the height of $\mathfrak{p}$ is $\geq i + 1$. This proves that we may carry out the induction as described.

When the induction terminates, we have an ideal $\mathfrak{q} = (x_1, \cdots, x_r) \subseteq \mathfrak{m}$ such that the only prime ideal of A containing $\mathfrak{q}$ is $\mathfrak{m}$. By Exercise 1.5 (ii), this implies that $\mathfrak{m} = r(\mathfrak{q})$. Thus by Exercise 5.4, $\mathfrak{q}$ is $\mathfrak{m}$-primary. Now by definition of $\delta(A)$, we have $\delta(A) \leq r \leq d = \dim(A)$. This completes the proof of Theorem 5.18. ∎

Proposition 5.19. *Let A be a Noetherian local ring, let $\mathfrak{m}$ be its maximal ideal, and let $K = A/\mathfrak{m}$. Then $\dim_K(\mathfrak{m}/\mathfrak{m}^2) \geq \dim(A)$.*

Proof. Let $\dim_K(\mathfrak{m}/\mathfrak{m}^2) = r$. It follows from Nakayama's Lemma (Proposition 5.5 (ii)) that the $\mathfrak{m}$-primary ideal $\mathfrak{m}$ may be generated by r elements. Consequently $\delta(A) \leq r$. ∎

If $\dim(\mathfrak{m}/\mathfrak{m}^2) = \dim(A)$, the Noetherian local ring A is called a *regular local ring*. If X is an affine variety, and $x \in X$, x is called a *smooth point* or *nonsingular point* if $\mathcal{O}_x$ is a regular local ring. We see that $\dim T_x(X) \geq \dim(X)$, with equality if and only if x is a smooth point. A point which is not smooth is called a *singularity* or *singular point*. We call a variety *smooth* or *nonsingular* if it has no singularities.

Let $G = \bigoplus G_n$ be a graded ring. An ideal $\mathfrak{a}$ of G is called *homogeneous* if $\mathfrak{a} = \bigoplus \mathfrak{a}_n$, where $\mathfrak{a}_n = \mathfrak{a} \cap G_n$. It is clear that if $\phi : G \to G'$ is a homomorphism of graded rings, so that $\phi(G_n) \subseteq G'_n$, then $\ker(\phi)$ is a homogeneous ideal.

Proposition 5.20. *Let A be a Noetherian local ring with maximal ideal $\mathfrak{m}$, and residue field $K = A/\mathfrak{m}$. Then A is regular if and only if the graded ring $G_{\mathfrak{m}}(A)$ is isomorphic to a polynomial ring $K[t_1, \cdots, t_d]$, graded by degree. If this is the case, then $d = \dim(A)$.*

Proof. Suppose that $G_{\mathfrak{m}}(A)$ is isomorphic as a graded ring to $K[t_1, \cdots, t_d]$. This implies that $\dim_K(\mathfrak{m}^n/\mathfrak{m}^{n+1})$ is equal to the dimension of the space of homogeneous polynomials of degree n in $t_1, \cdots, t_d$, i.e. $\dim(\mathfrak{m}^n/\mathfrak{m}^{n+1}) = \binom{d+n-1}{n}$, and so

$$l(A/\mathfrak{m}^n) = \sum_{i=0}^{n-1} \binom{d+i-1}{i} = \binom{d+n-1}{n-1}.$$

This is a polynomial in n of degree d, and so $\dim(A) = d(A) = d$. Since $\dim(\mathfrak{m}/\mathfrak{m}^2) = d$, this implies that A is a regular local ring.

    Conversely, assume that A is a regular local ring. Let $d = \dim(A)$, and suppose that $x_1, \cdots, x_d \in \mathfrak{m}$ such that the images $\overline{x}_1, \cdots, \overline{x}_d$ generate $\mathfrak{m}/\mathfrak{m}^2$ as a vector space over K. By Nakayama's Lemma (Proposition 5.5), $x_1, \cdots, x_d$ generate $\mathfrak{m}$ as an ideal. Consequently the homogeneous monomials of degree n in $x_1, \cdots, x_d$ generate $\mathfrak{m}^n$ as an ideal, and so the homogeneous monomials of degree n in $\overline{x}_1, \cdots, \overline{x}_d$ generate $\mathfrak{m}^n/\mathfrak{m}^{n+1}$ as a vector space. We define a homomorphism $\phi : K[t_1, \cdots, t_d] \to G_{\mathfrak{m}}(A)$ by $\phi(f(t_1, \cdots, t_d)) = f(\overline{x}_1, \cdots, \overline{x}_d)$. We have shown ϕ to be surjective. To prove that it is injective, we may argue as follows. Suppose that the kernel is a nontrivial ideal $\mathfrak{a}$, which is homogeneous. Suppose $g \in \mathfrak{a}$ is a nonzero homogeneous element of degree m. We will write $K[t_1, \cdots, t_d] = \oplus H_n$, where H_n is the homogeneous part of degree n. Then $\dim H_n = \binom{d+n-1}{n}$. Now if $n > m$, the kernel of $\phi|_{H_n}$ contains gH_{n-m}, which has dimension $\binom{d+n-m-1}{n-m}$. Consequently,

$$\dim(\mathfrak{m}^n/\mathfrak{m}^{n+1}) \leq \binom{d+n-1}{n} - \binom{d+n-m-1}{n-m}.$$

This is a polynomial of degree $d - 2$ which, for sufficiently large n, equals $\chi_{\mathfrak{m}}(n+1) - \chi_{\mathfrak{m}}(n)$. But $\chi_{\mathfrak{m}}(n)$ is a polynomial of degree d in n, so $\chi_{\mathfrak{m}}(n+1) - \chi_{\mathfrak{m}}(n)$ is a polynomial of degree $d-1$, which is a contradiction, since we have shown it to be bounded above by a polynomial of degree $d - 2$. ∎

Proposition 5.21. *Let F be a field, and let A be a subring. Let $x \in F$, and let M be a finitely generated nonzero A-submodule of F. If $xM \subseteq M$, then x is integral over A.*

Proof. Let R be any valuation ring of F containing A. By Exercise 2.5 (i), it is sufficient to show that $x \in R$. Let $x_1, \cdots, x_n$ generate M. By Proposition 4.16, there exists $\lambda \in F$ such that $\lambda x_1, \cdots, \lambda x_n \in R$, but not all λx_i lie in the maximal ideal $\mathfrak{m}$ of R. Thus replacing M by λM, we may assume that $M \subseteq R$ but $M \not\subseteq \mathfrak{m}$. Now if $x \notin R$, then $x^{-1} \in \mathfrak{m}$, and $M \subseteq x^{-1}M$ implies that $M \subseteq \mathfrak{m}$, which is a contradiction. ∎

Proposition 5.22. *Let A be an Noetherian integral domain, and let x be an element of its field of fractions. Then x is integral over A if and only if there exists $c \in A$ such that $cx^r \in A$ for all $r > 0$.*

Proof. First suppose that x is integral over A. Write $x = y/z$ where $y, z \in A$. We have a relation of integral dependence $x^n + a_{n-1}x^{n-1} + \ldots + a_0 = 0$ over A. This implies that $x^n \in A + Ax + Ax^2 + \ldots + Ax^{n-1}$. Now by induction we see that $x^r \in A + Ax + Ax^2 + \ldots + Ax^{n-1}$ for all r. Consequently, if $c = z^{n-1}$, we have $cx^r \in A$ for all $r > 0$.

On the other hand, suppose that $cx^r \in A$ for all $r > 0$. Then $A[x] \subseteq c^{-1}A$. Since $c^{-1}A$ is a finitely generated module over the Noetherian ring A, so is $A[x]$ by the Hilbert Basis Theorem (Exercise 1.4). Now it follows that x is integral over A by Proposition 5.21 with $M = A[x]$. ∎

Proposition 5.23. *Let A be a Noetherian local ring with maximal ideal $\mathfrak{m}$, and let $\mathfrak{a} \subseteq \mathfrak{m}$ be a proper ideal of A. Then $\bigcap_{n=0}^{\infty}(\mathfrak{a} + \mathfrak{m}^n) = \mathfrak{a}$.*

Proof. If $\mathfrak{a}$ is the zero ideal, this is Exercise 5.7. The general case follows from this special case by the following argument. Let $\overline{A} = A/\mathfrak{a}$. This is also a Noetherian local ring. Let $\overline{\mathfrak{m}}$ be the image of $\mathfrak{m}$ in $\overline{A}$. Then by Exercise 5.7, we have $\bigcap_{n=0}^{\infty} \overline{\mathfrak{m}}^n = 0$ in $\overline{A}$. Now the preimage of $\bigcap_{n=0}^{\infty} \overline{\mathfrak{m}}^n$ in A is $\bigcap_{n=0}^{\infty}(\mathfrak{a} + \mathfrak{m}^n)$, so this equals $\mathfrak{a}$. ∎

We now return to the graded ring $G_{\mathfrak{m}}(A)$ associated with a Noetherian local ring A with maximal ideal $\mathfrak{m}$. By Exercise 5.7, $\bigcap_{n=0}^{\infty} \mathfrak{m}^n = 0$. This implies that if $0 \neq a \in A$, there is an integer k such that $a \in \mathfrak{m}^k$ but $a \notin \mathfrak{m}^{k+1}$. We will denote $k = v(a)$. If $a = 0$, we will denote $v(a) = \infty$. It is our convention that $\infty + n = n + \infty = \infty$ for $n \in \mathbb{Z} \cup \{\infty\}$, and that $\infty > n$ for all $n \in \mathbb{Z}$. It is obvious that $v(x + y) \geq \min\bigl(v(x), v(y)\bigr)$.

Also, if $a \in A$ let $G(a)$ be the element of $G_{\mathfrak{m}}(A)$ which is defined as follows: let $k = v(a)$, and define $G(a)$ to be the image of a in $\mathfrak{m}^k/\mathfrak{m}^{k+1}$, which is homogeneous of degree k in $G(a)$.

Proposition 5.24. *Let A be a Noetherian local ring with maximal ideal* $\mathfrak{m}$.

(i) The graded ring $G_\mathfrak{m}(A)$ is Noetherian.

(ii) If $G_\mathfrak{m}(A)$ is an integral domain, then so is A. In this case, if $x, y \in A$, then $v(xy) = v(x) + v(y)$, and $G(xy) = G(x)\, G(y)$.

(iii) If $G_\mathfrak{m}(A)$ is an integrally closed integral domain, then so is A.

Proof. To see that $G_\mathfrak{m}(A)$ is Noetherian, let $x_1, \cdots, x_d$ generate $\mathfrak{m}$ as an ideal. Let $\overline{x}_i$ be the image of x_i in $\mathfrak{m}/\mathfrak{m}^2$. Then it is clear that the $\overline{x}_i$ generate $G_\mathfrak{m}(A)$ as an algebra over the field $A/\mathfrak{m}$, which is the homogeneous part of degree zero. Hence $G_\mathfrak{m}(A)$ is Noetherian by the Hilbert Basis Theorem (Exercise 1.4). This proves (i).

Suppose that $G_\mathfrak{m}(A)$ is an integral domain. Let $x, y \in A$, and suppose that neither x nor y is 0. As we have pointed out, since $\bigcap_{n=1}^{\infty} \mathfrak{m}^n = 0$ by Exercise 5.7, there exist integers $n = v(x)$ and $m = v(y)$ such that $x \in \mathfrak{m}^n$ but $x \notin \mathfrak{m}^{n+1}$, and $y \in \mathfrak{m}^m$ but $y \notin \mathfrak{m}^{m+1}$. Thus the images $G(x)$ and $G(y)$ of x and y in $\mathfrak{m}^n/\mathfrak{m}^{n+1}$ and $\mathfrak{m}^m/\mathfrak{m}^{m+1}$ are nonzero. Since $G_\mathfrak{m}(A)$ is an integral domain, this implies that the image $G(xy)$ of xy in $\mathfrak{m}^{n+m}/\mathfrak{m}^{n+m+1}$ is nonzero, so that $xy \notin \mathfrak{m}^{n+m+1}$. In particular, $xy \neq 0$, and $v(xy) = k + l$. This proves that A is an integral domain, and that $v(xy) = v(x) + v(y)$. It is obvious that this implies $G(xy) = G(x)\, G(y)$. This proves (ii).

Now suppose that $G_\mathfrak{m}(A)$ is an integrally closed integral domain. By part (ii) which we have already established, A is an integral domain. Since A and $G_\mathfrak{m}(A)$ are both Noetherian integral domains, we may use the criterion of Proposition 5.22 to test integral closure. Thus let x lie in the field of fractions of A. Suppose that x is integral over A. Then there exists $c \in A$ such that $cx^n \in A$ for all n. We must prove that $x \in A$.

Let $x = y/z$ where $y, z \in A$. We will prove that *if there exists $\mu \in A$ such that $v\big(z(x - \mu)\big) = k$, then $y \in (z) + \mathfrak{m}^k$, and moreover, there exists a $\mu' \in A$ such that $v\big(z(x - \mu')\big) > k$.* Of course if we can prove this, we may start the induction with $k = v(y)$ and $\mu = 0$. We will then have $y \in (z) + \mathfrak{m}^k$ for all k, which by Proposition 5.23 implies that $y \in (z)$, so that $x \in A$.

Suppose that $v\big(z(x - \mu)\big) = k$. since $z(x - \mu) = y - z\mu$, this implies that $y \in (z) + \mathfrak{m}^k$. We must show that we can find μ' so that $v\big(z(x - \mu')\big) > k$.

Since $cx^n \in A$ for all n, expanding $c(x - \mu)^n$ by the Binomial Theorem shows that $c(x - \mu)^n \in A$ for all n, and so $z^n|c(y - \mu z)^n$ for all n. Since $G(ab) = G(a)\, G(b)$, we have $G(z)^n|G(c)\, G(y - \mu z)^n$ in $G_\mathfrak{m}(A)$ for all n. Since $G_\mathfrak{m}(A)$ is an integrally closed integral domain, Proposition 5.22 implies that $G(y - \mu z)/G(z) \in G_\mathfrak{m}(A)$, so $G(z)|G(y - \mu z)$. Thus there exists $w \in A$ such that $G(y - \mu z) = G(w)\, G(z) = G(wz)$. Since $v(y - \mu z) = k$, this means that $wz \in \mathfrak{m}^k$, and so $y - \mu z - wz \in \mathfrak{m}^{k+1}$. Thus we may take $\mu' = \mu + w$. ∎

Theorem 5.25. *Let A be a regular local ring. Then A is an integrally closed integral domain.*

Proof. Since $G_\mathfrak{m}(A)$ is a polynomial ring by Proposition 5.20, it is a Noetherian integrally closed domain by Proposition 4.9. The Theorem thus follows from Proposition 5.24. ∎

The converse to this is not true by Exercise 5.8. Nevertheless, there is a converse for local rings of dimension 1. This is a particularly important special case.

Proposition 5.26. *Let A be a Noetherian local domain of dimension 1. If A is integrally closed, then it is a regular local ring.*

Proof. Let $\mathfrak{m}$ be the maximal ideal of A. We must show that $\mathfrak{m}/\mathfrak{m}^2$ is one dimensional over $A/\mathfrak{m}$. It is enough to show that $\mathfrak{m}$ is principal, since any generator of $\mathfrak{m}$ as an ideal clearly generates $\mathfrak{m}/\mathfrak{m}^2$ as a vector space. By Nakayama's Lemma (Proposition 5.5), $\mathfrak{m} \neq \mathfrak{m}^2$. Let $x \in \mathfrak{m} - \mathfrak{m}^2$. We will prove that $\mathfrak{m} = (x)$.

Since $\dim(A) = 1$, the only prime ideals in A are 0 and $\mathfrak{m}$. Thus $r((x)) = \mathfrak{m}$, and so (x) is an $\mathfrak{m}$-primary ideal. It follows that $\mathfrak{m}^n \subseteq (x)$ for some n. We will assume that n is minimal, so that $\mathfrak{m}^{n-1} \not\subseteq (x)$. We must show that $n = 1$.

We may choose $y \in \mathfrak{m}^{n-1}$ such that $y \notin (x)$. Let $z = y/x$. We will show that $z\mathfrak{m} \not\subseteq \mathfrak{m}$. Indeed, if $z\mathfrak{m} \subseteq \mathfrak{m}$, then by Proposition 5.21, z would be integral over A, and since A is integrally closed, we would have $z \in A$, which is a contradiction since $y \notin (x)$. On the other hand, $z\mathfrak{m} \subseteq A$ because $y\mathfrak{m} \subseteq \mathfrak{m}^n \subseteq (x)$. Thus $z\mathfrak{m}$ is an A-submodule of A, i.e. an ideal, which is not contained in $\mathfrak{m}$, and hence $z\mathfrak{m} = A$. Thus $y\mathfrak{m} = (x)$. Since $x \notin \mathfrak{m}^2$, this implies that $y \notin \mathfrak{m}$, and hence $n = 1$. This proves that $\mathfrak{m} = (x)$ is principal, and that A is a regular local ring. ∎

Let F be a field. By a *discrete valuation* on F, we mean a surjective function $v : F \to \mathbb{Z} \cup \{\infty\}$ with the properties that $v(xy) = v(x) + v(y)$; such that $v(x) = \infty$ if and only if $x = 0$; and such that $v(x + y) \geq \min(v(x), v(y))$. In this case, it is clear that $A = \{x \in F | v(x) \geq 0\}$ is a discrete valuation ring of F, with maximal ideal $\mathfrak{m} = \{x \in F | v(x) \geq 1\}$. A ring is called a *discrete valuation ring* if it arises in this way from a discrete valuation on its field of fractions.

The archetypal discrete valuation, which one should bear in mind, is the following example. Let Ω be a connected domain in $\mathbb{C}$, and let F be the field of meromorphic functions on Ω. Let $a \in \Omega$. Define $v_a(f)$ to be the order of vanishing of $f \in F$ at a, i.e. the unique integer such that $(x - a)^{-v(f)} f(x)$ is a nonzero holomorphic function near a. The corresponding discrete valuation ring is the ring of functions meromorphic on Ω which are holomorphic at a.

Thus one should think of the discrete valuation as the *order of vanishing* of a function.

Proposition 5.27. *Let $\mathfrak{o}$ be a discrete valuation ring, and let v be the associated valuation of its field F of fractions. Let $\mu \in \mathfrak{o}$ be such that $v(\mu) = 1$. Then the maximal ideal $\mathfrak{m}$ of $\mathfrak{o}$ is the principal ideal generated by μ. In fact, every ideal of $\mathfrak{o}$ is principal, and every nonzero ideal equals $\mathfrak{m}^n$ for some n. Indeed, if $\mathfrak{a}$ is any ideal of $\mathfrak{o}$, and if n is the smallest integer such that $v(x) = n$ for some $x \in \mathfrak{a}$, then $\mathfrak{a} = \mathfrak{m}^n$. Conversely, if $0 \neq x \in \mathfrak{o}$, then $v(x)$ is the largest integer n such that x is in the ideal $\mathfrak{m}^n$.*

Such a μ always exists, because by definition the valuation $v : F^\times \to \mathbb{Z}$ is surjective. Because $v(f/g) = v(f) - v(g)$, the valuation is determined by its values on $\mathfrak{o}$. Thus this proposition shows that if two discrete valuations give rise to the same valuation ring, they are equal. It also shows that a discrete valuation ring is a principal ideal domain.

Proof. If $g \in \mathfrak{m}$, let $n = v(g) \geq 1$. Then $v(\mu^{-1}g) \geq 0$, so $\mu^{-1}g \in \mathfrak{o}$. Therefore $g \in (\mu)$. This proves that $\mathfrak{m} \subseteq (\mu)$. Since $\mu \in \mathfrak{m}$, we have $\mathfrak{m} \subseteq (\mu)$, so $\mathfrak{m} = \mu$. A similar argument shows if $\mathfrak{a}$ is any nonzero ideal of $\mathfrak{o}$, and if n is the smallest integer such that $v(x) = n$ for some $x \in \mathfrak{a}$, then $\mathfrak{a} = (\mu^n) = \mathfrak{m}^n$. The final assertion is similarly clear. ∎

Proposition 5.28. *A ring is a discrete valuation ring if and only if it is a regular local ring of dimension 1.*

Proof. Let A be a discrete valuation ring. It is a principal ideal domain by Proposition 5.27. Hence every ideal is finitely generated, so it is a Noetherian local ring. If m is an element with $v(m) = 1$, then m generates the maximal ideal $\mathfrak{m} = \{x \in A | v(x) > 0\}$. Thus $\delta(A) = 1$. A generator of $\mathfrak{m}$ also generates $\mathfrak{m}/\mathfrak{m}^2$ as a vector space over $A/\mathfrak{m}$, and so dim $\mathfrak{m}/\mathfrak{m}^2 = 1$. This proves that A is regular.

On the other hand, suppose that A is a regular local ring of dimension 1. We may define a function $v : A \to \mathbb{Z} \cup \{\infty\}$ as in the discussion preceding Proposition 5.24, and it follows from that Proposition that $v(xy) = v(x)+v(y)$. The property $v(x + y) \geq \min(v(x), v(y))$ is clear.

By Theorem 5.25, A is an integral domain. To show that it is a discrete valuation ring, we must show that v may be extended to the field of fractions F of A in such a way that the properties $v(xy) = v(x) + v(y)$ and $v(x + y) \geq \min(v(x), v(y))$ remain true. If $0 \neq a \in F$, write $a = b/c$ with $b, c \in A$. Define $v(a) = v(b) - v(c)$. It follows from the fact that $v(xy) = v(x)+v(y)$ for $x, y \in A$ that this is well defined, and that this property remains true for $x, y \in F$. We must show that the property $v(x + y) \geq \min(v(x), v(y))$, which is obvious for

$x, y \in A$, remains true for $x, y \in F$. Find $w \in A$ such that $wx, wy \in A$. We have $v(x + y) = v(wx + wy) - v(w) \geq \min\big(v(wx), v(wy)\big) - v(w) = \min\big(v(wx) - v(w), v(wy) - v(w)\big) = \min\big(v(x), v(y)\big)$, as required. ∎

We will state without proof two further results about regular local rings. Although these theorems are very important, we will not use them. The proofs use homological algebra, whose use in commutative ring theory was developed independently by Serre and by Auslander and Buchsbaum, leading to the solution of several unsolved problems.

Theorem 5.29 (Auslander and Buchsbaum). *A regular local ring is a unique factorization ring.*

See Auslander and Buchsbaum, [**4**], Serre, [**24**], IV-39, Balcerzyk and Józefiak, Theorem 2.2.5 on p. 84, Eisenbud [**8**], Theorem 19.19 on p. 483, or Zariski and Samuel [**36**], Appendix 7 to Volume II.

Theorem 5.30 (Serre). *Let A be a regular local ring, and let $\mathfrak{p}$ be a prime ideal of A. Then $A_{\mathfrak{p}}$ is also a regular local ring.*

See Serre, [**24**], IV-41 or Balcerzyk and Józefiak [**5**], Corollary 2.1.6 on p. 81.

Although we do not discuss the intersection theory, the material in this Chapter is very closely connected with this topic in the form put forth by Samuel [**22**].

Exercises

Exercise 5.1. Let X be an affine variety, and Y a closed subvariety. Prove that the Krull dimension of the local ring $\mathcal{O}_{Y,X}$ is $\dim(X) - \dim(Y)$.

Exercise 5.2. Let $\mathfrak{m}$ be the ideal (p) in $\mathbb{Z}$, where p is a prime number. Prove that the associated graded ring $G_{\mathfrak{m}}(\mathbb{Z})$ is isomorphic to the polynomial ring in one variable over the field with p elements.

A *primary ideal* of a ring A is an ideal $\mathfrak{q}$ with the property that if $xy \in \mathfrak{q}$ then either $x \in \mathfrak{q}$ or $y^n \in \mathfrak{q}$ for some n.

Exercise 5.3. Let $\mathfrak{q}$ be a primary ideal, and let $\mathfrak{p} = r(\mathfrak{q})$. Prove that $\mathfrak{p}$ is prime. Prove that if $\mathfrak{p}'$ is any other prime containing $\mathfrak{q}$, then $\mathfrak{p} \subseteq \mathfrak{p}'$. Thus $\mathfrak{p}$ is the unique smallest prime ideal containing $\mathfrak{q}$. For this reason we say that $\mathfrak{q}$ is $\mathfrak{p}$-*primary*.

Exercise 5.4. Prove that if $\mathfrak{q}$ is any ideal of A such that $r(\mathfrak{q}) = \mathfrak{m}$ is *maximal*, then $\mathfrak{q}$ is primary. If A is Noetherian, prove that this is the case if and only if $\mathfrak{m} \supseteq \mathfrak{q} \supseteq \mathfrak{m}^n$ for some n.

Hint: Show that $A/\mathfrak{q}$ has a unique prime ideal, so every element is either a unit, or nilpotent.

Exercise 5.5. Let A be the coordinate ring of the quadric surface $x^2 = yz$ in $\mathbf{A}^3$. Prove that the ideal $\mathfrak{p} = (x, z)$ is prime, and that the ideal $\mathfrak{a} = \mathfrak{p}^2$ satisfies $r(\mathfrak{a}) = \mathfrak{p}$, but $\mathfrak{a}$ is not primary.

By a *numerical polynomial* we mean a polynomial $f(X)$ (with complex coefficients) such that $f(n) \in \mathbb{Z}$ for all sufficiently large n.

Exercise 5.6. Let $f(X)$ be a numerical polynomial of degree d. Prove that $f(n)$ is an integer for all n, and that the leading coefficient of f is of the form $e/d!$, where e is an integer.

Hint: Note that the binomial coefficient $\binom{X}{r}$ is a polynomial of degree r. Thus f may be written

$$f(X) = \sum_{k=0}^{d} a_k \binom{X}{k}.$$

Argue by induction on d that $a_k \in \mathbb{Z}$, noting that

$$f(X+1) - f(X) = \sum_{k=1}^{d} a_k \binom{X}{k-1}.$$

Now $e = a_d$.

Exercise 5.7. Prove the following theorem of Krull. Let A be a Noetherian local ring with maximal ideal m. Then $\bigcap_{n=0}^{\infty} \mathfrak{m}^n = 0$.

Hint: Let $M = \bigcap_{n=0}^{\infty} \mathfrak{m}^n$. Use the Artin-Rees Lemma to prove that $\mathfrak{m}M = M$. Now apply Nakayama's Lemma to conclude that $M = 0$.

Exercise 5.8. Let X be the quadratic cone in $\mathbf{A}^3$, defined by the equations $x^2 + y^2 + z^2 = 0$. Let A be the local ring of X at the origin. Prove that although A is not a regular local ring, nevertheless A is integrally closed.

Exercise 5.9. Let A be the local ring of the curve $y^2 = x^3$ at the origin, and let m be its maximal ideal. Show that A is an integral domain, but that $G_\mathfrak{m}(A)$ is not, and in fact has nonzero nilpotent elements.

Exercise 5.10. Let p be a prime number. Define a function $v : \mathbb{Q} \to \mathbb{Z} \cup \infty$ as follows. If $0 \neq x \in \mathbb{Q}$, we may write $x = p^k a/b$, where $a, b \in \mathbb{Z}$, and $p \nmid a, b$. Then we define $v(x) = k$. On the other hand we define $v(0) = \infty$. Prove that v is a discrete valuation.

Exercise 5.11. Let v be a discrete valuation. Show that if $v(x) \neq v(y)$ then $v(x+y) = \min(v(x), v(y))$.

Exercise 5.12. Let A be a Noetherian integral domain, and let $0 \neq f \in A$. Deduce from the Krull Dimension Theorem (Theorem 5.18) that if $\mathfrak{p}$ is a prime ideal of A which is minimal among those containing f, then $\mathfrak{p}$ is a minimal prime ideal of A. Thus obtain another proof of Theorem 4.12.

Hint: Show that (f) is primary in $A_\mathfrak{p}$, and hence $\dim(A_\mathfrak{p}) = 1$. Note that Theorem 4.12, and its consequences Proposition 4.15 and Proposition 4.13, have not been used in this chapter. So there is no circularity in this alternative proof.

6. Properties of Affine Varieties

We begin with a discussion of the fibers of a morphism. This chapter is strongly influenced by Mumford [21].

Let $\phi : X \to Y$ be a morphism of affine varieties. Let A and B be the respective coordinate rings of X and Y. Recall that ϕ is called *dominant* if $\phi(X)$ is dense in Y. It is easy to see that this is equivalent to the induced map $\phi^* : B \to A$ being injective. Therefore $\dim(X) \geq \dim(Y)$. We are interested in knowing what we can say about the *fibers* of a dominant map. Thus if $y \in Y$, the *fiber* over y is $\phi^{-1}(y)$, which is a closed subset of Y. As such, it may be decomposed into a finite union of irreducible components, $Y_1 \cup \cdots \cup Y_n$. In general, we expect that $\dim(Y_i) = \dim(X) - \dim(Y)$. However, this may not be true without qualification.

Example 6.1. Let $X = Y = \mathbf{A}^2$. Let $\phi : X \to Y$ be the map $\phi(\xi, \eta) = (\xi, \xi\eta)$. Clearly if $y = (\xi, \eta)$, and if $\xi \neq 0$, the fiber over y consists of a single point, as it should. If $\xi = 0$ and $\eta \neq 0$, then the fiber is empty. This is still consistent with the hope that every irreducible component of $\phi^{-1}(y)$ has dimension $\dim(X) - \dim(Y)$, since if the fiber is empty there are no irreducible components, i.e. $n = 0$ in the decomposition $\phi^{-1}(y) = \bigcup Y_i$. Alas, however, if $\xi = \eta = 0$, then $\phi^{-1}(y)$ is one-dimensional.

Even though this example shows that the fibers of a mapping may have dimension strictly greater than $\dim(X) - \dim(Y)$, at least we have an *inequality*. To prove this, we introduce the *graph* of a morphism. We associate with $\phi : X \to Y$ the closed subvariety $\Gamma_\phi \subseteq X \times Y$, where $\Gamma_\phi = \{(x,y)|y = \phi(x)\}$. It is clearly a closed set.

Proposition 6.2. *The graph Γ_ϕ of a morphism $\phi : X \to Y$ is a closed affine subvariety of $X \times Y$ isomorphic to X.*

Proof. Indeed, it is easy to give morphisms $X \to \Gamma_\phi$ and $\Gamma_\phi \to X$ which are inverses of each other, namely $x \to (x, \phi(x))$, and $(x,y) \to x$. ∎

Proposition 6.3. *Let $\phi : X \to Y$ be a dominant morphism of affine varieties, and let $W \subseteq Y$ be a closed subvariety of dimension r. Let Z be any component of $\phi^{-1}(W)$ such that $\phi(Z)$ is dense in W. Then $\dim(Z) \geq \dim(W) + \dim(X) - \dim(Y)$.*

Proof. Let d be the codimension of W in Y. By Proposition 4.15, there exist $f_1, \cdots, f_d$ in the coordinate ring $\mathcal{O}(Y)$ such that W is a component in the variety $V((f_1, \cdots, f_d))$. We claim that Z is a component in $V((f_1\phi, \cdots, f_d\phi))$. Indeed, since Z is a closed subset of $V((f_1\phi, \cdots, f_d\phi))$, we have $Z \subseteq Z_1 \subseteq V((f_1\phi, \cdots, f_d\phi))$ for some component Z_1 of $V((f_1\phi, \cdots, f_d\phi))$. Now $\phi(Z) \subseteq \phi(Z_1) \subseteq V((f_1, \cdots, f_d))$ and so $W = \overline{\phi(Z)} \subseteq \overline{\phi(Z_1)} \subseteq V((f_1, \cdots, f_d))$. Since W is a component of $V((f_1, \cdots, f_d))$, and since $\overline{\phi(Z_1)}$ is irreducible by Exercise 1.7, this implies that $W = \overline{\phi(Z_1)}$. Therefore $Z \subseteq Z_1 \subseteq \phi^{-1}(W)$, and since Z is a component of $\phi^{-1}(W)$, this implies that $Z_1 = Z$. This proves that Z is a component of $V((f_1\phi, \cdots, f_d\phi))$.

It follows that the codimension of Z in X is at most d by Proposition 4.13. ■

If we take $W = \{y\}$ to be a single point in Proposition 6.3, we see that the dimension of the components of the fiber are $\geq \dim(X) - \dim(Y)$, and Example 6.1 shows that this may be a strict inequality. We would like to say however that we *usually* have equality. The simplest case is when ϕ is a finite mapping. Recall from Exercise 4.3 that a dominant morphism $\phi : X \to Y$ is called *finite* if A is integral over the subring B, where A and B are the coordinate rings of X and Y respectively, and where we use ϕ^* to identify B as a subring of A.

Proposition 6.4. *Let $\phi : X \to Y$ be a finite dominant morphism of affine varieties. Then ϕ is surjective. The fibers of ϕ are all finite. If Z is a closed subset of X, then $\phi(Z)$ is closed, $\dim(Z) = \dim(\phi(Z))$, and the restriction of ϕ to Z is a finite dominant map onto $\phi(Z)$. If W is a closed subvariety of Y, and Z is any irreducible component of $\phi^{-1}(W)$, then $\phi(Z) = W$, and $\dim(Z) = \dim(W)$.*

Proof. Let us begin by showing that if Z is a closed subset of X, then $\phi(Z)$ is closed. Let A and B be the respective coordinate rings of X and Y, so A is integral over its subring B. Let $\mathfrak{p}$ be the prime ideal of A consisting of functions which vanish on Z. Let $y \in \overline{\phi(Z)}$. We will show that $y \in \phi(Z)$. Let $\mathfrak{m}$ be the maximal ideal of B consisting of all functions which vanish at y. We will first establish that $B \cap \mathfrak{p} \subseteq \mathfrak{m}$. Indeed, let $f \in B \cap \mathfrak{p}$. Since we are identifying $f \in B$ with $\phi^* f \in A$, this means that $\phi^* f \in \mathfrak{p}$, so that $\phi^* f$ vanishes on Z, which is the same as saying that f vanishes on $\phi(Z)$. Since

$y \in \overline{\phi(Z)}$, this implies that $f(y) = 0$, so $f \in \mathfrak{m}$. Therefore $B \cap \mathfrak{p} \subseteq \mathfrak{m}$. Now the "Going Up" Theorem (Theorem 4.5), there exists a prime ideal $\mathfrak{M}$ of A such that $\mathfrak{M} \supseteq \mathfrak{p}$, and $\mathfrak{M} \cap B = \mathfrak{m}$. By the Nullstellensatz, there exists an $x \in X$ such that $f(x) = 0$ for all $f \in \mathfrak{M}$. Since $\mathfrak{p} \subseteq \mathfrak{M}$, this implies that $x \in Z$. Now $\phi(x) = y$, because any function vanishing at y is in $\mathfrak{m}$, hence its image under ϕ^* is in $\mathfrak{M}$. Therefore $y \in \phi(Z)$. This shows that $\phi(Z)$ is closed.

We apply this in particular to $Z = X$. Since ϕ is dominant, $\phi(X)$ is dense in Y, and since it is closed, $\phi(X) = Y$. Thus ϕ is surjective.

Returning to general Z, the coordinate rings of Z and $f(Z)$ are $A/\mathfrak{p}$ and $B/(\mathfrak{p} \cap B)$, respectively. The map of coordinate rings induced by restriction $\phi|_Z : Z \to \phi(Z)$ is the canonical map $B/(\mathfrak{p} \cap B) \to A/\mathfrak{p}$ induced by inclusion. Clearly $A/\mathfrak{p}$ is integral over $B/(\mathfrak{p} \cap B)$ since A is integral over B, so $\phi|_Z : Z \to \phi(Z)$ is a finite dominant map. Because $A/\mathfrak{p}$ is integral over $B/(\mathfrak{p} \cap B)$, their fields of fractions have the same transcendence degree over k, and hence $\dim(Z) = \dim(\phi(Z))$.

Now let W be a closed subvariety of Y, and let Z be an irreducible component of $\phi^{-1}(W)$. Then, since $\dim(X) = \dim(Y)$, the previous proposition implies that $\dim(Z) \geq \dim(W)$. Now we have just proved that $\dim(\phi(Z)) = \dim(Z)$. Thus $\phi(Z)$ is a closed subset of W, of dimension $\geq \dim(W)$, and since W is irreducible, this implies that $\phi(Z) = W$, and that $\dim(Z) = \dim(\phi(Z)) = \dim(W)$.

In particular, if $W = \{y\}$ consists of a single point, the irreducible components of $\phi^{-1}(y)$ are zero dimensional affine varieties, i.e. points. This implies that the fiber $\phi^{-1}(y)$ is finite. ∎

Proposition 6.5. *Let $\phi : X \to Y$ be a dominant morphism of affine varieties. Then there exists a nonempty open subset $U \subseteq Y$ and a finite dominant map $\psi : \phi^{-1}(U) \to U \times \mathbf{A}^d$, where $d = \dim(X) - \dim(Y)$, such that if $p : U \times \mathbf{A}^d \to U$ is the projection, then $p \circ \psi = \phi|_{\phi^{-1}U}$, the restriction of ϕ to the dense open set $\phi^{-1}U$.*

Proof. Let A and B be the coordinate rings of X and Y, respectively. We identify B with a subring of A by means of ϕ^*. Let K and L be the respective fields of fractions of A and B, and let $A' = \{a/b | a \in A, 0 \neq b \in B\} \subseteq K$. Then A' is the smallest ring containing L and A. Since A is finitely generated over k, A' is finitely generated over L, so by Noether's Normalization Lemma (Theorem 4.2), there exist $x_1, \cdots, x_d \in A'$ such that $x_1, \cdots, x_d$ are algebraically independent over K, and A' is integral over $A'_0 = L[x_1, \cdots, x_d]$. Let $y_1, \cdots, y_r \in A$ be a set of generators over B. Each y_i satisfies a condition of algebraic dependence

$$y_i^{k_i} + \lambda_{i,k_i-1}(x_1, \cdots, x_d) y_i^{k_i-1} + \ldots + \lambda_{i,0}(x_1, \cdots, x_d) = 0,$$

with each $\lambda_{i,j} \in L[x_1, \cdots, x_d]$. We may choose a common denominator $g \in B$ of the x_i and of the $\lambda_{i,j}$, so that $gx_i \in A$, and $g\lambda_{i,j}(x_1, \cdots, x_d) \in B[x_1, \cdots, x_d]$ for all i, j. We consider the localizations B_g and A_g. Firstly, each $x_i \in A_g$. And secondly, all the above relations of integral dependence have coefficients $\lambda_{i,j}(x_1, \cdots, x_d) \in B_g[x_1, \cdots, x_d]$. Therefore A_g is integral over $B_g[x_1, \cdots, x_d]$. Now let U be the principal open set $Y_g = \{y \in Y | g(y) \neq 0\}$ (Exercise 3.1). Then $\phi^{-1}U = X_g$. By Exercise 3.1, U may be identified with the affine variety whose coordinate ring is B_g, and $\phi^{-1}U$ may be identified with the affine variety whose coordinate ring is A_g. The affine variety $U \times \mathbf{A}^d$ has coordinate ring $B_g[x_1, \cdots, x_d]$. We take $\psi : \phi^{-1}U \to U \times \mathbf{A}^d$ to be the morphism such that $\psi^* : B_g[x_1, \cdots, x_d] \to A_g$ is the inclusion map. Since A_g is integral over $B_g[x_1, \cdots, x_d]$, this is a finite dominant map. ■

Proposition 6.6. *Let $\phi : X \to Y$ be a dominant morphism of affine varieties. Then there exists a nonempty open subset U of Y such that $U \subseteq \phi(X)$, and such that if W is any closed subvariety of Y such that $W \cap U \neq \varnothing$, and if Z is any irreducible component of $\phi^{-1}(W)$ such that $Z \cap \phi^{-1}U \neq \varnothing$, then $\dim(Z) = \dim(W) + \dim(X) - \dim(Y)$.*

Proof. Let U and ψ be as in Proposition 6.5. By Proposition 6.4, the finite dominant map $\psi : \phi^{-1}(U) \to U \times \mathbf{A}^d$ is surjective, and so $U \subseteq \phi(X)$. Now let W be a closed subvariety of Y such that $W \cap U \neq \varnothing$, and let Z be an irreducible component of $\phi^{-1}(W)$ such that $Z \cap \phi^{-1}U$ is nonempty. Then $Z \cap \phi^{-1}U$ is an irreducible component of $\psi^{-1}(W \times \mathbf{A}^d)$. By Proposition 6.4, $\psi(Z \cap \phi^{-1}U) = (W \cap U) \times \mathbf{A}^d$, and these have the same dimension, so $\dim(Z) = \dim(Z \cap \phi^{-1}U) = \dim((W \cap U) \times A^d) = \dim(W) + d = \dim(W) + \dim(X) - \dim(Y)$. ■

Proposition 6.7 ("upper semicontinuity of dimension"). *Let $\phi : X \to Y$ be a surjective morphism. Define a function $e : X \to \mathbb{Z}$, where if $x \in X$, then $e(x)$ is the maximum dimension of any irreducible component of the fiber $\phi^{-1}(\phi(x))$. Then for $n \in \mathbb{Z}$, the set $S_n(\phi) = \{x \in X | e(x) \geq n\}$ is a closed subset of X.*

Proof. We argue by induction on $\dim(Y)$.

By Proposition 6.3, every fiber of ϕ has dimension $\geq r$, where $r = \dim(X) - \dim(Y)$. Thus if $n \leq r$, $S_n(\phi) = X$. Hence we may assume $n > r$.

Let U be a nonempty open subset as in Proposition 6.6. The fiber over any $y \in U$ has dimension r, and so $S_n(\phi) \subseteq \phi^{-1}(Y - U)$. Let W_i be the irreducible components of $Y - U$, and let $Z_{i,j}$ be the irreducible components of $\phi^{-1}(W_i)$. Let $\phi_{i,j}$ be the restriction of ϕ to a map $Z_{i,j} \to W_i$. It is clear that $S_n(\phi) = \bigcup_{i,j} S_n(\phi_{i,j})$, and by induction, this is a finite union of closed sets. ■

We turn now to the construction of the *tangent bundle* TX of an affine variety X. As a set, this is the disjoint union of the tangent spaces at the points of X:

$$TX = \coprod_{x \in X} T_x(X).$$

We have maps $p : TX \to X$ and $i : X \to TX$, where p maps the entire tangent space $T_x(X)$ to x, and i maps the point $x \in X$ to the zero vector in $T_x(X)$.

We wish to show that TX may be given the structure of an affine algebraic set. We will not prove that it is irreducible, so we do not claim that it is an affine variety.

We begin with an observation: if X is an affine variety and $A = \mathcal{O}(X)$ is its coordinate ring, then the points of X are in a natural bijection with the set of k-algebra homomorphisms $A \to k$. Indeed, this is an immediate consequence of Proposition 3.1, since if $x \in X$, the maximal ideal $\mathfrak{m}_x$ determines a k-algebra homomorphism $A \to A/\mathfrak{m}_x \cong k$, and every such isomorphism arises uniquely in this way. We will give a similar parametrization of TX as a set of k-algebra homomorphisms of A. The target of these k-algebra homomorphisms is not k, but the algebra of *dual numbers*. The *dual numbers* are the two-dimensional algebra over k generated by one element δ subject to the relation $\delta^2 = 0$. We may think of δ as an "infinitesimal." A special role is played by the homomorphism $\epsilon : k[\delta] \to k$ which sends δ to zero.

Proposition 6.8. *Let $A = \mathcal{O}(X)$ be the coordinate ring of the affine variety X. Suppose that $\lambda : A \to k[\delta]$ is a k-algebra homomorphism. Then there exists a point $x \in X$ and a tangent vector $D \in T_x(X)$ (i.e. a k-derivation of $\mathcal{O}_x(X)$ satisfying Eq. (5.1)) such that for every $f \in A$ we have $\lambda(f) = f(x) + \delta\, Df$. Every such pair (x, D) arises uniquely in this fashion. Hence there is a bijection between the tangent bundle TX and the set of k-algebra homomorphisms $\lambda : A \to k[\delta]$.*

Proof. Given λ, define two functions λ_0 and D from A to k by $\lambda(f) = \lambda_0(f) + \delta\, D(f)$. Clearly $\lambda_0 = \epsilon \circ \lambda$ is a k-algebra homomorphism $A \to k$, and as we have noted, Proposition 3.1 implies that there is a uniquely determined point $x \in X$ such that $\lambda_0(f) = f(x)$.

Since λ is a k-algebra homomorphism, we must have $\lambda(fg) = \lambda(f)\lambda(g)$. Since

$$\lambda(fg) = f(x)g(x) + \delta\, D(fg),$$

$$\lambda(f)\,\lambda(g) = \big(f(x) + \delta\, D(f)\big)\big(g(x) + \delta\, D(g)\big) = f(x)g(x) + \big(f(x)Dg + g(x)Df\big).\delta.$$

These are the same if and only if $D(fg) = f(x)\, Dg + g(x)\, Df$. Thus $D : A \to k$ satisfies the derivation condition Eq. (5.1). We may now extend D to all of

$\mathcal{O}_x(X)$ by means of the quotient rule: any element of $\mathcal{O}_x(X)$ may be written as f/g where f and g are elements of A and $g(x) \neq 0$. We then define

$$D(f/g) = \frac{g(x)\,D(f) - f(x)\,D(g)}{g(x)^2},$$

and it may be checked that this is a well-defined k-derivation, so $D \in T_x(X)$.

Conversely, if $x \in X$ and $D \in T_x(X)$, we may define $\lambda(f) = f(x) + \delta\,D(f)$, and it is easy to check that $\lambda : A \to k[\delta]$ is a k-algebra homomorphism. ∎

The map $p : TX \to X$ has the following interpretation in terms of k-algebra homomorphisms: p maps the element of TX corresponding to the k-algebra homomorphism $\lambda : A \to k[\delta]$ by Proposition 6.8 to the element of $x \in X$ corresponding to the k-algebra homomorphism $\epsilon \circ \lambda : A \to k$, whose kernel is the maximal ideal $\mathfrak{m}_x$ corresponding to x by Proposition 3.1. If $\phi : X \to Y$ is a morphism of algebraic varieties, there is an induced map $T\phi : TX \to TY$, which we may also describe in terms of the correspondence in Proposition 6.8; for if $\phi_* : B = \mathcal{O}(Y) \to A = \mathcal{O}(X)$ is the induced map of coordinate rings, and then composition with ϕ_* takes k-algebra homomorphisms $A \to k[\delta]$ to k-algebra homomorphisms $B \to k[\delta]$. We have a commutative diagram:

$$
\begin{array}{ccc}
TX & \xrightarrow{\ T\phi\ } & TY \\
{\scriptstyle p}\big\downarrow & & {\scriptstyle p}\big\downarrow \\
X & \xrightarrow{\ \phi\ } & Y
\end{array}
$$

Thus far we have only constructed the tangent bundle as functor from the category of affine varieties to the category of sets. We would like to know that the tangent bundle is an affine variety.

Proposition 6.9. *Let X be an affine variety. Then TX may be given the structure of an affine algebraic set in such a way that:*

(i) The maps $p : TX \to X$ and $i : X \to TX$ are morphisms of affine algebraic sets;

(ii) If $x \in X$ and $D_1, \cdots, D_r$ is a basis of $T_x(X) = p^{-1}(x)$, then the linear map $T_x(X) \to \mathbf{A}^r$ sending $\sum_i a_i D_i \to (a_1, \cdots, a_r)$ $(a_i \in k)$ is a morphism.

Note that we do not claim that this structure is unique, so we have *not* constructed the tangent bundle as a functor from the category of affine varieties to affine algebraic sets. Also note that we do not assert that TX is irreducible. See Borel [6], AG.16 for a more functorial construction. Borel constructed the tangent bundle as an affine scheme, not necessarily reduced. The simple construction given below is sufficient for our purposes.

Proof. Let $A = \mathcal{O}(X)$ be the coordinate ring, and let us realize A as a quotient of a polynomial ring, $A = k[X_1, \cdots, X_n]/\mathfrak{p}$. We may parametrize the k-algebra homomorphisms $\lambda : A \to k[\delta]$ by elements of $\mathbf{A}^{2n}$ as follows. We regard λ as a k-algebra homomorphism $k[X_1, \cdots, X_n] \to k[\delta]$ which must annihilate $\mathfrak{p}$; if $\lambda(X_i) = a_i + b_i\delta$, we associate λ with the point $(a_1, a_2, \cdots, a_n, b_1, b_2, \cdots, b_n)$. Now observe that the condition for λ to annihilate $\mathfrak{p}$ is given by algebraic conditions on each a_i and b_i, for if $f_1, \cdots, f_k$ are generators of $\mathfrak{p}$, then substituting

$$f(a_1 + b_1\delta, \cdots, a_n + b_n\delta) = 0,$$

expanding the left-hand side and collecting the coefficients of 1 and δ give a set of algebraic relations whose satisfaction is necessary and sufficient for $\lambda(\mathfrak{p}) = 0$. Conditions (i) are evident—if we identify X with $V(\mathfrak{p}) \subset \mathbf{A}^n$, then

$$p : (a_1, \cdots, a_n, b_1, \cdots, b_n) \mapsto (a_1, \cdots, a_n),$$

$$i : (a_1, \cdots, a_n) \mapsto (a_1, \cdots, a_n, 0, \cdots, 0),$$

are clearly morphisms. Condition (ii) is equally clear. ∎

We illustrate the proof of Proposition 6.9 with an example:

Example 6.10. Let X be the curve $y^2 = x^2(x+1)$. Substituting $a_1 + b_1\delta$ for x and $a_2 + b_2\delta$ for y, we obtain

$$a_2^2 + 2a_2b_2\delta = a_1^2(a_1 + 1) + (2a_1 + 3a_1^2)b_1\delta.$$

Therefore

$$a_2^2 = a_1^2(a_1 + 1),$$

$$2a_2b_2 = (2a_1 + 3a_1^2)b_1.$$

The first equation states that (a_1, a_2) is a point on the curve $y^2 = x^2(x+1)$. The second states that, (a_1, a_2) being fixed, the point (b_1, b_2) is on the line through the origin in $\mathbf{A}^2$ which is parallel to the tangent line to the curve at the point (a_1, a_2). Observe that at the singular point $(a_1, a_2) = (0,0)$, the second condition is vacuous. This is consistent with our discovery in Example 5.7, that the tangent space to this curve at the singularity is two-dimensional.

We may now prove:

Proposition 6.11. *The set of singular points of X is a closed subset.*

Proof. Let $d = \dim(X)$. Recall that x is singular if and only if $\dim\big(T_x(X)\big) > d$. Let S be the set of singular points of X, and let $p : TX \to X$ be the projection map from the tangent bundle. We do not know that TX is irreducible, so assume that $TX = \bigcup T_i$ is its decomposition into irreducible components. Each T_i is an affine variety. Let $p_i : T_i \to X$ be the restriction of p to T_i.

Note that $p^{-1}(S) = \bigcup p_i^{-1}(S)$ is the union of the spaces $T_x(X)$ which have dimension $\geq d + 1$. If $x \in S$ then $\dim\, p^{-1}(x) \geq d + 1$, and this implies that $\dim\, p_i^{-1}(x) \geq d + 1$ for some i. Thus $S = \bigcup S_i$, where $S_i = \{x \in X | p_i^{-1}(x) \geq d + 1\}$, and it is sufficient to show that S_i is closed.

By the upper semicontinuity of dimension (Proposition 6.7) applied to p_i, $p_i^{-1}(S_i)$ is a closed subset of T_i and hence of TX. Now, we have a morphism $X \to TX$ which maps each $x \in X$ to the zero vector in $T_x(X)$, and so we may regard X as a closed subspace of TX. Clearly $S_i = p_i^{-1}(S_i) \cap X$. This is the intersection of two closed subspaces, hence closed. ∎

We would like to prove that the singular points form a *proper* closed subset of X. We begin by showing that every affine variety is birationally equivalent to a hypersurface.

Recall that a field K is called *perfect* if either the characteristic of K is zero, or the characteristic of K is $p > 0$, and $K^p = K$. Algebraically closed fields are perfect, and so are finite fields. If E/F is a finitely generated field extension, then a *separating transcendence basis* of E is a transcendence basis $x_1, \cdots, x_d$ of E over F such that $E/F(x_1, \cdots, x_d)$ is separable. If E/F has a separating transcendence basis, we say that E/F is *separably generated*.

Proposition 6.12. *Let K be a perfect field, and let F/K be a finitely generated extension, then F/K is separably generated.*

Proof. For a proof of this, see Lang, [18], Corollary VIII.4.4 on p. 365. ∎

Proposition 6.13. *Let X be an affine variety of dimension d. Then there exists an irreducible separable polynomial*

$$f(y) = a_n(x_1, \cdots, x_d)y^n + a_{n-1}(x_1, \cdots, x_d)y^{n-1} + \ldots + a_0(x_1, \cdots, x_d)$$

with coefficients a_i in the polynomial ring $k[x_1, \cdots, x_d]$ with the following property. Let $B = k[x_1, \cdots, x_d, y]/(f)$, and let Y be the affine variety with coordinate ring B. Then X and Y are birationally equivalent.

The separability of this polynomial means that its derivative

$$n\, a_n(x_1, \cdots, x_d)y^{n-1} + (n-1)\, a_{n-1}(x_1, \cdots, x_d)\, y^{n-2} + \ldots + a_1(x_1, \cdots, x_n)$$

is nonzero, i.e. $a_k \neq 0$ for some k not divisible by the characteristic.

Proof. By Proposition 6.12, since k is separable, the function field F of X has a transcendence basis, i.e. $x_1, \cdots, x_d$ such that $F/k(x_1, \cdots, x_d)$ is a separable algebraic extension.

By the "Theorem of the Primitive Element," (Lang [18], Theorem V.4.6 on p. 243) there exists a single element $y \in F$ such that $F = k(x_1, \cdots, x_d, y)$. Since y is separable, it satisfies an irreducible separable polynomial as stated. Now X and the hypersurface Y have isomorphic function fields, so they are birationally equivalent. ∎

Proposition 6.14. *Let X be the hypersurface determined by the polynomial*

$$f(x_1, \cdots, x_d, y) =$$
$$a_n(x_1, \cdots, x_d)y^n + a_{n-1}(x_1, \cdots, x_d)y^{n-1} + \ldots + a_0(x_1, \cdots, x_d),$$

so that the coordinate ring of X is $A = k[x_1, \cdots, x_d, y]/(f)$.

(i) Suppose that $\xi_1, \cdots, \xi_d, \eta \in k$ are such that $f(\xi_1, \cdots, \xi_d, \eta) = 0$, while

$$\frac{\partial f}{\partial y}(\xi_1, \cdots, \xi_d, \eta) \neq 0.$$

Then $x = (\xi_1, \cdots, \xi_d, \eta)$ is a smooth point on X. Moreover, the images of the functions $x - \xi_1, \cdots, x - \xi_d$ generate the maximal ideal of $\mathcal{O}_x(X)$.

(ii) Suppose that the image of y in A is separable over $k[x_1, \cdots, x_d] \subseteq A$. Then the smooth points in X are dense.

Proof. We prove (i). Let

$$g(x_1, \cdots, x_d, y) = \frac{\partial f}{\partial y}(y) =$$
$$n\, a_n(x_1, \cdots, x_d)y^{n-1} + (n-1)\, a_{n-1}(x_1, \cdots, x_d)\, y^{n-2} + \ldots + a_1(x_1, \cdots, x_d).$$

Our hypothesis is that $g(\xi_1, \cdots, \xi_d, \eta) \neq 0$. The maximal ideal of the local ring $\mathcal{O}_x$ is generated by $x_1 - \xi_1, \cdots, x_d - \xi_d, y - \eta$. (Indeed, these generate the maximal ideal of the local ring at x in $\mathbf{A}^{d+1}$, and the canonical map $\mathcal{O}_x(\mathbf{A}^{d+1}) \to \mathcal{O}_x(X)$ is surjective.) Suppose that D is a derivation of the local ring $\mathcal{O}_x$. We will show that the value of $D(y - \eta)$ is determined by the values of $D(x_i - \xi_i)$. Indeed, the identity $f(x_1, \cdots, x_d, y) = 0$ holds in the ring $\mathcal{O}_x$, and so applying D to this, we obtain

$$\sum_{i=1}^{d} \frac{\partial f}{\partial x_i}(\xi_1, \cdots, \xi_d, \eta)Dx_i + \frac{\partial f}{\partial y}(\xi_1, \cdots, \xi_d, \eta)Dy = 0.$$

Since $\xi_i, \eta \in k$ are constants, $D\xi_i = D\eta = 0$, and we may rewrite this

$$(6.1) \qquad \sum_{i=1}^{d} \frac{\partial f}{\partial x_i}(\xi_1, \cdots, \xi_d, \eta) D(x_i - \xi_i) + \frac{\partial f}{\partial y}(\xi_1, \cdots, \xi_d, \eta) D(y - \eta) = 0.$$

Now the coefficient of $D(y - \eta)$ is nonzero, hence its value is determined by the values of the $D(x - \xi_i)$. Since the $x - \xi_i$ and $y - \eta$ generate the maximal ideal of $\mathcal{O}_x$, it follows from Proposition 5.1 that the space of derivations D of $\mathcal{O}_x$ is at most d-dimensional. Thus x is nonsingular.

We must also show that the images of the $x_i - \xi_i$ generate the maximal ideal $\mathfrak{m}_x$ of $\mathcal{O}_x(X)$. To see this, it is sufficient by Nakayama's Lemma (Proposition 5.5 (ii)) to show that they generate $\mathfrak{m}_x/\mathfrak{m}_x^2$. Hence we must show that any k-linear functional on $\mathfrak{m}_x/\mathfrak{m}_x^2$ which vanishes on the images of the $x_i - \xi_i$ is zero. By Proposition 5.1, it is sufficient to show that any derivation D of $\mathcal{O}_x$ which is zero on $x_i - \xi_i$ is zero. Any derivation is zero on constants, and we only have to show that it is zero on $\mathfrak{m}_x$. By Eq. (6.1), we have $D(y - \eta) = 0$, and we have already remarked that $\mathfrak{m}_x$ is generated by the $x_i - \xi_i$ together with $y - \eta$, hence $D = 0$.

As for (ii), if the image of y in $k[x_1, \cdots, x_d]$ is separable, the irreducible polynomial f which it satisfies is separable. This implies that the image of the derivative $\partial f/\partial y = g(x_1, \cdots, x_d, y)$ in A is not zero. By (i), any element of the principal open set X_g is smooth, and X_g is dense. ∎

Theorem 6.15. *Let X be an affine variety. Then the set of singular points is a proper closed subset of X.*

Proof. By Proposition 6.11, the singular points form a closed subset. We only have to show that they form a *proper* subset.

By Proposition 6.14, the theorem is true when X is as in part (ii) of that Proposition. For arbitrary X, by Proposition 6.13, X is birationally equivalent to a hypersurface Y determined by a separable polynomial. X and Y have open sets U and V such that ϕ is defined on U, and ϕ^{-1} is defined on V, and $\phi(U) = V$ while $\phi^{-1}(V) = U$. On Y, the singular points form a closed subset, which we have shown is proper, and hence the smooth points are dense. Thus V contains smooth points, and since ϕ is biregular between U and V, so does U. ∎

We recall from general topology the notion of a *proper map*. If $f : Y \to X$ is a continuous map of Hausdorff spaces, it is called *proper* if the inverse image of a compact set is always compact. See Exercise 6.5, where it is shown that (assuming that X is locally compact) a necessary and sufficient condition for f to be proper is that it is *universally closed*—this means that if X' is a locally compact Hausdorff space, and if $g : X' \to X$ is a continuous map, and if Y' is the *fiber product*

$$\{(y, x') \in Y \times X' | f(y) = g(x')\}$$

(with the topology induced from the product topology in $Y \times X'$), then the induced mapping $Y' \to X'$ is closed.

The original definition of a proper map—that the inverse image of a compact set be compact—does not work in the setting of varieties and morphisms. Instead, we take the equivalent condition of being universally closed and adapt it to the Zariski topology. A morphism $f : Y \to X$ is *proper* if for any morphism $g : X' \to X$, taking $Y' = \{(y, x') \in Y \times X' | f(y) = g(x')\}$ (with the Zariski topology), the induced mapping $Y' \to X'$ is closed.

One of the most important questions in algebraic geometry is the *resolution of singularities.* Thus if X is a variety (affine or otherwise), one wishes to find a smooth variety Y together with a proper surjective morphism $f : Y \to X$ which is a birational equivalence, and such that f^{-1} is regular on the set of smooth points of X. We call Y a *desingularization* of X, or a *resolution* of its singularities. Desingularizations always exist, at least characteristic zero, and conjecturally also in positive characteristic. The existence of a desingularization was proved for curves by Kronecker and independently by Max Noether in the 19-th century. For surfaces in characteristic zero, the resolution of singularities was investigated by the Italian geometers Segre, del Pezzo, Severi, Albanese and others, and by Beppo Levi before an analytic proof was found by Walker. Several algebraic proofs were given by Zariski, who also proved that 3-folds could be desingularized in characteristic zero. For algebraic varieties of arbitrary dimension in characteristic zero, Hironaka [15] proved the existence of a desingularization in 1964. Abhyankar also made fundamental contributions, especially in characteristic p. We refer to Lipman [20] for a survey of this topic, and to Zariski [35] and Abhyankar [1] for deeper studies.

Except for varieties of dimension 1, even if X is affine, the desingularization Y may not be, so the study of the resolution of singularities may not be carried out in the context of affine varieties.

Although, as we have just remarked, desingularization can take us out of the category of affine varieties, there is nevertheless one important tool which we can discuss in this context. This is *normalization*, which is our next topic.

Recall that if V is a finite-dimensional vector space over a field F, a bilinear form, i.e. an F-bilinear mapping $T : V \times V \to F$, is *nondegenerate* if there exists no nonzero element $v \in V$ such that $T(v, w) = 0$ for all $w \in V$.

For the next Proposition, we recall some facts about the norm and trace. For further details, any standard work on field theory may be consulted, for example Lang [18], Section VI.5. Let E/F be finite extension. The *trace* $\mathrm{tr} = \mathrm{tr}_{E,F} : E \to F$ is often defined by $\mathrm{tr}(y) = [E : F]_i \sum \sigma(y)$, where the summation is over all embeddings σ of E over F into an algebraic closure of F, and where $[E : F]_i$ is the degree of inseparability of E/F, equal to 1 if E/F is separable. Since $[E : F]_i$ is a multiple of the characteristic if E/F is not separable, the trace map is identically zero in this case. On the other hand, if E/F is separable, the trace mapping is nonzero (Lang, [**18**], Theorem VI.5.2 on p. 286). Similarly, the *norm* $N = N_{E/F} : E \to F$ is the map $N(y) = \left(\prod \sigma(y) \right)^{[E:F]_i}$.

There is another way of defining the norm and trace which is sometimes more convenient. Let $\lambda_y : E \to E$ be multiplication by y, so $\lambda_y(x) = xy$. Thus λ_y is an F-linear endomorphism of E, regarded as a vector space over F. Then $\mathrm{tr}_{E/F}(y)$ is the trace of this linear transformation, and $N_{E/F}(y)$ is its determinant. The equivalence of the two definitions is Lang [**18**], Proposition VI.5.6 on p. 288.

Proposition 6.16. *Let E/F be a finite separable extension, and let* $\mathrm{tr} : E \to F$ *be the trace. Then the trace bilinear form* $(x, y) \to \mathrm{tr}(xy)$ *is nondegenerate.*

Proof. This is a standard fact from field theory. See Lang [**18**] Theorem IV.5.2 on p. 286. ∎

Proposition 6.17. *Let A be an integrally closed integral domain, F its field of fractions, and B the integral closure of A in a finite separable extension E/F.*

(i) Let $N : E \to F$ and $\mathrm{tr} : E \to F$ *be the norm and trace. If $x \in B$, then $N(x) \in A$ and $\mathrm{tr}(x) \in A$.*

Assume furthermore that A is Noetherian.

(ii) Let D be defined as the set of $x \in E$ such that $\mathrm{tr}(xB) \subseteq A$. Then D is a finitely generated A-module.

(iii) We have $B \subseteq D$, and B is finitely generated as an A-module.

Proof. To prove (i), note that every conjugate of x satisfies the same equation of integral dependence over A as does x, hence is integral over A. Thus $N(x)$ and $\mathrm{tr}(x)$, being products and sums of elements of A, are elements of F which are integral over A. Since A is integrally closed, $N(x)$ and $\mathrm{tr}(x) \in A$.

To prove (ii), let $u_1, \cdots, u_n$ be a basis of E/F. Each satisfies an algebraic equation over F, so after multiplying each u_i by an element of A, we may assume that $u_1, \cdots, u_n$ are integral over A, hence are elements of B. Now let $v_1, \cdots, v_n$ be the dual basis with respect to the trace bilinear form, which is nondegenerate by Proposition 6.16. Let $V = Av_1 + \ldots + Av_n$. We will show that $D \subseteq V$. Indeed, if $d \in D$, write $d = \sum a_i v_i$ where $a_i \in F$. Then $a_i = \mathrm{tr}(du_i)$, which is in A by definition of D. Thus $D \subseteq V$. Now since D is an A-submodule of the finitely generated module V, and since A is Noetherian, D is finitely generated by Exercise 1.3.

It follows from (i) that $B \subseteq D$. Since A is Noetherian, B is finitely generated by Exercise 1.3. ■

Proposition 6.18. *Let K be a perfect field, L/K a finitely generated extension, and let $A \subseteq L$ be a ring such that A is finitely generated as a K-algebra, and such that L is separable over the field of fractions of A. Let B be the integral closure of A in L. Then B is finitely generated as a K-algebra, hence Noetherian.*

Proof. By the Noether Normalization Lemma (Theorem 4.2), we may find $x_1, \cdots, x_d$ in A which are algebraically independent, and such that A is integral over $A' = K[x_1, \cdots, x_d]$, and such that the field of fractions of A is separable over the field of fractions F of A'. Now it is clear that B is the integral closure of A' in L, and that L is separable over F. By Proposition 6.17 (iii), B is finitely generated as an A'-module, hence as an algebra over K. By the Hilbert Basis Theorem (Exercise 1.4), B is Noetherian. ■

Let X be an affine variety, and let A be its coordinate ring. X is called *normal* if A is integrally closed. Let F be the field of fractions of A. Whether or not X is normal, we may consider the integral closure B of A in F, which is an affine algebra by Proposition 6.18. Then B is the coordinate ring of an affine variety Y which is normal. The inclusion $A \to B$ induces a morphism $Y \to X$. It is clearly a finite dominant map, and a birational equivalence. It is also clear that if Y is a normal affine variety, and $f : Y \to X$ is a finite dominant map which is a birational equivalence, then the coordinate ring of Y is isomorphic to the integral closure of A in its field of fractions, so Y is uniquely characterized. It is called the *normalization* of X.

Proposition 6.19. *Let F be a field, let A be a subring, and let S be a multiplicative subset of A. Let B be the integral closure of A in F. Then $S^{-1}B$ is the integral closure of $S^{-1}A$ in F.*

Proof. It is obvious that the integral closure of $S^{-1}A$ in F contains $S^{-1}B$.

Conversely, to prove that the integral closure if $S^{-1}A$ in F is contained in $S^{-1}B$, assume that $x \in F$ is integral over $S^{-1}A$. Then

$$x^n + (a_{n-1}/s_{n-1})x^{n-1} + \ldots + a_0/s_0 = 0,$$

where $a_i \in A$ and $s_i \in S$. Let $t = \prod s_i$. Then

$$(tx)^n + b_{n-1}(tx)^{n-1} + \ldots + b_0 = 0,$$

where $b_i = t^{n-i}a_i/s_i \in A$. Hence tx is integral over A, and is therefore in B. Consequently, $x \in S^{-1}B$. ■

Proposition 6.20. *Let A be an integral domain. Then*

$$A = \bigcap_{\mathfrak{p} \text{ prime}} A_{\mathfrak{p}} = \bigcap_{\mathfrak{m} \text{ maximal}} A_{\mathfrak{m}}.$$

Thus A equals the intersection of the localizations of A at all prime (resp. maximal) ideals.

Proof. It is evident that $A \subseteq \bigcap A_{\mathfrak{p}} \subseteq \bigcap A_{\mathfrak{m}}$. On the other hand, let $x \in \bigcap A_{\mathfrak{m}}$. We will show that $x \in A$. Let $\mathfrak{a} = \{a \in A | ax \in A\}$. It is clearly sufficient to show that $1 \in \mathfrak{a}$. Suppose not. Then the ideal $\mathfrak{a}$ is proper, hence contained in a maximal ideal $\mathfrak{m}$. Now $x \in A_{\mathfrak{m}}$, so we may write $x = a/s$ where $a \in A$ and $s \in A - \mathfrak{m}$. Since $sx = a \in A$, $s \in \mathfrak{a}$, so $s \in \mathfrak{m}$, which is a contradiction. ■

Proposition 6.21. *Let A be an integral domain. The following conditions are equivalent:*

(i) A is integrally closed;

(ii) $A_{\mathfrak{p}}$ is integrally closed for all prime ideals $\mathfrak{p}$;

(iii) $A_{\mathfrak{m}}$ is integrally closed for all maximal ideals $\mathfrak{m}$.

Proof. Clearly (ii) implies (iii), and by Proposition 6.20, (iii) implies (i). We must therefore show that (i) implies (ii). Indeed, this follows from Proposition 6.19, taking F to be the field of fractions of A, so that $B = A$ if A is integrally closed, and $S^{-1}B = S^{-1}A$. If $S = A - \mathfrak{p}$, this means that $A_{\mathfrak{p}}$ is integrally closed. ■

If $x \in X$, we say that the point x is *normal* if the local ring $\mathcal{O}_x(X)$ is integrally closed. Proposition 6.21 shows that X is normal if and only if every point of X is normal. Hence normality is a local property.

Proposition 6.22. *Let X be an affine variety, and let S be the closed subset of singular points of X. Let Y be a closed subvariety of X such that $\mathcal{O}_{Y,X}$ is a regular local ring. Then Y is not contained in S.*

Proof. Let $d = \dim(X)$, and $r = \dim(Y)$, and let $s = d - r$. Let A be the coordinate ring of X, and let $\mathfrak{p}$ be the ideal of Y in A. Then (Exercise 5.1) the dimension of the local ring $\mathcal{O}_{Y,X} = A_{\mathfrak{p}}$ is s. Since by assumption $A_{\mathfrak{p}}$ is regular, we may find s elements $f_1, \cdots, f_s$ which generate its maximal ideal $\mathfrak{p}A_{\mathfrak{p}}$. We will show that there exists $h \in A - \mathfrak{p}$ such that $f_1, \cdots, f_s \in A_h$, and that $f_1, \cdots, f_s$ generate the ideal $\mathfrak{p}A_h$ in A_h. Let $x_1, \cdots, x_n$ generate the ideal $\mathfrak{p}$ of A. Then they also generate the ideal $\mathfrak{p}A_{\mathfrak{p}}$ over $A_{\mathfrak{p}}$, so there exist $c_{ij} \in A_{\mathfrak{p}}$ such that $f_i = \sum_j c_{ij} x_j$. On the other hand, the f_i also generate the ideal $\mathfrak{p}A_{\mathfrak{p}}$ over $A_{\mathfrak{p}}$, and so there exist $d_{ij} \in A_{\mathfrak{p}}$ such that $x_j = \sum_i d_{ji} f_i$. The f_j, c_{ij} and

d_{ij} are all fractions whose denominators are in $S = A - \mathfrak{p}$. We may choose h to be a common denominator for the f_j, c_{ij} and d_{ij}, so that the f_j, c_{ij} and d_{ij} are in A_h. Then it is clear that the f_j and x_i generate the same ideal in A_h, so that $f_1, \cdots, f_s$ generate $\mathfrak{p}A_h$.

We may replace A by A_h, and X by the principal open set X_h. Then Y is replaced by the principal open set $Y_{\overline{h}}$, and its coordinate ring $\overline{A} = A/\mathfrak{p}$ is replaced by $\overline{A}_{\overline{h}}$, where $\overline{h}$ is the image of h in $\overline{A}$. This does not change the local ring $\mathcal{O}_{Y,X}$. Thus, we may assume that the ideal $\mathfrak{p}$ may be generated by s elements $f_1, \cdots, f_s$.

Now by Theorem 6.15, Y contains a point y which is smooth as a point of Y. We will show that y is smooth as a point of X. Thus $y \in Y - S$, which proves the Proposition.

Let $\mathfrak{m}$ be the maximal ideal of y in A. Since $y \in Y$, we have $\mathfrak{p} \subseteq \mathfrak{m}$. The maximal ideal of y in the coordinate ring $\overline{A} = A/\mathfrak{p}$ of Y is the image $\overline{\mathfrak{m}} = \mathfrak{m}/\mathfrak{p}$ of $\mathfrak{m}$ in $\overline{A}$. Since y is smooth as a point of Y, $\overline{A}_{\overline{\mathfrak{m}}}$ is a regular local ring of dimension r. Hence there exist $g_1, \cdots, g_r \in \mathfrak{m}$ such that the images $\overline{g}_1, \cdots, \overline{g}_r \in \overline{\mathfrak{m}}$ generate the maximal ideal $\overline{\mathfrak{m}}\overline{A}_{\overline{\mathfrak{m}}}$ in $\overline{A}_{\overline{\mathfrak{m}}}$. Note that $\overline{A}_{\overline{\mathfrak{m}}} = A_{\mathfrak{m}}/\mathfrak{p}A_{\mathfrak{m}}$. It follows that the ideal $(g_1, \cdots, g_r)A_{\mathfrak{m}}$ generated by the g_i in $A_{\mathfrak{m}}$ satisfies $(g_1, \cdots, g_r)A_{\mathfrak{m}} + \mathfrak{p}A_{\mathfrak{m}} = \mathfrak{m}A_{\mathfrak{m}}$. Now since $\mathfrak{p}$ is generated by s elements $f_1, \cdots, f_s$, $\mathfrak{p}A_{\mathfrak{m}}$ is generated over $A_{\mathfrak{m}}$ by these same s elements, and consequently $\mathfrak{m}A_{\mathfrak{m}}$ is generated by $f_1, \cdots, f_s, g_1, \cdots, g_r$. This shows that $\mathfrak{m}A_{\mathfrak{m}}$ is generated by $r + s = \dim(A_{\mathfrak{m}})$ elements, hence is a regular local ring. This proves that y is smooth as a point of X. ∎

Proposition 6.23. *Let X be an affine variety, Y a closed subvariety. Let S be the closed set of singular points of X. Then $\mathcal{O}_{Y,X}$ is a regular local ring if and only if $Y \not\subseteq S$.*

The proof of this depends on Theorem 5.30, which we have not proved. We will only use the weaker Proposition 6.22 in the sequel.

Proof. We have proved in Proposition 6.22 that if $\mathcal{O}_{Y,X}$ is regular, then $Y \not\subseteq S$. Conversely, suppose that $y \in Y - S$. Then $\mathcal{O}_y$ is a regular local ring. Now $\mathcal{O}_{Y,X}$ is a localization of $\mathcal{O}_y$, namely, if $\mathfrak{m}$ is the maximal ideal of y, and $\mathfrak{p}$ is the prime ideal of Y, then $\mathcal{O}_y = A_{\mathfrak{m}}$ and $\mathcal{O}_{Y,X} = A_{\mathfrak{p}}$, and it is not hard to see that $A_{\mathfrak{p}}$ is the localization of $A_{\mathfrak{m}}$ at the prime ideal $\mathfrak{p}A_{\mathfrak{m}}$. Hence by Theorem 5.30, $\mathcal{O}_{Y,X}$ is a regular local ring. ∎

Theorem 6.24. *Let X be a normal affine variety of dimension d, and let S be the closed subset of singular points. Then every irreducible component of S has dimension $\leq d - 2$.*

Proof. Let Y be an irreducible component of S. Then by Theorem 6.15, $\dim(Y) \leq d-1$. Suppose that $\dim(Y) = d-1$. Then $\mathcal{O}_{Y,X}$ is a Noetherian local ring of dimension 1 by Exercise 5.1. It is a localization of the coordinate ring of X, which is integrally closed because X is normal, so by Proposition 6.21, $\mathcal{O}_{Y,X}$ is integrally closed. Now by Proposition 5.26, it is a regular local ring. This contradicts Proposition 6.22. ∎

Proposition 6.25. *Let X be an affine curve, and let A be its coordinate ring. A necessary and sufficient condition for X to be nonsingular is that A be integrally closed.*

Proof. A is integrally closed if and only if each $A_{\mathfrak{m}}$ is integrally closed for each maximal ideal $\mathfrak{m}$ of A, by Proposition 6.21. These are precisely the local rings of X. A necessary and sufficient condition for $A_{\mathfrak{m}}$ to be integrally closed is that it is a regular local ring, by Theorem 5.25 and Proposition 5.26, i.e. if and only if the corresponding point of X is nonsingular. ∎

Theorem 6.26. *Let X be an affine curve, and let X' be the normalization of X. Then X' is nonsingular.*

Proof. This is clear from Theorem 6.24. ∎

Proposition 6.27. *Let A be a discrete valuation ring, F its field of fractions. Let B be any subring of F containing A. Then either $B = F$ or $B = A$.*

Proof. Suppose that $B \neq A$. We will prove that $B = F$. By assumption, there exists an element $a \in B$ which is not in A. Since $A = \{x \in F | v(x) \geq 0\}$, where v is a discrete valuation, we have $v(a) < 0$. Now let $x \in F$. For sufficiently large n, $v(xa^{-n}) = v(x) - nv(a) > 0$, so $xa^{-n} \in A \subseteq B$. Therefore $x = xa^{-n}.a^n \in B$. ∎

Proposition 6.28. *Let A be an integral domain, F its field of fractions, and let B be a subring of F containing A, which is finitely generated as an A-algebra. Let $\mathfrak{q}$ be a prime ideal of B, and let $\mathfrak{p} = A \cap \mathfrak{q}$. Suppose that $A_{\mathfrak{p}}$ is a discrete valuation ring. Then there exists an element $g \in A - \mathfrak{p}$ such that $A_g = B_g$.*

Proof. Since $A_{\mathfrak{p}}$ is not a field, $\mathfrak{p}$ is not 0, and so $\mathfrak{q}$ is not zero and therefore $B_{\mathfrak{q}}$ is not a field. We have $A_{\mathfrak{p}} \subseteq B_{\mathfrak{q}}$, and so by Proposition 6.27, $A_{\mathfrak{p}} = B_{\mathfrak{q}}$. In particular, $B \subseteq A_{\mathfrak{p}}$. Now let $x_1, \cdots, x_n$ generate B as an A-algebra. Each x_i can be written a_i/s_i where $s \in A - \mathfrak{p}$, and if $g = \prod s_i$, then clearly $B \subseteq A_g$, so that $B_g = A_g$. ∎

Recall from Chapter 3 the notion of a *birational* morphism. If X and Y are affine varieties with coordinate rings A and B respectively, and if $f : Y \to X$ is a dominant morphism, then there is induced an injection $f^* : A \to B$, and we may regard A as a subring of B. Then f is called *birational* if the field of fractions of A is equal to the field of fractions of B.

The following result is a special case of Zariski's Main Theorem. (The general case for higher dimensional varieties is much deeper.)

Proposition 6.29. *Let X and Y be affine curves, and let $f : Y \to X$ be a birational morphism. Let $y \in Y$, and let $x = f(y) \in X$. Assume that x is a normal point of X. Then there exists a open set U containing x such that f induces an isomorphism $f^{-1}U \to U$.*

Proof. Let A and B be the coordinate rings of X and Y respectively. Identify A as a subring of B by the injection $f^* : A \to B$. Let $\mathfrak{q}$ and $\mathfrak{p}$ be the maximal ideals of y and x in B and A respectively, so that $\mathfrak{p} = A \cap \mathfrak{q}$. Since $A_{\mathfrak{p}}$ is integrally closed, it is a discrete valuation ring by Proposition 5.26 and Proposition 5.28. Therefore by Proposition 6.28, there exists $g \in A - \mathfrak{p}$ such that $A_g = B_g$. This means that the principal open sets $U = X_g$ and $Y_g = f^{-1}(U)$ are isomorphic. ■

Exercises

Exercise 6.1. Let X be a topological space. A subset of X is called *locally closed* if it is the intersection of an open set and a closed set. A subset is called *constructible* if it is the union of a finite number of locally closed subsets.

(i) Prove that the set Σ of constructible subsets of X forms a *Boolean algebra*. That is, Σ is closed under finite unions, intersections, and formation of set theoretic complements. In fact, Σ is the smallest Boolean algebra containing all the open sets in X.

(ii) Prove that if X and Y are affine varieties, and $\phi : X \to Y$ is a morphism, then $\phi(X)$ is a constructible subset of Y.

Hint: Show that one may assume ϕ dominant. Let $U \subseteq Y$ be as in Proposition 6.6. Observe that $\phi(X) = U \cup \bigcup \phi(Z_{ij})$, where Z_{ij} are as in the proof of Proposition 6.7. Use induction to show that $\phi(Z_{ij})$ are constructible, while U is open, hence constructible.

(iii) Let X and Y be affine varieties, let $\phi : X \to Y$ be a morphism, and let $S \subseteq X$ be constructible. Prove that $\phi(S)$ is constructible.

Exercise 6.2. Let A be the coordinate ring of the affine curve X whose equation is $y^2 = x^2(x + 1)$. Show that the integral closure of $A = k[x, y]$ is the ring $k[y/x]$, and that if Y is the normalization of X, then the fiber over the singular point at the origin consists of two points, corresponding to the values $t = 1$ and $t = -1$.

Exercise 6.3. Do the same for the curve $y^2 = x^3$.

Hint: The integral closure is $k[y/x]$. The fiber over the singularity in this case consists of a single point.

Exercise 6.4. Use dual numbers to obtain equations for the tangent bundle over the curve $y^2 = x^3$.

The purpose of the next exercise is to motivate the algebro-geometric notion of a *proper mapping*.

Exercise 6.5. Proper Mappings. (i) Let X and Y be Hausdorff spaces. Assume that X is locally compact. Let $f : Y \to X$ be a proper mapping, i.e. a continuous map such that the inverse image of a compact set is compact. Prove that f is a closed mapping.

(ii) Let X, Y and X' be Hausdorff topological spaces, and let $f : Y \to X$ and $g : X' \to X$ be continuous mappings. Let Y' be the "fibered product" $\{(y, x') \in Y \times X' | f(y) = g(x')\}$. There are obvious induced maps $g' : Y' \to Y$ and $f' : Y' \to X'$ such that $f \circ g' = g \circ f'$. Prove that if Z is any topological space, and if $p : Z \to Y$, $q : Z \to X'$ are continuous maps such that $f \circ p = g \circ q$, then there exists a unique continuous map $r : Z \to Y'$ such that $p = g' \circ r$ and $q = f' \circ r$.

(iii) In the context of (ii), show that if $f : Y \to X$ is proper, then so is $f' : Y' \to X'$.

(iv) Show that if $f : Y \to X$ is a proper mapping, where X is locally compact, and if $g : X' \to X$ is any continuous map, where X' is locally compact, then the induced map $f' : Y' \to X'$ described in (iii) is closed.

If X is a Hausdorff space, a *compactification* of X consists of a compact space $\widehat{X}$ and a homeomorphism of X onto a dense subset of $\widehat{X}$. In practice we identify X with its image so $X \subset \widehat{X}$. Assuming X is locally compact, the *one point compactification* X_∞ is the set obtained by adjoining to X a single point $\{\infty\}$. The open subsets of X_∞ are the open subsets of X, together with the complements in X_∞ of the compact subsets of X. Our assumption that X is locally compact is needed in order for this topology on X_∞ be Hausdorff.

(v) Let X be a locally compact Hausdorff space, and let $f : X \to W$ be a map into a compact space. Prove that there exists a compactification $\widehat{X}$ such that f may be extended to a continuous map on $\widehat{X}$.

Hint: Let $\widehat{X}$ be the closure of the graph of f in $X_\infty \times W$.

(vi) Prove conversely that if X and Y are Hausdorff spaces, and Y is locally compact, and if $f : Y \to X$ is a continuous mapping, and if for any locally compact Hausdorff space X', and any mapping $g : X' \to X$, the induced map $f' : X' \to Y'$ is closed, then f is proper.

Hint: If W is a compact subset of Y, take X' to be the compactification of $f^{-1}(W)$ as in (v). Show that $f^{-1}(W)$ is closed in the compact set X'.

(vii) Prove that if Y is a locally compact Hausdorff space, then Y is compact if and only if for any locally compact Hausdorff space Z the projection map $Y \times Z \to Z$ is a closed mapping.

7. Varieties

Varieties which are not affine arise in a number of ways. The first examples which one encounters are *projective spaces*. Recall that *projective n-space* $\mathbf{P}^n$ may be constructed as follows. Let V be a finite-dimensional vector space over k. Then $\mathbf{P}(V)$ is the set of lines through the origin in V. If $\dim(V) = n + 1$, then we will interpret $\mathbf{P}(V)$ as an n-dimensional variety. Since any two $n + 1$-dimensional vector spaces are isomorphic, we may as well take $V = k^{n+1}$, and we denote $\mathbf{P}(V) = \mathbf{P}^n$.

Although $\mathbf{P}^n$ is not an affine variety, nevertheless it admits a covering by open sets, each of which may be regarded as a copy of $\mathbf{A}^n$. To see this, let H be any $n - 1$-dimensional linear subspace through the origin in V. The set $\mathbf{A}_H$ of lines through the origin in V which are not contained in H may be identified with $\mathbf{A}^n$ as follows. Choose any vector $v \notin H$. Then each line $l \in \mathbf{A}_H$ intersects $v + H$ in a unique point. Since (after choosing a system of coordinates) $v + H$ is an n-dimensional affine space, the correspondence $l \to l \cap (v + H)$ identifies $\mathbf{A}_H$ with $\mathbf{A}^n$. Thus we get a bijection $\phi_H : \mathbf{A}^n \to \mathbf{A}_H$, which we will call a *chart*. (See Proposition 7.7 below for a general definition of this term.) We give $\mathbf{P}^n$ the *Zariski topology*. A subset of $\mathbf{P}^n$ is Zariski-closed if its intersection with each of these affine sets is closed in the Zariski topology on $\mathbf{A}^n$.

Projective spaces and their closed subsets are examples of *varieties*. Different foundations for algebraic geometry varieties have been given, by Weil, Serre and Grothendieck. We will take the viewpoint of Serre [23], in which a variety is a space, together with a sheaf of rings (basically specifying the regular functions on each open set), which is locally isomorphic to affine varieties. This is also the point of view taken in the first chapter of Mumford [21]. We should think of a variety as a space which is modeled on affine varieties in the same way that a manifold is a space which is modeled on Euclidean space.

If X is a topological space, a *presheaf* $\mathcal{F}$ on X is a rule which associates to every open set U an abelian group $\mathcal{F}(X)$; it is assumed that $\mathcal{F}(\varnothing) = 0$. If $U \supseteq V$ are a pair of nonempty open sets then there is a *restriction map* $\rho_{U,V} : \mathcal{F}(U) \to \mathcal{F}(V)$. These are assumed compatible in that if $U \supseteq V \supseteq W$, then $\rho_{V,W} \circ \rho_{U,V} = \rho_{U,W}$. Moreover $\rho_{U,U}$ is the identity map on U. A *morphism* of presheaves $\phi : \mathcal{F} \to \mathcal{G}$ consists of a homomorphism $\phi(U) : \mathcal{F}(U) \to \mathcal{G}(U)$ for every open set U such that if $U \supseteq V$ then $\rho_{U,V} \circ \phi(U) = \phi(V) \circ \rho_{U,V}$.

If the following *sheaf axiom* is satisfied, then the presheaf $\mathcal{F}$ is called a *sheaf*:

Sheaf Axiom 7.1. *Let U is an open set in X, and let $\{U_i | i \in I\}$ be a covering of U by nonempty open sets. Suppose $s_i \in U_i$ are given such that $\rho_{U_i, U_i \cap U_j}(s_i) = \rho_{U_j, U_i \cap U_j}(s_j)$ for each $i, j \in I$. Then there exists a unique $s \in \mathcal{F}(U)$ such that $s_i = \rho_{U, U_i}(s)$.*

If I is a partially ordered set, it is called a *directed set* if for each $i, j \in \mathcal{X}$ there exists a k such that $k \geq i, j$. If $\mathcal{C}$ is a category, a *directed family* in the category $\mathcal{C}$ consists of a directed set I, for each $i \in I$ an object A_i of $\mathcal{C}$, and, when $i \leq j$, a morphism $\rho_{i,j} : A_i \to A_j$. The *direct limit* $\varinjlim A_i$ (if it exists) consists of an object A, together with morphisms $\rho_i : A_i \to A$ satisfying $\rho_j \circ \rho_{i,j} = \rho_i$ when $i \leq j$, having the following universal property: if B is any object together with morphisms $\beta_i : A_i \to B$ such that $\beta_j \circ \rho_{i,j} = \beta_i$ when $i \leq j$, then there is a unique morphism $\beta : A \to B$ such that $\beta_i = \beta \circ \rho_i$. As usual, the universal property characterizes the direct limit up to isomorphism if it exists. Direct limits exist in the categories of sets, abelian groups, modules over a fixed ring, and (the category that interests us) k-algebras. One approach is to prove the existence of the direct limit in the category of abelian groups as in Theorem III.10.1 on p. 128 of Lang [18], then to note that if each of the A_i is an abelian group which has the extra structure of a k-algebra, and if the ρ_i are k-algebra homomorphisms, then the direct limit in the category of abelian groups has an induced k-algebra structure. We call a directed subset J of I *cofinal* if every element of I is majorized by an element of J. In this case it is easy to show that $\varinjlim_{i \in J} A_i = \varinjlim_{i \in I} A_i$.

Proposition 7.2. *Let A_i $(i \in I)$ be a direct system of abelian groups, with maps $\rho_{i,j} : A_i \to A_j$, $i \leq j$, and let $A = \varinjlim A_i$ be the direct limit, with maps $\rho_i : A_i \to A$. If each of the maps $\rho_{i,j}$ is injective, then each of the maps ρ_i is injective.*

Proof. The direct limit may be constructed as the quotient of the direct sum $\bigoplus_{i \in I} A_i$, modulo the subgroup R_I generated by elements of the form $\beta_i - \rho_{i,j}\beta_i$, where $\beta_i \in A_i$ and $i \leq j$. We call such a generator of R_I a *relation*. If $\alpha_i \in A_i$ is in the kernel of ρ_i, this means that $\alpha_i \in R_I$. Now in writing α_i as a finite sum of relations, there will be a $k \in I$ greater than or equal to every index involved in these relations. Let $J = \{j \in I | j \leq k\}$. Then J is itself a directed set, and by construction $\alpha_i \in R_J$, so the image of α_i in $\varinjlim_{j \in J} A_j$ is zero. But since J has a maximal element k, $\varinjlim_{j \in J} A_j$ is just A_k, for A_k clearly has the universal property of the direct limit. Hence $\rho_{i,k}(\alpha_i) = 0$. Because we are assuming that $\rho_{i,k}$ is injective, this shows that $\alpha_i = 0$. ■

If $\mathcal{F}$ is a presheaf on X, and U is an open subset, then $\mathcal{F}$ induces a presheaf, denoted $\mathcal{F}|_U$ on U. If $\mathcal{F}$ is a sheaf, so is $\mathcal{F}|_U$. Also, if $x \in X$, then the open sets containing x form a directed set, ordered by reverse inclusion (i.e. $U < V$ if $U \supset V$). The sets $\mathcal{F}(U)$ with $x \in U$ form a direct system, with respect to the restriction morphisms, so we may form the direct limit

$$\mathcal{F}_x = \varinjlim_{U \ni x} \mathcal{F}(U).$$

The group $\mathcal{F}_x$ is known as the *stalk* of $\mathcal{F}$ at x. We have an induced map, also known as restriction, $\rho_x : \mathcal{F}(U) \to \mathcal{F}_x$ when $x \in U$.

The sheaf $\mathcal{O}$ is a *sheaf of rings* if each $\mathcal{O}(U)$ has the structure of a ring, and the $\rho_{U,V}$ are ring homomorphisms. A *morphism* in the category of sheaves of rings is a morphism of sheaves in which each homomorphism is a ring homomorphism. A *ringed space* is an ordered pair $(X, \mathcal{O})$, where X is a topological space, and $\mathcal{O}$ is a sheaf of rings on X, sometimes known as the *structure sheaf*. A *morphism of ringed spaces* $(X, \mathcal{O}_X) \to (Y, \mathcal{O}_Y)$ consists of the following data. First, a continuous map $\phi : X \to Y$ of the underlying spaces, called the *underlying mapping*. Second, for each open subset U of Y, a ring homomorphism $\phi^* : \mathcal{O}_Y(U) \to \mathcal{O}_X(\phi^{-1}U)$ which is compatible with restriction: if $U \supseteq V$, then we have a commutative diagram:

$$
\begin{array}{ccc}
\mathcal{O}_Y(U) & \xrightarrow{\phi^*} & \mathcal{O}_X(\phi^{-1}U) \\
{\scriptstyle \rho_{U,V}} \downarrow & & \downarrow {\scriptstyle \rho_{\phi^{-1}U, \phi^{-1}V}} \\
\mathcal{O}_Y(V) & \xrightarrow{\phi^*} & \mathcal{O}_X(\phi^{-1}V)
\end{array}
$$

For the class of examples which most interest us—prevarieties—we will see below in Proposition 7.5 that the morphism is determined by its underlying map.

For example if X is an affine variety, for every nonempty subset $U \subseteq X$, let $\mathcal{O}_X(U)$ denote the ring of regular functions on U. This is a ringed space. The stalk $\mathcal{O}_x$ is just the local ring, as in Chapter 3. A morphism $\phi : X \to Y$ induces a morphism of ringed spaces in an obvious way—each map ϕ^* is induced by composition with ϕ. We recall that if X is an affine variety, then each $\mathcal{O}(U)$ and each $\mathcal{O}_x$ may be regarded as a subring of the function field F, which is the field of fractions of any one of these rings. The restriction maps are injective, meaning a regular function on U is determined by its values on any smaller open set, or by its image in the local ring at any point.

Remark 7.3. *It might seem that there is some loss of information in regarding an affine variety as a ringed space in this way—originally, the elements of $\mathcal{O}(U)$ were interpreted as functions $U \to k$, whereas in the ringed space, $\mathcal{O}(U)$ is an abstract ring. However the value $f(x)$ of an element $f \in \mathcal{O}(U)$ may be recovered as follows from the "abstract" data. The value $f(x)$ is the unique constant $a \in k$ such that the restriction $\rho_x(f - a)$ lies in the maximal ideal of the local ring $\mathcal{O}_x$. Hence there is no harm in thinking of the elements of $\mathcal{O}(U)$ as functions, when $(X, \mathcal{O})$ is an affine variety.*

If $(X, \mathcal{O})$ is a ringed space, an *affine open set* $U \subseteq X$ is a nonempty open set such that the ringed space $(U, \mathcal{O}|_U)$ is isomorphic to an affine algebraic variety. A *prevariety* is a ringed space $(X, \mathcal{O}_X)$ such that X is irreducible, and X admits a finite open cover by affine open sets. We will denote $\mathcal{O}_X = \mathcal{O}$ when there is no chance for confusion. This is consistent with our notations in Chapter 3.

Proposition 7.4. *Let $(X, \mathcal{O})$ be a prevariety. There exists a field F containing k together with an injective map $i_U : \mathcal{O}(U) \to F$ for every nonempty open set U. If $U \supseteq V$ are nonempty open sets, then the restriction map $\rho_{U,V} : \mathcal{O}(U) \to \mathcal{O}(V)$ is injective, and satisfies $i_V \circ \rho_{U,V} = i_U$. Therefore if we identify each $\mathcal{O}(U)$ with its image in F by means of the injection i_U, the restriction maps become inclusions. The local rings $\mathcal{O}_x$ $(x \in X)$ may also be identified with their images in F, and the restriction maps $\rho_x : \mathcal{O}(U) \to \mathcal{O}_x$ $(x \in U)$ become inclusions. With these identifications, F is the field of fractions of any $\mathcal{O}_x$, or any $\mathcal{O}(U)$ where U is affine.*

Proof. Let us show that if $U \supseteq V$ are any nonempty open sets then $\rho_{U,V}$ is injective. Let $s \in \mathcal{O}(U)$ be in the kernel. Let U_i be affine open sets which cover X. We note that

$$\rho_{U \cap U_i, V \cap U_i}\big(\rho_{U, U \cap U_i}(s)\big) = \rho_{V, V \cap U_i}\big(\rho_{U,V}(s)\big) = 0.$$

Since $U \cap U_i$ and $V \cap V_i$ are contained in the affine variety U_i, and since we know that the restriction maps are injective for affine varieties, this implies that $\rho_{U, U \cap U_i}(s) = 0$. Now the uniqueness assertion from the Sheaf Axiom 7.1 implies that $s = 0$.

We consider the set $\mathcal{X}$ of nonempty open sets. Since X is irreducible, any two nonempty open sets intersect. Thus $\mathcal{X}$ forms a directed set in which $U \leq V$ if $U \supseteq V$. We may therefore form the direct limit

$$F = \varinjlim \mathcal{O}(U),$$

and since we have proved that each $\rho_{U,V}$ is injective for $U, V \in \mathcal{X}$, the induced maps $i_U \to F$ are injective by Proposition 7.2. Hence denoting $F = \varinjlim \mathcal{O}(U)$, we obtain a ring $F(X)$ into which each $\mathcal{O}(U)$ is injected. Clearly we may regard every $\mathcal{O}(U)$ as a subset of F.

If U is any affine open set, then the set of nonempty open subsets of U are cofinal in X, so $F(X) = F(U)$. From Chapter 3, we know that $F(U)$ is the field of fractions of $\mathcal{O}(U)$, and of any $\mathcal{O}_x$ with $x \in U$. ■

It is a consequence of the sheaf axiom that $\mathcal{O}(U)$ is the intersection of the $\mathcal{O}_P$ over all $P \in U$ (Exercise 7.1). Since by Proposition 7.4 the restriction maps $\rho_{U,V} : \mathcal{O}(U) \to \mathcal{O}(V)$ and $\rho_x : \mathcal{O}(U) \to \mathcal{O}_x$ are injective, even if U is not affine, we may still identify $\mathcal{O}(U)$ with a subring of F. However if U is not affine, F will no longer be the field of fractions of $\mathcal{O}(U)$. Identifying $\mathcal{O}(U)$ with its image in $\mathcal{O}(V)$ when $U \subseteq V$, the union of the $\mathcal{O}(U)$ is the function field F (Exercise 7.2).

We will say that $f \in F$ is *regular* at $P \in X$ if $f \in \mathcal{O}_P$. The set of all points where f is regular forms an open set. If f is regular at x, then Remark 7.3 gives us a prescription for assigning a value $f(x)$—it is the unique $a \in k$ such that $f - a$ lies in the maximal ideal of $\mathcal{O}_x$. We will therefore regard the ring $\mathcal{O}(U)$ (U open) as the ring of functions regular on U, which may be thought of as functions in the usual sense $U \to k$.

Proposition 7.5. *Let $(X, \mathcal{O}_X)$ and $(Y, \mathcal{O}_Y)$ be prevarieties, and let $f : X \to Y$ be a continuous map. There exists at most one morphism of prevarieties with underlying map f.*

Proof. We must show that if $U \subset Y$ is an open set, the map $f^* : \mathcal{O}(U) \to \mathcal{O}(f^{-1}U)$ is determined by f. As we have pointed out, we may regard elements of $\mathcal{O}(U)$ as functions on U taking values in k. Regarding the elements of $\mathcal{O}(U)$ as functions in this concrete way, the map $f^* : \mathcal{O}(U) \to \mathcal{O}(f^{-1}U)$ is just composition with f. Hence for prevarieties the morphism of ringed spaces is completely determined by the underlying map f. ■

A closed irreducible subset Y of a prevariety X is a prevariety. We may define, as in Chapter 3, a local ring $\mathcal{O}_{Y,X}$ associated with Y. This is the set of all $f \in F$ such that $f \in \mathcal{O}_P$ for some $P \in Y$. (This is consistent with the definition given in Chapter 3 for affine varieties.) The local maximal ideal $\mathfrak{M}_{Y,X}$ of $\mathcal{O}_{Y,X}$ then consists of all $f \in \mathcal{O}_{Y,X}$ such that $f(P) = 0$ for all $P \in Y$ where f is regular. It is not hard to see that the function field of Y is isomorphic to $\mathcal{O}_{Y,X}/\mathfrak{M}_{Y,X}$ (Exercise 7.11).

An open set in a prevariety also inherits the structure of a prevariety. There is a slight nuance here: it is possible that an open set in an affine variety is a prevariety, but is not isomorphic to an affine variety.

Example 7.6. Let $X = \mathbf{A}^2$, and let U be the open set which is the complement of the origin. Then $\mathcal{O}(U) = \mathcal{O}(\mathbf{A}^2) = k[x, y]$ is the ring of polynomials in two variables. According to our definitions, the open set U is itself a prevariety, but it is not isomorphic to any affine variety. It admits a cover by affine sets which are affine varieties. Two will suffice: consider the two complements of the coordinate axes. These are principal open sets in $\mathbf{A}^2$, hence are affine varieties (Exercise 3.1). For example, the coordinate ring of the complement of the x-axis is the ring $k[x, y, y^{-1}]$.

Because of such examples, we define a *quasi-affine* prevariety to be a prevariety which is isomorphic to an open set in an affine variety.

Observe that to specify the structure of a prevariety on a space, it is sufficient to specify the topology, the function field F and the local rings $\mathcal{O}_P \subseteq F$, since if the local rings are known, the rings $\mathcal{O}(U)$ are determined by the requirement that $\mathcal{O}(U) = \bigcap_{P \in U} \mathcal{O}_P$. Moreover, if a cover of X by open sets U_i is given, and each of the latter given the structure of an affine variety, then the local rings $\mathcal{O}_P$ are determined, and so these data, if consistent, are sufficient to specify the prevariety structure. One formulation of the consistency requirement resembles the definition of a manifold as a space locally modeled on Euclidean space:

Proposition 7.7 (Gluing Lemma). *Let X be an irreducible Noetherian topological space, and let $\{U_i \mid i \in I\}$ be a finite open cover of X. Suppose that there is given for each $i \in I$ a bijection $\phi_i : V_i \to U_i$, where V_i is an affine variety. Suppose that for each $i, j \in I$ there is given a birational map $\phi_{i,j} : V_i \to V_j$ such that $\phi_{i,j}$ is regular on $\phi_i^{-1}(U_i \cap U_j)$, and agrees with $\phi_j^{-1} \circ \phi_i$ on this set. Then X admits a unique structure of a prevariety such that the ϕ_i are morphisms.*

We call the maps $\phi_i : V_i \to U_i$ *charts*.

Proof. We leave it to the reader to complete the following sketch (Exercise 7.3). One defines X to be the quotient disjoint union $\coprod V_i$ modulo the relation $\phi_i^{-1}(x) \sim \phi_j^{-1}(x)$ for $x \in U_i \cap U_j$. We define $\mathcal{O}(U)$, for $U \subset X$ to be the ring of functions $f : U \to k$ such that $f \circ \phi_i \in \mathcal{O}(\phi_i^{-1} U_i)$ for each $I \in I$. It must be checked that $\mathcal{O}$ is a sheaf. ∎

Proposition 7.8. *Let X and Y be prevarieties. Suppose that $\{U_i \mid i \in I\}$ and $\{V_i \mid i \in I\}$ are affine open covers of X and Y, and $f : X \to Y$ a continuous map. Suppose that $f(U_i) \subseteq V_i$, and that the restriction of f to U_i is a morphism of affine varieties $U_i \to V_i$. Then f is a morphism of prevarieties.*

There is an abuse of language here, since strictly speaking a homomorphism of prevarieties is a morphism of ringed spaces, not just a continuous map of the underlying spaces. This abuse of language is justified by Proposition 7.5, since the morphism of ringed spaces (if it exists) is uniquely determined by its underlying map.

Proof. Let V be an open set in Y. We regard elements of $\mathcal{O}_Y(V)$ and $\mathcal{O}_X(f^{-1}V)$ as k-valued functions on V and $f^{-1}V$, respectively. We will show that composition with f defines a map $\mathcal{O}_Y(V) \to \mathcal{O}_X(f^{-1}V)$. Indeed, if $\phi : V \to k$ is a regular function, then $\phi \circ f : f^{-1}V \to k$ is in the local ring $\mathcal{O}_x(X)$ for each $x \in f^{-1}V$, because x is in some U_i, and by assumption, f induces a morphism $U_i \to V_i$ of affine varieties. Since $\phi \circ f \in \mathcal{O}_x(X)$ for each $x \in f^{-1}V$, $\phi \circ f \in \mathcal{O}_X(f^{-1}U)$. Composition with f therefore induces a map $\mathcal{O}_Y(V) \to \mathcal{O}_X(f^{-1}V)$, which is a morphism of ringed spaces. ∎

Proposition 7.9. *Projective space* $\mathbf{P}^n$ *is naturally a prevariety.*

Proof. In the opening paragraphs of this chapter, we pointed out that $\mathbf{P}^n$ could be covered by spaces which are naturally regarded as affine spaces. A finite number of these affine spaces ($n + 1$ to be exact) are sufficient to cover $\mathbf{P}^n$. Call these U_i, and let $\phi_i : \mathbf{A}^n \to U_i$ be the charts. It is easy to see that the maps $\phi_{i,j}$ in Proposition 7.7 are morphisms, and so $\mathbf{P}^n$ becomes a prevariety. ∎

Proposition 7.10. *Let X and Y be prevarieties. Then the Cartesian product $X \times Y$ admits the structure of a prevariety in such a way that if $U \subseteq X$ and $V \subseteq Y$ are affine open sets, then the inclusion map $U \times V \to X \times Y$ is a morphism.*

Proof. To topologize $X \times Y$, we say that $W \subset X \times Y$ is open if and only if $W \cap (U \times V)$ is open for every pair of affine open subsets $U \subseteq X$ and $V \subseteq Y$. If $\{U_i\}$ is a finite open cover of X and $\{V_j\}$ is a finite open cover of Y, then we may take $\{U_i \times V_j\}$ to be a finite open cover of $X \times Y$. Picking charts, the hypotheses of Proposition 7.7 are easy to verify. ∎

Theorem 7.11. *Let X and Y be prevarieties. If $f : Z \to X$ and $g : Z \to Y$ are morphisms of prevarieties, the map $\phi : Z \to X \times Y$ given by $\phi(z) = \big(f(z), g(z)\big)$ is a morphism of prevarieties. The projection maps $p : X \times Y \to X$ and $q : X \times Y \to Y$ are morphisms. The prevariety $X \times Y$ is the product in the category of prevarieties.*

Proof. First let us check that $p : X \times Y \to X$ and $q : X \times Y \to Y$ defined by $p(x, y) = x$ and $q(x, y) = y$ are morphisms. Indeed, if $(x, y) \in X \times Y$, find affine open sets $U \subseteq X$ and $V \subseteq Y$ such that $x \in U$ and $y \in V$. Then p and q agree with the projection maps $U \times V \to U$ and $U \times V \to V$, which are morphisms of affine varieties. By Proposition 7.8, it follows that p and q are morphisms.

Let $P \in Z$. Since f and g are continuous, and since the affine open sets form a basis of the topology of any prevariety, we may find affine open sets

$W \subseteq Z$, $U \subseteq X$ and $V \subseteq Y$ such that $P \in W$, and $f(W) \subseteq U$, $g(W) \subseteq V$. It follows from the definition of a morphism that the restrictions of f and g to W are morphisms of affine varieties. According to Theorem 4.19, $U \times V$ is the product in the category of affine varieties, and the universal property of the product implies that f and g induce a map $\phi_W : W \to U \times V$ such that $p \circ \phi_W = f$ and $q \circ \phi_W = g$. This means that $\phi_W(z) = \big(f(z), g(z)\big)$ for $z \in W$, so ϕ_W is the restriction of ϕ to $U \times V$. Now $U \times V$ is an affine open set in $X \times Y$, and so we have verified that ϕ satisfies the definition of a morphism. Obviously ϕ is the unique map such that $p \circ \phi = f$ and $q \circ \phi = g$, and so we have verified that $X \times Y$ satisfies the universal property of the product in the category of prevarieties. ■

We say a prevariety *separated* if the diagonal is closed in $X \times X$. A *variety* is a separated prevariety. This requirement is imposed to exclude certain pathology (Exercise 7.4). According to Exercise 7.4 an affine variety is separated. A map $\phi : X \to Y$ between varieties is called a *morphism* if and only if it is a morphism of prevarieties. A variety of dimension one is called a *curve*, and a variety of dimension two is called a *surface*.

Proposition 7.12. *(i) Projective space* $\mathbf{P}^n$ *and affine space* $\mathbf{A}^n$ *are separated. (ii) If* X *is separated, then any open or closed subspace* $Y \subseteq X$ *is separated.*

Proof. To prove (i), affine varieties are separated by Exercise 4.4. Hence $\mathbf{A}^n$ is separated. As for $\mathbf{P}^n$, suppose that $(P, Q) \in \mathbf{P}^n \times \mathbf{P}^n$ is in the closure of the diagonal. It is possible to find an open affine set U isomorphic to $\mathbf{A}^n$ such that P, Q are both in U. Since the diagonal is closed in U, we have $P = Q$. This proves that the diagonal is closed in $\mathbf{P}^n$.

To prove (ii), observe that the diagonal in $Y \times Y$ is the intersection of $Y \times Y$ with the diagonal in $X \times X$. The latter is closed by assumption, hence the diagonal is closed in $Y \times Y$. ■

We see that any irreducible closed subset of $\mathbf{P}^n$ is a variety. If a variety is isomorphic to a closed subset of $\mathbf{P}^n$ we call it a *projective variety*. A variety which is isomorphic to an open set in a projective variety is called *quasi-projective*.

Proposition 7.13. *If* X *and* Y *are separated, then so is* $X \times Y$.

Proof. The diagonals $\Delta_X \subseteq X \times X$ and $\Delta_Y \subseteq Y \times Y$ are closed, and so, by Exercise 7.5 is $\Delta_X \times \Delta_Y \subseteq X \times X \times Y \times Y$. The diagonal $\Delta_{X \times Y} \subseteq X \times Y \times X \times Y$ is the image of this set under the isomorphism $X \times X \times Y \times Y \to X \times Y \times X \times Y$ which interchanges the middle factors. ■

We define a variety X to be *complete* if for every variety Y, the morphism $p : X \times Y \to Y$ defined by $p(x, y) = y$ is a closed mapping. According to Exercise 6.5 (vi), the corresponding property for Hausdorff spaces is compactness, so the role of completeness in algebraic geometry is very analogous to the role of compactness in topology.

Proposition 7.14. *Let X be a variety. Suppose the projection $X \times U \to U$ is closed for every affine variety Y. Then X is complete.*

Proof. We must show that the projection $p : X \times Y \to Y$ is a closed map for an arbitrary variety Y. Let $V \subseteq X \times Y$ be a closed set. We will show that $p(V) \subseteq Y$ is closed. Let $\{U_i\}$ be a covering of Y by affine open sets, and let $p_i : X \times U_i \to U_i$ be the projections. Obviously $p(V) \cap U_i = p_i(V \cap U_i)$. By hypothesis p_i is a closed map, so $p(V) \cap U_i$ is closed for every i. It follows that $p(V)$ is closed. ∎

Proposition 7.15. *Let X and Y be prevarieties, and assume that Y is separated. Let $\phi : X \to Y$ be a morphism. Let Γ_ϕ be the graph $\{(x, \phi(x)) \mid x \in X\} \subseteq X \times Y$ of ϕ. Then Γ_ϕ is closed subset of $X \times Y$.*

Proof. Define $\Phi : X \times Y \to Y \times Y$ by $\Phi(x, y) = (\phi(x), y)$. This is a morphism by Theorem 7.11. The graph Γ_ϕ is the preimage under Φ of the diagonal in $Y \times Y$ which is closed by hypothesis. Hence Γ_ϕ is closed. ∎

Proposition 7.16. *(i) If X and Y are complete varieties, then $X \times Y$ is complete;*

(ii) If X is a complete variety, and if Y is a closed subset, then each irreducible component of Y is a complete variety;

(iii) If X and Y are varieties, and X is complete, then for any morphism $\phi : X \to Y$, the image $\phi(X)$ is closed in Y, and is a complete variety; moreover, ϕ is a closed mapping;

(iv) If X is a complete affine variety, then X consists of a single point.

Proof. To prove (i), observe that the projection $X \times Y \times Z \to Z$ may be factored as $X \times Y \times Z \to Y \times Z \to Z$. The first map is closed since X is complete, and the second is closed since Y is complete. Hence the composite is closed, so $X \times Y$ is complete.

To prove (ii), let Y_i be an irreducible component of Y. Consider the projection $Y_i \times Z \to Z$. This factors as the composition $Y_i \times Z \to X \times Z \to Z$. The first map is the inclusion of a closed subset, hence is a closed map, while the second map is closed since X is complete. Hence Y_i is complete.

To prove (iii), consider the graph $\Gamma_\phi = \{(x, \phi(x)) \mid x \in X\} \subseteq X \times Y$. This is closed by Proposition 7.15. Now $\phi(X)$ is the image of Γ_ϕ under projection $X \times Y \to Y$, and is closed since X is complete. To show that $\phi(X)$ is complete, let $V \subseteq \phi(X) \times Z$ be a closed set. We will show that the projection of V onto Z is closed. Define a morphism $\Phi : X \times Z \to Y \times Z$ by $\Phi(x, z) = (\phi(x), z)$. Since we have just proved that $\phi(X)$ is closed in Y, and $\phi(X) \times Z$ is closed in $Y \times Z$. Since V is closed in $\phi(X) \times Z$, it follows that V is closed in $Y \times Z$. Hence $\Phi^{-1}(V)$ is closed in $X \times Z$, and since X is complete, its projection on Z is closed. But this is the same as the projection of $V \subset \phi(X) \times Z$ onto Z. This proves that $\phi(X)$ is complete. It is irreducible since X is irreducible, by Exercise 1.7. We have proved that $\phi(X)$ is a complete variety. Finally to show that ϕ is a closed map, let $Z \subseteq X$ be a closed subset. Then Z is complete by (ii), so by what we have just shown, $\phi(Z)$ is a complete closed subset of Y.

To prove (iv), let us first show that $\mathbf{A}^1$ is not complete. Indeed, we consider the hyperbola $xy = 1$ in $\mathbf{A}^1 \times \mathbf{A}^1$. This is closed. However, its projection on the second factor is the complement of the origin and this is not closed. This shows that the projection $\mathbf{A}^1 \times \mathbf{A}^1 \to \mathbf{A}^1$ on the second factor is not a closed map, and so $\mathbf{A}^1$ is not complete.

We may now settle the general case. Suppose that X is an affine variety of dimension greater than zero. Let A be its coordinate ring. Let $f \in A - k$. Then the inclusion $k[f] \to A$ gives rise to a morphism $X \to \mathbf{A}^1$. Now suppose that X is complete. Then the image has dimension greater than zero since f is not a constant function, and is closed by (iii). Hence the image is all of $\mathbf{A}^1$. This image is complete by (iii), so we see that $\mathbf{A}^1$ is complete. This is a contradiction. ∎

Many of the concepts developed in Chapters 3 and 6 apply to varieties, with the hard work being to do the affine case. *Rational maps* are defined just as in the affine case—a rational map $X \to Y$ is a morphism defined on a dense open set $U \subseteq X$. We identify two rational maps even if their domains are different if they agree on a nonempty open set. There is a maximal open set on which a given rational map is regular. A morphism or rational map is called *dominant* if there is an open set on which it is regular whose image is dense. A dominant rational map induces a map of function fields, and the category of varieties and dominant rational maps is equivalent to the (opposite category of) the category of function fields over k, i.e. Theorem 3.8 remains true for varieties. A dominant rational map is a *birational map* or *birational equivalence* if it induces an isomorphism of function fields.

Just as with affine varieties, we say that a point P in a variety X is *smooth* or *nonsingular* if $O_P(X)$ is a regular local ring. If P is not smooth, we say it is *singular*. X is a *smooth* or *nonsingular* variety if every point is nonsingular. Theorem 6.15, asserting that the singular points form a proper closed subset, is valid for varieties. The general case follows immediately from the affine case.

Just as in the affine case, we may define the *normalization* of a variety. If X is a variety, the *normalization* of X is a variety Y together with a surjective morphism $f : Y \to X$, which is a birational equivalence, and such that if $U \subseteq X$ is an affine, open set then $f^{-1}(U)$ is affine, and $\mathcal{O}\big(f^{-1}(U)\big)$ is the integral closure of $\mathcal{O}(U)$. The normalization may be constructed from the affine case by means of the Gluing Lemma (Proposition 7.7).

Now we give a *valuative criterion* for a variety to be complete. This approach to completeness is due to Chevalley.

Proposition 7.17 (Valuative Criterion for Completeness). *Let X be a variety with function field F, and suppose that for every irreducible closed subset $Z \subseteq X$, and for every valuation ring R of $F(Z)$ containing k, there is a point $P \in Z$ such that $\mathcal{O}_P(Z) \subseteq R$. Then X is complete.*

Proof. Let Y be an affine variety, and let $V \subseteq X \times Y$ be a closed subset. Let $p : X \times Y \to X$ and $q : X \times Y \to Y$ be the projection maps. By Proposition 7.14, it is sufficient to show that $q(V)$ is closed in Y. Clearly we may assume that V is irreducible. Let Z be the closure of $p(V) \subseteq X$. Then Z is irreducible by Exercise 1.7. By hypothesis Z has the property that every valuation ring of $F(Z)$ containing k contains the local ring $\mathcal{O}_P(Z)$ for some $P \in Z$.

We may replace Y by the closure of $q(V)$ without changing the validity of the assertion which we are trying to prove (that $q(V)$ is closed). Therefore we may assume that the composite morphism $V \to Z \times Y \to Y$ is a dominant map. Thus $q(V)$ is dense in Y, and we must show that it is all of Y.

Let $F(Z)$ be the function field of Z, and let K be the function field of V. The composition $V \to Z \times Y \to Z$ is dominant, and so there is induced an injection $j : F(Z) \to K$ of function fields.

Now let $Q \in Y$. We will construct $P \in Z$ such that $(P, Q) \in V$. Note that this will show that $q(V) = Y$, as required. Let B be the coordinate ring of Y. Let $\phi : B \to k$ be the homomorphism $\phi(g) = g(Q)$. Since the map $V \to Z \times Y \to Y$ is dominant, B is naturally injected as a subring in K. By the Extension Theorem (Theorem 2.2), we may find a valuation ring R of K containing B and a homomorphism $\Phi : R \to k$ extending ϕ. Let us consider the inverse image of R under the homomorphism $j : F(Z) \to K$ just constructed. This is a valuation ring of $F(Z)$ containing k. By hypothesis, there exists $P \in Z$ such that $\mathcal{O}_P(Z) \subseteq j^{-1}(R)$. We will show that $(P, Q) \in V$.

Let U be an affine open neighborhood of P in Z. Let $V' = (U \times Y) \cap V$. Then V' is a nonempty affine open subset of V. Let A and C be the respective coordinate rings of the affine varieties U and V. Then C is a subring of K. We will show that $C \subseteq R$. Let $i : V' \to U \times Y$ be the inclusion, and let $i^* :

$A \otimes B \to C$ be the induced map of coordinate rings. We have homomorphisms $p^* : A \to A \otimes B$ and $q^* : B \to A \otimes B$ induced by the projection maps, so $p^*(a) = a \otimes 1$ and $q^*(b) = 1 \otimes b$. The map i^* is surjective since i is the inclusion of a closed set into an affine variety. We recognize now the restriction of $j : F(Z) \to K$ to the ring A as the composition $i^* \circ p^*$, and since the image of j is contained in R, this shows that $i^*(A \otimes 1) \subseteq R$. On the other hand, $i^*(1 \otimes B)$ is the image of B under the injection $B \to K$, which is contained in R by construction. Now $C = i^*(A \otimes B)$ is generated by $i^*(A \otimes 1)$ and $i^*(1 \otimes B)$, and we have thus proved that C is contained in R. Thus the homomorphism $\Phi : R \to k$ is defined on all of C, and its kernel is a maximal ideal $\mathfrak{m}$, which clearly corresponds to the point (P, Q) of V. The image of this point in Y is of course Q, which completes the proof that $q(V) = Y$. This shows that X is complete. ■

Proposition 7.18. *Let k be an algebraically closed field, F a field containing k. Let R be a valuation ring of F containing k, and let $\mathfrak{m}$ be the maximal ideal of R. Then there is a valuation ring R' of F containing with maximal ideal $\mathfrak{m}'$ such $R' \subseteq R$, $\mathfrak{m} \subseteq \mathfrak{m}'$, and such that the composition $k \to R' \to R'/\mathfrak{m}'$ is an isomorphism.*

Proof. Let $K = R/\mathfrak{m}$. We identify k with its image under the composition $k \to R \to K$. Since k is algebraically closed, we may apply the Extension Theorem (Theorem 2.2) to the identity homomorphism $k \to k$, and we see that there exists a valuation ring $R_1 \subseteq K$ containing k with maximal ideal $\mathfrak{m}_1$ such that the composition $k \to R_1 \to R_1/\mathfrak{m}_1$ is an isomorphism. Let R' be the preimage of R_1 in R, and let $\mathfrak{m}'$ be the preimage of $\mathfrak{m}_1$. The composition $k \to R' \to R'/\mathfrak{m}' \cong R_1/\mathfrak{m}_1$ is therefore an isomorphism. We only have to show that R' is a valuation ring of f. Suppose that $f \in F - R'$. We will show that $f^{-1} \in \mathfrak{m}'$. If $f \notin R$, then since R is a valuation ring, $f^{-1} \in \mathfrak{m} \subseteq \mathfrak{m}'$. Thus we may assume that $f \in R$. Now $f \notin \mathfrak{m}$ since $\mathfrak{m} \subseteq R'$, and therefore $f^{-1} \in R$ also. Let $\overline{f}$ be the image of f in K. By hypothesis $\overline{f} \notin R_1$, and since R_1 is a valuation ring, $\overline{f}^{-1} \in \mathfrak{m}_1$. This means that $f^{-1} \in \mathfrak{m}'$, as required. ■

By a *linear subspace* of $\mathbf{P}^n = \mathbf{P}(k^{n+1})$, we mean a subspace $\mathbf{P}(U)$, where U is a vector subspace of k^{n+1}. If the codimension of U is one, we may also describe $\mathbf{P}(U)$ as a *hyperplane*.

Proposition 7.19. *Projective varieties are complete.*

Proof. By Proposition 7.16 (ii), it is sufficient to show that $\mathbf{P}^n$ itself is complete. For this, we will use the valuative criterion of Proposition 7.17. Let $Z \subseteq \mathbf{P}^n$ be an irreducible closed subspace, and let R be a valuation ring of $F(Z)$, with maximal ideal $\mathfrak{m}$. We will show that $\mathcal{O}_P(Z) \subseteq R$ for some $P \in Z$.

We may assume that Z is not contained in any linear subspace of $\mathbf{P}^n$, since such a space is isomorphic to $\mathbf{P}^r$ for $r < n$, and by induction we may assume that this property of Z has been demonstrated for irreducible closed subspaces of $\mathbf{P}^r$. By Proposition 7.18, we may find a valuation ring $R' \subseteq R$ containing k, with maximal ideal $\mathfrak{m}'$, such that the composition $k \to R' \to R/\mathfrak{m}'$ is an isomorphism. Since $\mathcal{O}_P \subseteq R'$ implies $\mathcal{O}_P \subseteq R$, we may replace R with R'. Hence we may assume that the composition $k \to R \to R/\mathfrak{m}$ is an isomorphism. Therefore there exists a homomorphism $\phi : R \to k$ which is the identity on k, and whose kernel is precisely $\mathfrak{m}$.

Let $x_0, x_1, \cdots, x_n$ be the coordinate functions on k^{n+1}. These are not themselves functions in $F = F(\mathbf{P}^n)$, but after dividing by x_0, the functions $\xi_0 = 1, \xi_1 = x_1/x_0, \cdots, \xi_n = x_n/x_0 \in F$. Note that ξ_i is in $\mathcal{O}_{Z,X}$ because we have arranged that Z is not contained in any linear subspace of $\mathbf{P}^n$, so that in particular it is not in the linear subspace determined by the equation $x_0 = 0$, and therefore x_i/x_0 is regular at some point of Z. Let η_i be the image of ξ_i in $F(Y)$, for $i = 0, \cdots, n$. By Proposition 4.16, there exists $\lambda \in F$ such that the $\lambda\eta_0, \lambda\eta_1, \cdots, \lambda\eta_n$ all lie in R, but not all of them lie in $\mathfrak{m}$, so one of them is a unit. If it is $\lambda\eta_i$, then dividing by this, we see that $\eta_0/\eta_i, \eta_1/\eta_i, \cdots, \eta_n/\eta_i \in R$. Now η_j/η_i is the image of x_j/x_i in $\mathcal{O}_{Z,X}$, and after reordering the coordinates, we may assume that $\eta_1, \cdots, \eta_n \in R$, where η_i is the image of x_i/x_0 in $\mathcal{O}_{Z,X}$. Now let U be the affine open subset of $\mathbf{P}^n$ which is the complement of the hyperplane determined by the equation $x_0 = 0$ in k^{n+1}. Then U is isomorphic to $\mathbf{A}^n$, with coordinate functions $\xi_1, \cdots, \xi_n$. The closed subspace $Z \cap U$ of U is an affine variety with coordinate ring $B = k[\eta_1, \cdots, \eta_n] \subseteq R$. Let P be the point in U whose coordinates are $(a_1, \cdots, a_n)$ where $a_i = \phi(\eta_i)$. It is clear that $P \in Z \cap U$. We will show that $\mathcal{O}_P \subseteq R$. Any element of $\mathcal{O}_P$ may be written as f/g where $f, g \in B$, and $g(P) \neq 0$. Write $g = g(\eta_1, \cdots, \eta_n)$. Then $\phi(g) = g(a_1, \cdots, a_n) = g(P) \neq 0$, so $g \notin \mathfrak{m}$. Hence g is invertible in the valuation ring R, and therefore $f/g \in R$. This proves that $\mathcal{O}_P \subseteq R$. ∎

The following rigidity theorem says that a constant map from a complete variety into an arbitrary variety cannot be deformed into a nonconstant map.

Proposition 7.20. *Let X be a complete variety, and let Y and Z be arbitrary varieties. Suppose that $f : X \times Y \to Z$ is a morphism. Suppose that there exists a point $y_0 \in Y$ such that $f(X \times \{y_0\})$ consists of a single point $z_0 \in Z$. Then there exists a morphism $g : Y \to Z$ such that $f(x,y) = g(y)$ for all $(x,y) \in X \times Y$.*

Proof. Let U be any affine open set containing z_0. Let $W = Z - U$, and let

$$V = \{y \in Y \mid f(x,y) \notin W \text{ for all } x \in X\}.$$

Let us show that V is open in Y. Indeed, W is closed in Z, so $f^{-1}(W)$ is closed in $X \times Y$, and the complement of V is the image of $f^{-1}(W)$ under the projection $X \times Y \to Y$. Since X is complete, it follows that the complement of V is closed, so V is open. Now suppose that $y \in V$. Then $f(X \times \{y\})$ is the image of a complete variety, hence is complete by Proposition 7.16 (iii). However, $f(X \times \{y\})$ is contained the affine open set U by construction. This means that $f(X \times \{y\})$ is a complete, irreducible, and affine, hence consists of a single point by Proposition 7.16 (iv).

Now let x_0 be any fixed element of X, and define $g(y) = f(x_0, y)$. We have shown that if $y \in V$, then $f(X, y)$ consists of a single point. Thus $f(x, y) = g(y)$ if $y \in V$ and $x \in X$. Now $X \times V$ is open in $X \times Y$, and we have shown that the morphisms $(x, y) \mapsto f(x, y)$ and $(x, y) \mapsto g(y)$ agree on this open set. Hence these two morphisms are the same. ∎

We end this chapter with some remarks about *blowing up* and the resolution of singularities, a topic which we have already touched on in Chapter 6. There are two closely related questions, one global, the other local. The global question is, given a variety X, to find a complete nonsingular variety birationally equivalent to X. This problem may not have a unique solution. For example, we call a surface *rational* if it is birationally equivalent to $\mathbf{P}^2$, or equivalently if its function field is isomorphic to the *rational function field* $k(x, y)$, where x and y are algebraically independent. Thus $\mathbf{P}^2$ and $\mathbf{P}^1 \times \mathbf{P}^1$ are birationally equivalent, both being rational surfaces. Each is nonsingular. They are nonisomorphic, as may be seen by the following consideration: in $\mathbf{P}^1 \times \mathbf{P}^1$, there may be found two closed curves which do not intersect, for example $\mathbf{P}^1 \times \{0\}$ and $\mathbf{P}^1 \times \{\infty\}$. No two such nonintersecting closed curves may be found in $\mathbf{P}^2$.

The *local* question which presents itself is that of desingularization. If X is a (not necessarily complete) variety, we wish to find a surjective proper morphism $f : Y \to X$ which is a birational equivalence, where Y is a smooth variety, such that f^{-1} is regular on the set of smooth points of X. In characterizing this question as local, we have in mind the possibility of dealing with the singular subvarieties of Y one at a time.

We have already introduced one tool for desingularization, namely normalization. The normalization is always nonsingular in codimension one by Theorem 6.24. Thus for a curve, the normalization is nonsingular, and for a surface, the singularities of the normalization form a zero-dimensional closed set, i.e. a finite set of points. The process of normalization does not take one out of the category of affine varieties, i.e. if X is affine, so is its normalization.

The surface V which is the locus of $x^2 + y^2 + z^2 = 0$ in $\mathbf{A}^3$ is normal (Exercise 5.8), yet it has a singularity at the origin. This example shows that the process of normalization does not solve the desingularization problem for singularities of surfaces, though it is helpful in reducing the problem to that of zero dimensional singularities.

To resolve the singularity of $x^2 + y^2 + z^2 = 0$, we make use of another tool, namely *blowing up* of the singularity. Let X be a variety, and let $P \in X$ be a point, which may or may not be singular. By *blowing up* the point, we mean exhibiting a variety Y together with a proper surjective morphism $\phi : Y \to X$ which is a birational equivalence, such that ϕ^{-1} is an isomorphism of $X - P$ onto $\phi^{-1}(X - P)$. Since ϕ is proper, $\phi^{-1}(P)$ is complete (Exercise 7.12 (iii)). It may or may not be irreducible, and it may or may not be nonsingular.

Returning to the special case of the variety V which is the locus of $x^2 + y^2 + z^2 = 0$ in $\mathbf{A}^3$, we can blow up the singularity. We will exhibit a nonsingular surface Y together with a desingularization $\phi : Y \to V$. The map ϕ^{-1} will be biregular on $V - (0,0,0)$, while $\phi^{-1}(0,0,0)$ will be a curve isomorphic to $\mathbf{P}^1$. This example shows how desingularization can take us out of the category of affine varieties, since $\mathbf{P}^1$ cannot be embedded in affine space.

To accomplish the desingularization of V, we make use of the morphism $\psi : \mathbf{A}^3 - (0,0,0) \to \mathbf{P}^2$ which maps each point (x,y,z) with x,y,z not all zero to the line through the origin and (x,y,z). We interpret this line as a point in $\mathbf{P}^2$, which we will denote by its *homogeneous coordinates* $x : y : z$. Thus $\lambda x : \lambda y : \lambda z = x : y : z$ if $\lambda \neq 0$. Now $\psi\big(V - (0,0,0)\big)$ is the curve

$$C = \{x : y : z \mid x^2 + y^2 + z^2 = 0\}.$$

It may be checked that C is isomorphic to $\mathbf{P}^1$, namely the rational function

$$x : y : z \to \begin{cases} x/y & \text{if } y \neq 0 \\ \infty & \text{if } y = 0 \end{cases}$$

is an isomorphism $C \to \mathbf{P}^1$.

Now let Γ be the graph of ψ in $(V - (0,0,0)) \times C$, and let Y be the closure of Γ in $V \times C$. (Compare Exercise 6.5 (v).) Projection on V gives a map $Y \to V$. The fiber above $\xi = (x,y,z)$ consists of a single point $(\xi, \psi(\xi))$ if ξ is not the origin, or of $\{(0,0,0)\} \times C$ if $\xi = (0,0,0)$. To see that Y is nonsingular, we divide C up into three affine sets,

$$U_1 = \{(x : y : z) \in C \mid x \neq 0\},$$

$$U_2 = \{(x : y : z) \in C \mid y \neq 0\},$$

$$U_3 = \{(x : y : z) \in C \mid z \neq 0\}.$$

Let $Y_i = Y \cap V \times U_i$. It is sufficient to show that each of these affine open sets of Y is nonsingular, since their union is Y. Consider for example Y_3. We replace the homogeneous coordinates $x : y : z$ ($z \neq 0$) on the curve U_3 by the affine $u = x/z$ and $v = y/z$. Thus Y_3 is

$$\{(x,y,z,u,v) \in \mathbf{A}^5 \mid x^2 + y^2 + z^2 = 0, z \neq 0, x = uz, y = vz\}.$$

This is an affine variety, because it is a principal open set (Exercise 3.1) in the affine variety

$$\{(x, y, z, u, v) \in \mathbf{A}^5 \mid x^2 + y^2 + z^2 = 0, x = uz, y = vz\}.$$

We have a morphism $(x, y, z, u, v) \mapsto (z, u, v)$ from Y_3 in these coordinates to

$$\{(z, u, v) \in \mathbf{A}^3 \mid u^2 + v^2 + 1 = 0, z \neq 0\}.$$

Since this morphism has an obvious inverse $(z, u, v) \mapsto (uz, vz, z, u, v)$, it is an isomorphism. The latter variety is the Cartesian product of the principal open set $\{z \neq 0\}$ in $\mathbf{A}^1$ with the circle $u^2 + v^2 + 1 = 0$, which is nonsingular by Proposition 6.14—the polynomial $F(u, v) = u^2 + v^2 + 1$ cannot vanish simultaneously with both its partial derivatives $\partial F / \partial u = 2u$ and $\partial F / \partial v = 2v$. We see that Y_3 is nonsingular, and similarly Y_1 and Y_2. Hence Y is nonsingular.

This example does not do justice to the possibilities for complexity in the resolution of singularities, even in the limited case of isolated (normal) singularities on surfaces. For the spectacular example $z^5 = x^2 + y^3$, see Laufer [**19**], p. 23–37, where other interesting examples will be found. The cusp singularities of Hilbert modular surfaces, first studied by Hirzebruch, are particularly interesting examples. For these see van der Geer [**28**], Chapter II.

We may also blow up smooth points on a surface. For example, let us blow up the origin in $\mathbf{A}^2$. We make use of the morphism $\theta : \mathbf{A}^2 - (0, 0) \to \mathbf{P}^1$ which maps $(x, y) \neq (0, 0)$ to the point with homogeneous coordinates $x : y$, corresponding to the line through the origin and (x, y). Let us consider the closure Y in $\mathbf{A}^2 \times \mathbf{P}^1$ of the graph of θ in $(\mathbf{A}^2 - (0, 0)) \times \mathbf{P}^1$, and let $p : Y \to \mathbf{A}^2$ be the projection map. It is not hard to see that p^{-1} is biregular on $\mathbf{A}^2 - (0, 0)$, while $p^{-1}(0, 0) \cong \mathbf{P}^1$. We have therefore blown up a smooth point of $\mathbf{A}^2$ into a $\mathbf{P}^1$.

We have already noted that $\mathbf{P}^2$ and $\mathbf{P}^1 \times \mathbf{P}^1$ are smooth, complete, birationally equivalent, but nonisomorphic surfaces. We may factor the birational map $\mathbf{P}^2 \to \mathbf{P}^1 \times \mathbf{P}^1$ as follows. First, let us pick two points P_1 and P_2 in $\mathbf{P}^2$. Let ℓ be the line connecting them. We blow them up, resulting in a surface Y. Let C_1, C_2 be the preimages of P_1 and P_2 in Y, and let ℓ' be the preimage of ℓ in Y. Then, we claim that ℓ' may be "blown down" to obtain a variety isomorphic to $\mathbf{P}^1 \times \mathbf{P}^1$. That is, we may find birational maps $Y \to \mathbf{P}^2$ and $Y \to \mathbf{P}^1 \times \mathbf{P}^1$, the first consisting of blowing up two points, and the second consisting of blowing up a single point.

To see this, we realize Y as a subvariety of $\mathbf{P}^2 \times \mathbf{P}^1 \times \mathbf{P}^1$. If we take $P_1 = 0 : 0 : 1$ and $P_2 = 0 : 1 : 0$, then Y is consists of $(x : y : z, t : u, v : w) \in \mathbf{P}^2 \times \mathbf{P}^1 \times \mathbf{P}^1$ such that $xu = yt$ and $xw = zv$. The projection onto $\mathbf{P}^2$ realizes Y as the blowup of the two points P_1 and P_2. The projection on $\mathbf{P}^1 \times \mathbf{P}^1$ blows down the line $\ell' = (0 : y : z, 0 : 1, 0 : 1)$ into the single point $(0 : 1, 0 : 1)$ of $\mathbf{P}^1 \times \mathbf{P}^1$.

Exercises

Exercise 7.1. Show that if X is a prevariety and $U \subseteq X$ an open set, then $\mathcal{O}(U)$ is the intersection of all $\mathcal{O}_P$ such that $P \in U$.

Hint: If U is affine, this is the definition of $\mathcal{O}(U)$. For the general case, let U_i be an affine open cover of U. It is enough to show that $\mathcal{O}(U) = \bigcap \mathcal{O}(U_i)$. Deduce this from the sheaf axiom.

Exercise 7.2. Show that if X is a prevariety, then the union over all nonempty open sets U of $\mathcal{O}(U)$ is the function field F.

Exercise 7.3. Verify Proposition 7.7.

Exercise 7.4. Construct a prevariety which is not separated as follows. Start with two copies X_1 and X_2 of $\mathbf{A}^1$. Identify $a \in X_1$ and $b \in X_2$ if the coordinates a and b are equal but nonzero. In other words, X is to be a line with a single point doubled. If O_1 and O_2 are the two distinct images of the origin, show that (O_1, O_2) is in the closure of the diagonal, and hence this prevariety is not separated.

Exercise 7.5. Prove that if X and Y are prevarieties, and $X_1 \subseteq X$, $Y_1 \subseteq Y$ are closed subsets, then $X_1 \times Y_1$ is a closed subset of $X \times Y$.

Hint: By the Gluing Lemma (Proposition 7.7), it is sufficient to verify this for X and Y affine.

Exercise 7.6. Prove that a morphism of prevarieties is continuous. (Use the Gluing Lemma (Proposition 7.7), and the fact that this is true for affine varieties.)

Exercise 7.7. The Normalization of a Variety. Let X be a variety. Prove that there exists a variety X', and a surjective morphism $n : X' \to X$ such that if U is any affine open subset of X, then $n^{-1}(U)$ is the normalization of the affine variety U, as described in Chapter 6. (Use the Gluing Lemma Proposition 7.7.)

If X is a variety, then any affine open set is an affine variety, hence has a dimension. Since X is irreducible, any two affine open sets have nonempty intersection. It follows that the dimension of any two affine open sets are equal, and so we may define the dimension of X to be the dimension of any affine open set. It is easy to see that this is the transcendence degree over k of the function field of F. A morphism $\phi : X \to Y$ of varieties is called *dominant* if the image is dense.

Exercise 7.8. Let $\phi : X \to Y$ be a dominant morphism of varieties. Show that there exists a nonempty open subset U of Y such that $U \subseteq \phi(X)$, and such that if W is any closed subvariety of Y such that $W \cap U \neq \varnothing$ is nonempty, and if Z is any irreducible component of $\phi^{-1}(W)$ such that $Z \cap \phi^{-1}U \neq \varnothing$, then $\dim(Z) = \dim(W) + \dim(X) - \dim(Y)$.

Hint: If X and Y are affine, this is Proposition 6.6. You must show that the general case follows from this special case.

Exercise 7.9. Similarly generalize Proposition 6.7 to varieties which are not necessarily affine.

Exercise 7.10. Let X be a prevariety, Y an irreducible closed subspace. Show that the function field $F(Y)$ is isomorphic to $\mathcal{O}_{Y,X}/\mathfrak{M}_{Y,X}$. (Check this first for X affine.)

Exercise 7.11. (i) Let X be a prevariety, U an open subset, and Y an irreducible closed subset of X. Show that $\mathcal{O}_{Y,X} \cap \mathcal{O}(U) = \mathcal{O}_{U \cap Y, U}$.

(ii) Let X be a variety, $P \in X$, and let Y be an irreducible closed subset of X. Show that $P \in Y$ if and only if $\mathcal{O}_P \subseteq \mathcal{O}_{Y,X}$.

Hint: Using (i), you may reduce to the case where X is affine.

Exercise 7.12 (i) Prove that a variety X is complete if and only if the morphism mapping X onto a point is proper.

(ii) Prove that if $\phi : Y \to X$ is proper and X is complete, then Y is complete.

(iii) Prove that if $\phi : Y \to X$ is proper, and $Z \subseteq X$ is complete, then $\phi^{-1}(Z)$ is complete.

8. Complete Nonsingular Curves

By a *function field of dimension one* over our fixed algebraically closed field k, we mean a finitely generated extension field F of transcendence degree one over k. The function field of a curve is such a field; two curves are birationally equivalent if they have isomorphic function fields.

In this chapter, we will establish that every function field of dimension one is the function field of a unique complete nonsingular curve. This is different from the situation for surfaces, where in Chapter 7 we saw that two nonisomorphic complete nonsingular surfaces could be birationally equivalent. We will also show that morphisms of complete nonsingular curves correspond exactly to field homomorphisms, so the study of curves and their mappings is intimately related to the Galois theory of fields.

If F is a field, by a *nontrivial* valuation ring $\mathfrak{o}$ of F, we mean a valuation ring $\mathfrak{o}$ of F such that $\mathfrak{o} \neq F$.

Proposition 8.1. *Let A be the coordinate ring of a nonsingular affine curve X, and let F be the function field of X, which is the field of fractions of A. The nontrivial valuation rings of F which contain A are precisely the local rings $\mathcal{O}_P$ of $P \in X$. They are discrete valuation rings.*

Proof. By Proposition 5.28, every local ring of $A_{\mathfrak{p}}$ with $\mathfrak{p}$ a nonzero prime ideal is a discrete valuation ring. It is clearly nontrivial and contains A.

Conversely, suppose that $\mathfrak{o}$ is a nontrivial valuation ring of F containing A. Let $\mathfrak{m}$ be its maximal ideal. Then $\mathfrak{m} \cap A$ is a prime ideal in A, and we will show that $\mathfrak{m} \cap A \neq 0$. Indeed, if $h \in F$, write $h = f/g$ with $f, g \in A$ and $g \neq 0$. If $\mathfrak{m} \cap A = 0$, then $g \notin \mathfrak{m}$, so $g^{-1} \in \mathfrak{o}$ and so $h \in \mathfrak{o}$. Thus $F = \mathfrak{o}$, contradicting the assumed nontriviality of $\mathfrak{o}$.

Now $P = V(\mathfrak{m} \cap A)$ is a nontrivial irreducible subspace of X. Since X is one-dimensional, by Theorem 4.11, P is zero dimensional, hence a point. We see that $\mathfrak{m} \cap A$ is the ideal of functions $f \in A$ which vanish at P. Let us show that $\mathfrak{o} = \mathcal{O}_P$. Any element of $\mathcal{O}_P$ may be written as f/g where $f \in A \subseteq \mathfrak{o}$, and $g \in A - \mathfrak{m} \subseteq \mathfrak{o} - \mathfrak{m}$. Thus g is a unit in $\mathfrak{o}$, and so $f/g \in \mathfrak{o}$. Thus $\mathcal{O}_P \subseteq \mathfrak{o}$. Now $\mathcal{O}_P$ is a discrete valuation ring by Proposition 5.28. It follows from Proposition 6.27 that $\mathfrak{o} = \mathcal{O}_P$. ∎

Proposition 8.2. *Every function field of dimension one is the function field of an affine curve X. If $\mathfrak{o}_1, \cdots, \mathfrak{o}_n$ are nontrivial valuation rings of F, then we may choose X so that X is nonsingular, and the coordinate ring A of X is contained in $\bigcap \mathfrak{o}_i$. Then there exist points $P_1, \cdots, P_n$ of X with each $\mathfrak{o}_i$ equal to the local ring $\mathcal{O}_{P_i}(X)$.*

Proof. Let us show first that if $x \in F$, then there exists $y \in F$ such that $k(x) = k(y)$, and $y \in \mathfrak{o}_j$ for $j = 1, \cdots, n$. We construct inductively $y_1, \cdots, y_n$ such that $y_j \in \mathfrak{o}_i$ for $i \leq j$, and $k(x) = k(y_j)$. Suppose that y_{j-1} is constructed. If $y_{j-1} \in \mathfrak{o}_j$, then we take $y_j = y_{j-1}$. Otherwise, we proceed as follows. Since k is infinite, we may find $a \in k$ such that the image of $y_{j-1} - a$ does not lie in the maximal ideal of k_i for $i = 1, \cdots, j-1$. Therefore we have $(y_{j-1} - a)^{-1} \in \mathfrak{o}_i$ for $i < j$. Since $y_{j-1} - a \notin \mathfrak{o}_j$, and since the latter is a valuation ring, $(y_{j-1} - a)^{-1} \in \mathfrak{o}_j$. Thus we may take $y_j = (y_{j-1} - a)^{-1}$. Now let $y = y_n$.

Let $x_1, \cdots, x_r$ generate F as a field over k. By what we have just shown, we may assume that $x_i \in \mathfrak{o}_j$ for each i, j. Now let A be the integral closure of $k[x_1, \cdots, x_r]$ in F. This is an affine algebra by Proposition 6.18. Let X be the affine variety whose coordinate ring is A. Then X is nonsingular since it is normal, by Theorem 6.26. The existence of points P_i such that $\mathfrak{o}_i = \mathcal{O}_{P_i}(X)$ now follows from Proposition 8.1. ∎

Proposition 8.3. *Let $A = k[x]$ be the polynomial ring which is the coordinate ring of the affine variety $\mathbf{A}^1$, and let $F = k(x)$ be its field of fractions. There is bijection between the points of $\mathbf{P}^1$ and the set of valuation rings of F which contain k. To the point $P \in \mathbf{A}^1$ there corresponds its local ring $\mathcal{O}_P$. To ∞ there corresponds the valuation ring*

$$\mathcal{O}_\infty = \{ x^{-r} f(x^{-1})/g(x^{-1}) \in F | f, g \text{ polynomials}, r \geq 0, f(0), g(0) \neq 0 \}.$$

Proof. $\mathcal{O}_\infty$ is also a discrete valuation ring, with the valuation

$$v_\infty \left(x^{-r} f(x^{-1})/g(x^{-1}) \right) = r.$$

We must show that every valuation ring $\mathfrak{o}$ of F which contains k is one of these. There are two possibilities. If $x \in \mathfrak{o}$, then $A \subseteq \mathfrak{o}$, and so by Proposition 8.1, $\mathfrak{o} = \mathcal{O}_P$ for some point $P \in \mathbf{A}^1$. The other possibility is that $x \notin \mathfrak{o}$. Then x^{-1} is in the maximal ideal $\mathfrak{m}$ of $\mathfrak{o}$. Now let σ be the field automorphism of F over k such that $\sigma(x) = x^{-1}$. Then x is in the maximal ideal $\sigma(\mathfrak{m})$ of the valuation ring $\sigma(\mathfrak{o})$, and so $A \subseteq \sigma(\mathfrak{o})$. By Proposition 8.1, $\sigma(\mathfrak{o}) = \mathcal{O}_P$ for some $P \in \mathbf{A}^1$. Now since x is in the maximal ideal of $\mathcal{O}_P$, it vanishes at P, which of course implies that $P = O$ is the origin. Since $\sigma(\mathfrak{o}) = \mathcal{O}_O$, we see that $\mathfrak{o}$ is the ring described in the Proposition. ∎

If E/F is a finite extension of function fields, and if $\mathfrak{o}$ is a valuation ring of F and $\mathcal{O}$ is a valuation ring of E, we say that $\mathcal{O}$ *lies above* $\mathfrak{o}$ if $\mathcal{O} \cap F = \mathfrak{o}$. Since $\mathcal{O} \cap F$ is clearly a valuation ring of F, every valuation ring of E lies above a unique valuation ring of F.

Proposition 8.4. *Let F be a function field of dimension one over k. Suppose $x \in F$ such that F is a finite separable extension of $k(x)$, and let $\mathfrak{o}$ be a nontrivial valuation ring of $k(x)$. Then there are only finitely many valuation rings of F lying above $\mathfrak{o}$.*

Proof. Replacing x with x^{-1} if necessary, we may assume that $x \in \mathfrak{o}$. Thus $\mathfrak{o} \supset k[x]$. Let A be the integral closure of $k[x]$ in F. Let X and $\mathbf{A}^1$ denote the affine varieties with coordinate rings A and $k[x]$, respectively; corresponding to the inclusion $k[x] \to A$ we have a morphism $X \to \mathbf{A}^1$, which is a finite dominant morphism. By Proposition 6.4, its fibers are finite. By Proposition 8.3, the valuation ring $\mathfrak{o}$ is the local ring in $k[x]$ of some point P in $\mathbf{A}^1$. If $\mathcal{O}$ is a valuation ring of F containing $\mathfrak{o}$, then it contains $k[x]$, and so it contains its integral closure A by Exercise 2.5 (i). By Proposition 8.3, the nontrivial valuation rings of F containing A correspond to points of X, and those lying above $\mathcal{O}$ correspond to points in the finite fiber of the finite dominant morphism $X \to \mathbf{A}^1$. There are thus only finitely many. ∎

Proposition 8.5. *Let X be an affine curve, let A be its coordinate ring, and let F be its function field. There are only finitely many nontrivial valuation rings of F not containing A.*

Proof. By the Noether Normalization Lemma (Theorem 4.2), there exists $x \in F$ such that A is integral over x. Any valuation ring of F containing $k[x]$ contains A by Exercise 2.5 (i). Hence the intersection of any valuation ring of F not containing A with $k(x)$ must be the valuation ring $\mathcal{O}_\infty$ of Proposition 8.3. Since by Proposition 8.4 there are only finitely many valuation rings of F above this valuation ring of $k(x)$, the Proposition follows. ∎

Theorem 8.6. *Every function field F of dimension one is the function field of a complete nonsingular curve X. The local rings of X are precisely the nontrivial valuation rings of F. If Y is any nonsingular curve with function field F, then Y is isomorphic to a subvariety of X. The complete nonsingular curve X is determined up to isomorphism by its function field F.*

Proof. Let X be a set which is in bijection with the nontrival valuation rings of F. If $x \in X$, we denote by $\mathfrak{o}_x$ the corresponding valuation ring. We give X the cofinite topology, in which a set is open if it is either empty, or the complement of a finite set. If $U \subseteq X$ is a nonempty open set, we define $\mathcal{O}_X(U) = \bigcap_{x \in U} \mathcal{O}_x$.

We will show that this is a sheaf of rings, and that $(X, \mathcal{O}_X)$ is a variety, in fact a complete nonsingular curve.

It is straightforward that X is a sheaf. We must show that X has a finite cover by affine open sets. Note that if U is an affine variety with function field F, its coordinate ring A is an affine algebra inside A, and U since the set of valuation rings of F not containing A is finite by Proposition 8.5, U may be identified with the complement of this finite set, and it is open in the cofinite topology. Any finite set of points of X may be embedded in such an affine open set by Proposition 8.2, and the complement of this affine set (being finite by Proposition 8.5) can be embedded in another by the same Proposition. So two affine open sets suffice to cover X, which is therefore a variety.

Every local ring of X is a discrete valuation ring, hence a regular local ring, so X is smooth. It is complete by the valuative criterion Proposition 7.17, as is clear since X being one-dimensional, we only have to test this criterion for $Z = X$, and it is clear in this case since every valuation ring of F is a local ring of X.

To prove that any other complete nonsingular curve Y with function field F is isomorphic to X, we first assume that Y is affine, before treating the general case. Let B be the function field of Y, so that we may identify F with the function field of B. Because Y is smooth, all its local rings are discrete valuation rings of F, and their intersection is B. We identify Y with a subset of X by matching up the local rings. If $U \subseteq Y$ is an open set, then

$$\mathcal{O}_Y(U) = \bigcap_{U \ni x} \mathcal{O}_x = \mathcal{O}_X(U),$$

so this injection of Y in X is a morphism of varieties, and so Y is a subvariety of X.

Now if Y is not necessarily affine, we can cover it with affine sets, each of which is injected into X. It follows from Proposition 7.8 that Y is injected into X.

Now we may show that any complete nonsingular variety with function field F is isomorphic to X. Indeed, if Y is such a variety, we have just shown there exists an injection $Y \to X$. The image is closed because Y is complete. Since the only nonfinite closed set in X is X itself, this embedding of Y is an isomorphism. ■

Theorem 8.7. *(i) Let E and F be the function fields of the complete nonsingular curves X and Y. If $\phi : X \to Y$ is a nonconstant morphism, then, since ϕ is a dominant rational map, there is induced a field injection $\phi^* : F \to E$ of F into E over k, and every field injection of F into E over k arises uniquely in this way.*

(ii) The categories of function fields of dimension one and complete nonsingular curves are contravariantly equivalent.

Proof. To prove (i), we must show that every isomorphism $F \to E$ arises uniquely as ϕ^* for some ϕ. The map ϕ may be described explicitly as follows: if $x \in X$, by Theorem 8.6, the local ring $\mathfrak{o}_x$ is a nontrivial valuation ring of E. Its preimage in F is a nontrival valuation ring of F, which again by Theorem 8.6 is the local ring of a point $\phi(x) \in Y$. We have described a map $\phi : X \to Y$, and it is not hard to see that it is a morphism.

By Exercise 3.6, (ii) is a paraphrase of (i). ∎

Exercises

Exercise 8.1. Let U be the affine curve $y^2 = (x - e_1)(x - e_1)(x - e_3)$, where the e_i are distinct elements of k. Let X be the complete nonsingular curve containing U as an affine open set. Show that $X - U$ consists of a single point.

Exercise 8.2. Generalize Exercise 8.1 as follows. Let U be the affine curve $y^2 = f(x)$, where $f \in k[x]$ is a polynomial of degree n with no repeated roots. Let X be the complete nonsingular curve containing U. Show that $X - U$ consists of a single point if the degree of f is odd, and two points if the degree of f is even.

Exercise 8.3. Suppose that the characteristic of k is p. Let $X = \mathbf{P}^1$, and let $k(x)$ be the function field of X, the field of rational functions in one variable. Let $F : X \to X$ be the "Frobenius" morphism induced by the mapping $x \to x^p$ from $k(x^p) \to k(x)$. Show that F is a bijection, but that F^{-1} is not a morphism.

9. Ramification

In this chapter, we will develop some algebra having to do with the behavior of prime ideals when the ground field is extended. This has to do with the fibers of a finite morphism. The algebraic facts which we prove, in addition to the applications to algebraic geometry, are the first theorems which must be proved when one studies algebraic number theory.

Let us consider the fibers of a nonconstant morphism $f : Y \to X$ of complete nonsingular curves. This corresponds by Theorem 8.7 to an embedding of the function field F of X into the function field E of Y. In this situation, we may regard E as an extension field of F. In view of the close relation between morphisms of curves and homomorphisms of fields which we saw in Theorem 8.7, we will call the morphism of curves *separable* or *purely inseparable* if the corresponding map of fields is separable or purely inseparable. Similarly, if E/F is Galois, we call the morphism *Galois*, and so forth. We call $[E : F]$ the *degree* of the morphism f.

It is proved in field theory that if E/F is a finite extension, then there exists a subfield E_0 such that E/E_0 is purely inseparable, and E_0/F is separable. Let Y_0 be the complete nonsingular curve associated with E_0. Then f is factored into $f' \circ f_0$, where $f_0 : Y \to Y_0$ is purely inseparable, and $f' : Y_0 \to X$ is separable. We will study purely inseparable morphisms later in Proposition 12.3. At the moment, our aim is to study a nonconstant separable morphism $f : X \to Y$.

Example 9.1. Assume that the characteristic of k is not equal to 2. Let Y be the elliptic curve $y^2 = (x - e_1)(x - e_2)(x - e_3)$, where the e_i are distinct elements of k. Let $X = \mathbf{P}^1$ be the rational curve whose function field is $k(x)$. Let $A = k[x]$ and $B = k[x, y]$ be the affine algebras which are the coordinate rings of the affine open sets $\mathbf{A}^1 \subset \mathbf{P}^1$ and $U = f^{-1}(\mathbf{A}^1) \subseteq Y$. Corresponding to the embedding $k(x) \to k(x, y)$, we have a morphism $f : Y \to X$. Let $a \in X$. Suppose first that $a \in \mathbf{A}^1$, and let $\mathfrak{m}$ be the corresponding maximal ideal of A, which is the principal ideal generated by $(x - a)$. If a is not one of e_1, e_2, e_3, then the fiber over a consists of two points. We may see this as follows. The points in the fiber over a correspond to the maximal ideals $\mathfrak{M}$ of B lying above $\mathfrak{m}$. We have a k-algebra homomorphism $\rho : A \to k$ which is evaluation at a, namely $\rho(f) = f(a)$. This homomorphism has kernel $\mathfrak{m}$. If $\mathfrak{M}$ is an ideal of B lying above $\mathfrak{m}$, then $B \to B/\mathfrak{M} \cong k$ will be a k-algebra homomorphism which extends ρ, and which we will also denote as ρ. Now since $\rho(x) = a$, $\rho(y)^2 = (a - e_1)(a - e_2)(a - e_3) \neq 0$. There are two possibilities for $\rho(y)$, namely $\pm\sqrt{(a - e_1)(a - e_2)(a - e_3)}$. On the other hand, if $a = e_1$, e_2 or e_3, there is only one choice for $\rho(y)$, and so the fiber over these three points consists of just a single element. Also, the fiber over $\infty \in \mathbf{P}^1$ may also be seen to consist of just a single element (Exercise 8.1).

In this example, we see that the fiber over a typical point $P \in X$ consists of n points, where n is the degree of the field extension F/E. The points $P \in X$ where this fails are called *points of ramification*. If $Q \in f^{-1}(P)$, we would like to define a *ramification index* $e(Q|P)$ which is one if P is not ramified, and which is to be thought of as the "multiplicity" with which it occurs in the fiber. We should have $\sum_{Q \in f^{-1}P} e(Q|P) = n$ for every point $P \in X$. And finally, we should be able to prove that there are only a finite number of points of ramification. The proofs of these facts will occupy this chapter. We will base our proofs on ideal theory, and in the process develop another important chapter in commutative algebra, namely ramification theory for prime ideal in *Dedekind domains*. These results are of importance in algebraic number theory as well as algebraic geometry, and for this reason we will work in somewhat more generality than is needed for the applications at hand.

A *Dedekind domain* is an integrally closed Noetherian domain of dimension one.

Proposition 9.2. *A principal ideal domain is a Dedekind domain. In particular, a discrete valuation ring is a Dedekind domain.*

Proof. A principal ideal domain is a unique factorization ring, hence integrally closed by Proposition 4.9. It is well known that a principal ideal domain is Noetherian, and that every nonzero prime ideal is maximal. ■

However, there are examples, such as $\mathbb{Z}[\sqrt{-5}]$, of Dedekind domains which are not principal ideal domains.

Proposition 9.3. *Let A be a Dedekind domain, and let F be its field of fractions. Let E/F be a finite separable extension, and let B be the integral closure of A in E. Then B is a Dedekind domain.*

The conclusion is true if E/F is merely assumed to be a finite algebraic extension. However, the case where E/F is purely inseparable must be treated separately, and we omit it. See Zariski and Samuel [**36**], Theorem 19 in Chapter V (p. 281 of Volume I) for a proof which does not assume E/F to be separable.

Proof. Clearly B is an integrally closed integral domain. It is Noetherian, because it is finitely generated as an A-module by Proposition 6.17 (iii), hence as an A-algebra, and hence Noetherian by the Hilbert basis theorem (Exercise 1.4). That it has dimension one is a consequence of Proposition 4.4 (ii) and assumption that A has dimension one. ■

The situation of Proposition 9.3 will be the main topic of study in this Chapter. Let us take a moment to show how this situation can arise geometrically. (It also arises nongeometrically in algebraic number theory.)

Proposition 9.4. *(i) Let U be a nonsingular affine curve, A its coordinate ring. Then A is a Dedekind domain.*

(ii) Let Y and X be complete nonsingular curves with function fields E and F respectively. Let $\phi : Y \to X$ be a nonconstant morphism. Identify F with its image in E under the induced field injection $\phi^* : F \to E$ (see Theorem 8.7). Let $U \subset X$ be an affine open set, and let $V = \phi^{-1}(U) \subset Y$. Then the coordinate rings A and B of U and V, respectively, are Dedekind domains with fields of fractions F and E, respectively, and B is the integral closure of A in E.

Proof. For (i), A is integrally closed by Proposition 6.25; it is Noetherian by the Hilbert Basis Theorem and of dimension one because U is a curve. So A is a Dedekind domain.

For (ii), B is the intersection of the local rings of V by Proposition 3.7. We claim that these valuation rings are precisely the valuation rings of E containing A. Note that this implies that B is the integral closure of A by Exercise 2.5. Indeed, every valuation ring of E is of the form $\mathcal{O}_y(Y)$ for some $y \in Y$ by Theorem 8.6, and if $x = \phi(y)$, then $\mathcal{O}_x(X) = F \cap \mathcal{O}_x(Y)$, since a function f on X is obviously regular at x if and only if $\phi^*(f) = f \circ \phi$ is regular at y. It follows that $\mathcal{O}_y(Y)$ contains A if and only if $\mathcal{O}_x(Y)$ contains A. By Proposition 8.1, a necessary and sufficient condition for this is that $x \in U$, i.e. $y \in V$. Thus the valuation rings of E containing A are precisely the local rings of V, as required. ■

Proposition 9.5. *Let A be a Dedekind domain, and $\mathfrak{p}$ a nonzero prime ideal. Then $A_{\mathfrak{p}}$ is a discrete valuation ring. Its unique maximal ideal is $\mathfrak{p}A_{\mathfrak{p}}$, and $A_{\mathfrak{p}}/\mathfrak{p}A_{\mathfrak{p}} \cong A/\mathfrak{p}$.*

Proof. $A_{\mathfrak{p}}$ has precisely two prime ideals, namely 0 and $\mathfrak{p}A_{\mathfrak{p}}$, so it has dimension one. It is integrally closed by Proposition 6.21, and so by Proposition 5.26 and Proposition 5.28, it is a discrete valuation ring. By Exercise 3.4 (iv), $A_{\mathfrak{p}}/\mathfrak{p}A_{\mathfrak{p}}$ is isomorphic to the field of fractions of $A/\mathfrak{p}$. However, $\mathfrak{p}$ is maximal, and so $A/\mathfrak{p}$ is itself a field, and therefore $A_{\mathfrak{p}}/\mathfrak{p}A_{\mathfrak{p}} \cong A/\mathfrak{p}$. ■

It is proved in Exercise 9.1 that in a Dedekind domain, every nonzero ideal may be factored uniquely into prime ideals. (However, it may or may not be the case that a prime ideal is principal, so this does not imply that a Dedekind domain is a unique factorization domain.)

We recall the basic construction of *extension of scalars*, whereby a module over a ring A gives rise to a module over an A-algebra. Let A be a ring. An *A-algebra* is a ring B together with an A-module homomorphism $i : A \to B$, called the *structure map*, which may or may not be injective. Then B becomes an A-module, with the multiplication $A \times B \to B$ given by $(a, b) \to i(a) b$. (Note: sometimes the notion of an algebra is expanded to allow B to be a noncommutative ring. In that case, the it is assumed in the definition that $i(A)$ be contained in the center of B. We will only consider *commutative* A-algebras.)

If M is an A-module, and B an A-algebra, then $B \otimes_A M$ is naturally a B-module, with multiplication $b \cdot (b' \otimes m) = bb' \otimes m$. $B \otimes M$ is called the B-module obtained from M by *extension of scalars*.

Proposition 9.6. *Let A be an integral domain, and let F be its field of fractions. Let V be a finite-dimensional vector space over F, and let M be an A-submodule of V such that the F-span of M is V. Let $\mathfrak{p}$ be a maximal ideal of A, and let M' be the $A_{\mathfrak{p}}$-submodule $A_{\mathfrak{p}}M$ of V. Then $M'/\mathfrak{p}M'$ is isomorphic as an A-module to $M/\mathfrak{p}M$.*

Proof. Recall that by Exercise 3.4 (iv) the canonical map $A/\mathfrak{p} \to A_{\mathfrak{p}}/\mathfrak{p}A_{\mathfrak{p}}$ is an isomorphism. We have

$$
\begin{aligned}
M'/\mathfrak{p}M' = M'/(\mathfrak{p}A_{\mathfrak{p}})M' &\cong (A_{\mathfrak{p}}/\mathfrak{p}A_{\mathfrak{p}}) \otimes_{A_{\mathfrak{p}}} M' && \text{[Exercise 9.7]} \\
&\cong (A_{\mathfrak{p}}/\mathfrak{p}A_{\mathfrak{p}}) \otimes_{A_{\mathfrak{p}}} (A_{\mathfrak{p}} \otimes M) && \text{[Ex. 9.12, 9.11]} \\
&\cong (A_{\mathfrak{p}}/\mathfrak{p}A_{\mathfrak{p}}) \otimes_A M && \text{[Exercise 9.4 (ii)]} \\
&\cong (A/\mathfrak{p}) \otimes_A M && \text{[Exercise 3.4 (iv)]} \\
&\cong M/\mathfrak{p}M && \text{[Exercise 9.7]} \quad \blacksquare
\end{aligned}
$$

Let A be a ring. An A-module M is called *free* if it is isomorphic to a direct sum of copies of A. Thus there exists an indexing set X such that M is isomorphic to the set $A^{(X)}$ of all mappings from $X \to A$, are zero for all but finitely many elements of X. $A^{(X)}$ is an A-module with componentwise operations: if $f, g \in A^{(X)}$, $a, b \in A$, then for $x \in X$, the A-module structure on $A^{(X)}$ is given by $(af + bg)(x) = af(x) + bg(x)$. Thus $A^{(X)}$ is the direct sum of copies of A, indexed by X. If X is a finite set $\{1, \cdots, n\}$, then we may interpret $A^{(X)}$ as the set of ordered n-tuples of elements of A, and denote it as usual by A^n. The *rank* of a free A-module M is the cardinality of X where $M \cong A^{(X)}$. This is well-defined by Exercise 9.8.

We recall that if A is an integral domain and M is an A-module, then $m \in M$ is *torsion* if there exists a nonzero $a \in A$ such that $am = 0$. The torsion elements of M form a submodule T of M, called the *torsion submodule*. M is called *torsion* if $T = M$, and *torsion-free* if $T = 0$.

Proposition 9.7. *Let A be a principal ideal domain, and let F be its field of fractions. Let M be a finitely generated A-submodule of an n-dimensional vector space V. Assume that $FM = V$. Then M is a free module of rank n, and in fact contains a basis $x_1, \cdots, x_n$ of F such that $M = Ax_1 \oplus \ldots \oplus Ax_n$.*

Proof. We recall the well-known structure theorem for finitely generated modules over a principal ideal domain, which is proved in Lang [**18**], Section III.7, as well as in many other sources. This theorem asserts that any finitely generated module has the form $A^r \oplus T$ for some r, where T is the torsion submodule, and T may be further identified as a direct product of "cyclic" submodules, each of which has the form $A/\mathfrak{a}$ for some ideal $\mathfrak{a}$ of A. In the case at hand, M is contained in an F-vector space, so M is torsion-free. This implies that the torsion submodule $T = 0$, and so $M \cong A^r$ for some r. Now the F-span of M in V is isomorphic to $F \otimes_A M \cong F^r$ (Exercise 9.6), which implies that $r = n$. Thus M is a free module of rank n, and a basis $x_1, \cdots, x_n$ for M will clearly have the required properties. ■

Proposition 9.8. *Let A be a Dedekind domain and let F be its field of fractions. Let M be a finitely generated A-submodule of an n-dimensional vector space V. Assume that M contains an F-basis of V. Let $\mathfrak{p}$ be a nonzero prime ideal of A. Then $M/\mathfrak{p}M$ is naturally a vector space over $A/\mathfrak{p}$, and its dimension is n.*

Proof. First consider the case where A is a principal ideal domain. By Proposition 9.7, $M \cong A^n$. Now $M/\mathfrak{p}M \cong (A/\mathfrak{p}) \otimes_A A^n \cong (A/\mathfrak{p})^n$ (Exercise 9.8), so $M/\mathfrak{p}M$ is indeed n-dimensional as an $A/\mathfrak{p}$-module.

If A is not a principal ideal domain, we may still proceed as follows. Let $M' = A_{\mathfrak{p}}M \subseteq V$. This is a module over $A_{\mathfrak{p}}$, which is a principal ideal domain by Proposition 9.5 and Proposition 5.27, and so by the special case just established, the dimension of $M'/\mathfrak{p}M'$ is n. But it follows from Proposition 9.5 and Proposition 9.6 that this $A/\mathfrak{p}$-module is isomorphic to $M'/\mathfrak{p}M'$. ■

Recall that if A is a ring, ideals $\mathfrak{a}$ and $\mathfrak{b}$ are called *coprime* if $\mathfrak{a} + \mathfrak{b} = A$.

Proposition 9.9 (the Chinese Remainder Theorem). *Let A be a ring, and let $\mathfrak{a}_1, \cdots, \mathfrak{a}_n$ be ideals which are coprime in pairs.*

(i) Let $x_1, \cdots, x_n \in A$. Then there exists an $x \in A$ such that $x - x_i \in \mathfrak{a}_i$ for $i = 1, \cdots, n$.

(ii) We have an isomorphism of A-modules

$$A / \bigcap_i \mathfrak{a}_i \cong \bigoplus_i A/\mathfrak{a}_i.$$

Proof. Let us prove (i). Assume by induction that the assertion is true for $n-1$. Thus we may find y such that $y - x_i \in \mathfrak{a}_i$ for $i = 1, \cdots, n-1$. By Exercise 9.1 (i) and (ii), $\mathfrak{a}_n$ is coprime to $\mathfrak{a}_1 \cap \cdots \cap \mathfrak{a}_{n-1}$, and so we may write $1 = \lambda + \mu$ where $\lambda \in \mathfrak{a}_1 \cap \cdots \cap \mathfrak{a}_{n-1}$, and $\mu \in \mathfrak{a}_n$. Now we let $x = \mu y + \lambda x_n$. Then $x - x_i = \mu(y - x_i) + \lambda(x_n - x_i) \in \mathfrak{a}_i$ for all i.

To prove (ii), note that we have for each i a canonical map $A \to A/\mathfrak{a}_i$, and hence a map

$$A \to \bigoplus_i A/\mathfrak{a}_i$$

whose kernel is obviously $\bigcap \mathfrak{a}_i$. It follows from part (i) that this map is surjective, whence (ii). ∎

Let E/F be a finite separable extension of degree n, let A be a Dedekind domain with field of fractions F, and let B be the integral closure of A in E. Let $\mathfrak{p}$ be a prime ideal of A. We may factor $\mathfrak{p}B$ into prime ideals of B (Exercise 9.1). Let $\mathfrak{p}B = \prod_i \mathfrak{P}_i^{e_i}$. The primes of $\mathfrak{P}_i$ of B occurring in this decomposition clearly satisfy $\mathfrak{P}_i \cap A \supseteq \mathfrak{p}$. Since $\mathfrak{p}$ is maximal, this implies that $\mathfrak{P}_i$ is one of the primes of B lying above A. Conversely, if $\mathfrak{P}$ is a prime of B lying above A, $\mathfrak{P} \supseteq \mathfrak{p}B$, and since $\mathfrak{p}B = \prod \mathfrak{P}_i^{e_i} = \bigcap \mathfrak{P}_i^{e_i}$ is a primary decomposition (Exercise 9.1), any prime containing $\mathfrak{p}B$ must contain one of the $\mathfrak{P}_i$ (Proposition 5.17). The latter are maximal ideals since the Dedekind domain B has dimension one, so the $\mathfrak{P}_i$ are the only prime ideals of B containing $\mathfrak{p}B$. Hence the primes lying above $\mathfrak{p}$ are precisely the $\mathfrak{P}_i$. Now $B/\mathfrak{P}_i$ is a finite extension of the field $A/\mathfrak{p}$ by Proposition 6.17 (iii). Let f_i be the degree of this extension. We say e_i is the *ramification index*, and f_i the *residue class degree* of $\mathfrak{P}_i$ over $\mathfrak{p}$. We will also use the notation $e(\mathfrak{P}|\mathfrak{p})$ and $f(\mathfrak{P}|\mathfrak{p})$ for the ramification index and residue class degree of a prime $\mathfrak{P}$ over $\mathfrak{p}$, so in our previous notation $e_i = e(\mathfrak{P}_i|\mathfrak{p})$ and $f_i = f(\mathfrak{P}_i|\mathfrak{p})$. If $e(\mathfrak{P}|\mathfrak{p}) > 1$, we say that $\mathfrak{P}$ is *ramified* over A, or that the prime $\mathfrak{p}$ *ramifies* in B. If there is a unique prime $\mathfrak{P}$ lying above $\mathfrak{p}$, and if $f(\mathfrak{P}|\mathfrak{p}) = 1$, then we say that $\mathfrak{p}$ (or $\mathfrak{P}$) is *totally ramified*. It follows from the following Theorem that this implies that $e(\mathfrak{P}|f) = n$.

Theorem 9.10. *Let E/F be a finite separable extension of degree n, let A be a Dedekind domain with field of fractions F, and let B be the integral closure of A in E. Let $\mathfrak{p}$ be a prime ideal of A, and let $\mathfrak{p}B = \prod_i \mathfrak{P}_i^{e_i}$ be the decomposition of $\mathfrak{p}B$ into prime powers. Let f_i be the residue class degree of $\mathfrak{P}_i$ over $\mathfrak{p}$. Let $K = A/\mathfrak{p}$, and let $K_i = B/\mathfrak{P}_i$.*

(i) We have $B/\mathfrak{p}B \cong \bigoplus (B/\mathfrak{P}_i^{e_i})$ as $A/\mathfrak{p}$-modules;

(ii) We have $\sum_i e_i f_i = n$.

(iii) Let $y \in B$. Let $\overline{y}_i$ denote the residue class of y in K_i. Then the residue class of $\mathrm{tr}_{E/F}(y)$ in K is

$$\sum_i e_i \, \mathrm{tr}_{K_i/K}(\overline{y}_i).$$

(iv) We have $\mathfrak{p}B_{\mathfrak{P}_i} = \mathfrak{P}_i^{e_i} B_{\mathfrak{P}_i}$.

Proof. First let us establish the isomorphism of (i). In this exercise dim means $\dim_{A/\mathfrak{p}}$. By Proposition 9.9, $B/\mathfrak{p}B \cong \bigoplus B/\mathfrak{P}_i^{e_i}$. Thus $\dim(B/\mathfrak{p}B) = \sum_i \dim(B/\mathfrak{P}_i^{e_i})$. Clearly

$$\dim(B/\mathfrak{P}_i^{e_i}) = \dim(B/\mathfrak{P}_i) + \dim(\mathfrak{P}_i/\mathfrak{P}_i^2) + \ldots + \dim(\mathfrak{P}_i^{e_i-1}/\mathfrak{P}_i^{e_i}).$$

We will now show that $\mathfrak{P}_i^r/\mathfrak{P}_i^{r+1} \cong B/\mathfrak{P}_i$ as B-modules. Indeed, let $\xi \in \mathfrak{P}_i^r - \mathfrak{P}_i^{r+1}$, and let (ξ) denote the principal ideal of B generated by ξ. Consider the ideal $(\xi) + \mathfrak{P}_i^{r+1}$. This ideal is contained in $\mathfrak{P}_i^r$, and properly contains $\mathfrak{P}_i^{r+1}$. It follows from Exercise 9.1 (vii) that $(\xi) + \mathfrak{P}_i^{r+1} = \mathfrak{P}_i^r$. Therefore multiplication by ξ induces an isomorphism of $B/\mathfrak{P}_i$ onto $\mathfrak{P}_i^r/\mathfrak{P}_i^{r+1}$. We have therefore established the isomorphism of (i).

Note that this implies (ii), since the dimension of $B/\mathfrak{p}B$ as a vector space over $A/\mathfrak{p}$ is n by Proposition 9.8, while the dimension of $B/\mathfrak{P}_i$ over $A/\mathfrak{p}$ is f_i by definition. Thus $n = \sum e_i f_i$.

Regarding (iii), since $A_\mathfrak{p}$ is a discrete valuation ring, it is a principal ideal domain. We may replace A by $A_\mathfrak{p}$, and B by $S^{-1}B$, where $S = A - \mathfrak{p}$. This does not change E, F, K nor K_i. Thus we may assume that A is a discrete valuation ring. Thus by Proposition 9.7, we may find an F-basis $\xi_1, \cdots, \xi_n$ of E which is also an A-basis of B. Now let $\lambda_y : E \to E$ denote multiplication by y. If $y\xi_i = \sum a_{ij}\xi_j$ with $a_{ij} \in F$, then $\mathrm{tr}_{E/F}(y) = \sum_i a_{ii}$ is equal to the trace of the endomorphism λ_y calculated relative to this basis. Note that actually $a_{ij} \in B$. On the other hand, the images $\overline{\xi}_i$ of the ξ_i in $B/\mathfrak{p}B$ are a basis for $B/\mathfrak{p}B$ as a vector space over $A/\mathfrak{p}$, and if $\overline{a}_{ij}$ is the image of a_{ij} in A, then multiplication by y induces the endomorphism of $B/\mathfrak{p}B$ whose matrix with this basis is $(\overline{a}_{ij})$. Hence the image of $\mathrm{tr}_{E/F}(y)$ in $A/\mathfrak{p}$ is the residue class of $\sum \overline{a}_{ii}$, which is the trace of the endomorphism of $B/\mathfrak{p}B$ induced by multiplication by

y. Now by part (i), we have an isomorphism

$$B/\mathfrak{p}B \cong \bigoplus_i (B/\mathfrak{P}_i)^{e_i}.$$

This endomorphism induces on each of the spaces $B/\mathfrak{P}$ multiplication by $\overline{y}_i$, and the trace of this endomorphism is $\mathrm{tr}_{K_i/K}(\overline{y}_i)$. Since this module occurs e_i times, we obtain the formula of (iii).

For (iv), we have $\mathfrak{p}B_{\mathfrak{P}_i} = \prod_j \mathfrak{P}_j^{e_j} B_{\mathfrak{P}_i}$. If $j \neq i$, then $\mathfrak{P}_j$ meets the multiplicative set $B - \mathfrak{P}_i$, so $\mathfrak{P}_j B_{\mathfrak{P}_i} = B_{\mathfrak{P}_i}$ by Exercise 3.4 (i). Thus we obtain (iv). ■

Proposition 9.11. *In the setting of* Theorem 9.10, *let* $\mathfrak{P}$ *be one of the* $\mathfrak{P}_i$, *and let* $e = e_i$ *be the ramification index of* $\mathfrak{P}$ *over* $\mathfrak{p}$. *Let* w *be the discrete valuation of* E *corresponding to* $\mathfrak{P}$, *and let* v *be the discrete valuation of* F *corresponding to* $\mathfrak{p}$. *Then for* $x \in F$, *we have* $w(x) = e\, v(x)$.

By abuse of language, we say that the valuation w *extends* v.

Proof. Clearly the restriction $w|_F : F \to \mathbb{Z}$ satisfies the definition of a discrete valuation, except for one point: it may not be surjective. Its image is an ideal in $\mathbb{Z}$, so let e' be the positive generator. Then $x \to e'^{-1} w(x)$ is a discrete valuation of F. Its valuation ring $\{x \in F | w(x) \geq 0\}$ is $B_{\mathfrak{P}} \cap F$. Now we show that $B_{\mathfrak{P}} \cap F \neq F$. Indeed, if p is any nonzero element of $\mathfrak{p}$, then $1/p \in F$ cannot be an element of $B_{\mathfrak{P}}$, since if $1/p = b/s$ with $b \in B$ and $s \in B - \mathfrak{P}$, then $s = pb \in \mathfrak{p}B$, which is a contradiction since $\mathfrak{p}B \subseteq \mathfrak{P}$. Now $A_{\mathfrak{p}} \subseteq B_{\mathfrak{P}} \cap F$, and so by Proposition 6.27, $B_{\mathfrak{P}} \cap F = A_{\mathfrak{p}}$. Thus the two valuations v and $e'^{-1}w$ have the same valuation ring. These two valuations therefore coincide by the comment to Proposition 5.27.

We must therefore show that $e' = e$. Let $u \in A_{\mathfrak{p}}$ have $v(u) = 1$. Thus $u A_{\mathfrak{p}} = \mathfrak{p} A_{\mathfrak{p}}$ and $u B_{\mathfrak{P}} = u A_{\mathfrak{p}} B_{\mathfrak{P}} = \mathfrak{p} B_{\mathfrak{P}} = \mathfrak{P}^e$ by Theorem 9.10 (iv), so $w(u) = e$. It follows that $e' = e$. ■

Proposition 9.12. *Let* A *be a Dedekind domain,* F *its field of fractions. Let* $E \supseteq K \supseteq F$ *be finite separable extensions of* F. *Let* B *and* C *be the integral closures of* A *in* K *and* E *respectively. Let* $\mathfrak{p}$ *be a prime of* F, *let* $\mathfrak{q}$ *be a prime of* B *lying above* $\mathfrak{p}$, *and let* $\mathfrak{P}$ *be a prime of* C *lying above* $\mathfrak{q}$. *Then* $e(\mathfrak{P}|\mathfrak{p}) = e(\mathfrak{P}|\mathfrak{q})\, e(\mathfrak{q}|\mathfrak{p})$ *and* $f(\mathfrak{P}|\mathfrak{p}) = f(\mathfrak{P}|\mathfrak{q})\, f(\mathfrak{q}|\mathfrak{p})$.

Proof. The multiplicativity of the ramification index follows from the characterization of the ramification index in Proposition 9.11, while the multiplicativity of the residue class degree is clear from the definition. ■

Let $f : Y \to X$ be a nonconstant morphism of complete nonsingular curves, and let F, E be the function fields of X and Y, respectively. Then we have seen that f induces an embedding of F into E, and we may regard F as a subfield of E. Let U be an affine open subset of X, and let A be its coordinate ring. By Proposition 9.4, $f^{-1}U = V$ is an affine open set in Y, and if B is the integral closure of A in E, then B is the coordinate ring of Y. Theorem 9.10 is applicable to this situation. In this case, if $P \in U$, P corresponds to a prime $\mathfrak{p}$ of A. The elements Q_i of the fiber $f^{-1}(P)$ correspond to the primes $\mathfrak{P}_i$ of B lying above $\mathfrak{p}$. The residue class degree f_i is equal to 1, because $A/\mathfrak{p} \cong B/\mathfrak{P}_i \cong k$. Thus Theorem 9.10 associates with each point Q_i in $f^{-1}(Q)$ a *ramification index* $e_i = e(Q|P)$, and we have $\sum_i e_i = n$. We should think of $e(Q|P)$ as the multiplicity with which Q occurs in the fiber. Proposition 9.11 gives an algebraic interpretation of the ramification index, which is not, however, without geometric content, since the valuation $w(f)$ of $f \in F$ is to be thought of as the order of vanishing of a function f at P.

The fact that the residue class degrees equal 1 in this geometric situation is due to the fact that we are considering curves over an algebraically closed field k. In the study of curves over a field which is not algebraically closed, there can be a residue class degree $f_i > 1$. Curves over a field which is not algebraically closed are considered in Exercise 9.15 and in Chapter 14. Even over an algebraically closed field, a situation where the residue class degree is important is considered in Exercise 9.16.

Proposition 9.13. *Let X be an affine variety with coordinate ring A, and let $x \in X$. Let $\mathfrak{m}$ be the maximal ideal of A consisting of functions vanishing at x. Then the tangent space $T_x(X)$ is isomorphic to the dual space to the k-vector space $\mathfrak{m}/\mathfrak{m}^2$.*

Proof. Let $A_\mathfrak{m} = \mathcal{O}_x$ be the local ring at x, and let $\mathfrak{M} = \mathfrak{m}\mathcal{O}_x$ be the maximal ideal of $A_\mathfrak{m}$. By definition, $T_x(X)$ is the dual space to the k-vector space $\mathfrak{M}/\mathfrak{M}^2$, so what we really need to show is that $\mathfrak{M}/\mathfrak{M}^2 \cong \mathfrak{m}/\mathfrak{m}^2$, and this follows from Proposition 9.6. ■

Proposition 9.14. *Let $f(x,y)$ be an irreducible polynomial. Let X be the affine curve $f(x,y) = 0$, that is, the affine curve whose coordinate ring is $k[x,y]/(f)$. Let $\xi = (a,b)$ be a point on the curve, so that $f(a,b) = 0$. Then ξ is a singular point if and only if both partial derivatives*

$$\frac{\partial f}{\partial x}(a,b) = \frac{\partial f}{\partial y}(a,b) = 0.$$

Proof. Suppose first that $(\partial f/\partial y)(a,b) \neq 0$. In this case, the nonsingularity of (a,b) is a consequence of Proposition 6.14 (i). The case where $(\partial f/\partial x)(a,b) \neq 0$ is similar.

On the other hand, suppose that both partials vanish—we show that (a,b) is a singular point. Indeed, let us decompose the polynomial

$$f(x,y) = \sum_n f_n(x-a, y-b),$$

where f_n is homogeneous of degree n. Then since $f(a,b) = 0$, the constant term f_0 vanishes, and since

$$\frac{\partial f}{\partial x}(a,b) = \frac{\partial f}{\partial y}(a,b) = 0,$$

f_1 also vanishes. Now consider the images of $x-a$ and $y-b$ in the coordinate ring $k[x,y]/(f)$. It is clear that these are both in the maximal ideal $\mathfrak{m}$ of functions vanishing at (a,b). Since f only involves terms of degree greater than or equal to 2, it is clear that the images of these are linearly independent modulo $\mathfrak{m}^2$, so $\dim_k(\mathfrak{m}/\mathfrak{m}^2) > 1$. By Proposition 9.13, it follows that (a,b) is a singular point. ■

There is an important distinction between the nature of ramification when the field extension E/F is Galois, i.e. normal and separable, and when it is not. To see this clearly, let us have in mind an example of a morphism $f : X \to Y$ of complete nonsingular curves such that the corresponding injection of function fields is separable but not normal.

Example 9.15. Assume that char$(k) \neq 2, 3$. Let $F = k(x)$ be the rational function field, which is the function field of $X = \mathbf{P}^1$. Let $E = k(x,y)$ be obtained by adjoining y, which satisfies $y^3 - 3y + 2x = 0$. Let Y be the corresponding complete nonsingular curve. The inclusion $F \to E$ corresponds to a morphism $f : Y \to X$. We will study the ramification of this morphism. Let $U = \mathbf{A}^1 \subseteq X$, and let $V = f^{-1}U$. The coordinate ring of U is of course $A = k[x]$, and the coordinate ring of V is $k[x,y]$, which we will now show to be integrally closed. It is sufficient to show that the locus of $f(x,y) = 0$ in $\mathbf{A}^2$ is a smooth variety, and indeed it follows from the nonvanishing of the partial derivative with respect to x that this is the case. Hence $k[x,y]$ is integrally closed. Since the equation $y^3 - 3y + 2x = 0$ for y over $k[x]$ is integral, $k[x,y]$ is integral over $k[x]$, and it is therefore the integral closure of $k[x]$ in $k(x,y)$. Hence it is the coordinate ring of U.

If x is fixed, the discriminant of the cubic equation $y^3 - 3y + 2x = 0$ is $4(-3)^3 + 27(2x)^2$, so this polynomial has repeated roots if and only if $x = \pm 1$.

If $x \neq \pm 1$, then there are three roots, and hence the fiber over x consists of three distinct points, each with ramification index two. If $x = \pm 1$, then there are only two points in the fiber, and since $\sum e_i = n$ by Theorem 9.10, it is clear that one of which must have ramification index 1, and the other must have ramification index two. (See Exercise 9.13.)

Finally, let us compute the ramification at infinity. Let $z = 1/x$, and let U' be the affine set in $\mathbf{P}^1$ whose coordinate ring is $k[z]$. Thus $U' = \mathbf{P}^1 - \{0\}$. We may use this affine set to compute with, because $\infty \in U'$. Let $V' = f^{-1}(U')$. The coordinate ring of V' is the integral closure of $k[z]$ in $k(x,y)$, and we must determine the latter ring. First note that if $w = y/x$, then w satisfies the polynomial equation

$$w^3 - 3wz^2 + 2z^2 = 0.$$

Hence w is integral over $k[z]$, and so the integral closure of $k[z]$ contains $k[z,w]$. However, we see that the partial derivatives with respect to both z and w vanish when $z = w = 0$, and so $(z,w) = (0,0)$ is a singular point on the curve with locus $w^3 - 3wz^2 + 2z^2 = 0$. By Proposition 6.25, the presence of this singularity shows that the ring $k[z,w]$ is not integrally closed. We look therefore for another element of $k(x,y)$ which is integral over this ring. Let $u = z/w$. We have

$$u^2 - (3w - 2)^{-1}w = 0.$$

Note that expanding the numerator on the right in

$$z^2 = \frac{w^3}{3w-2} = \frac{\left(\frac{1}{3}(3w-2) + \frac{2}{3}\right)^3}{3w-2}$$

and rearranging shows that $(3w - 2)^{-1} \in k[z,w]$, and so u is integral over $k[z,w]$. Moreover, $z = uw \in k[u,w]$, and so $k[u,w]$ is an integral extension of $k[z,w]$. We are therefore led to hope that $k[u,w]$ is integrally closed. We have

$$w - 3wu^2 + 2u^2 = 0,$$

and so what we need is for this affine curve to be nonsingular. It is easily checked that the three equations

$$g(u,w) = w - 3wu^2 + 2u^2 = 0, \qquad \frac{\partial g}{\partial w}(u,w) = 1 - 3u^2 = 0,$$

$$\frac{\partial g}{\partial u}(u,w) = 2u(2 - 3w)$$

do not have any simultaneous solution $(u,w) = (a,b)$, and so this affine curve is nonsingular. Therefore its coordinate ring is integrally closed. Now let $\mathfrak{p}$ be the maximal ideal of $k[z]$ corresponding to $\infty \in \mathbf{P}^1$. Then $z \in \mathfrak{p}$. Let $\mathfrak{P}$ be a maximal ideal of $k[u,w]$ lying above $\mathfrak{p}$. Then $uw = z \in \mathfrak{P}$, so either $u \in \mathfrak{P}$ or $w \in \mathfrak{P}$. However, since $w - 3wu^2 + 2u^2 = 0$, this implies that both u and $w \in \mathfrak{P}$. Now let $\mathfrak{P}'$ be the ideal of $k[u,w]$ generated by u and w. Obviously $k[u,w]/\mathfrak{P}' \cong k$, so $\mathfrak{P}'$ is maximal, and $\mathfrak{P}' \subseteq \mathfrak{P}$. Thus $\mathfrak{P}' = \mathfrak{P}$. We have shown that $\mathfrak{P}'$ is the unique maximal ideal of $k[u,w]$ lying above $\mathfrak{p}$, and so $\infty \in \mathbf{P}^1$ is totally ramified in Y.

Hence there are three ramified points P in $\mathbf{P}^1$. For one, ∞, there is a unique point in the fiber above P having ramification index three. For each of the other two ramified points, there are two points in the fiber, one of which has ramification index two, and the other of which has ramification index one.

In this example, the function field of Y is a separable extension of the function field of X, but it is not a normal extension. As we will see shortly, the case of a Galois extension is somewhat simpler.

Proposition 9.16. *Let A be an integrally closed integral domain, and F its field of fractions. Let E/F be a finite Galois extension, and let B be the integral closure of A in E. Let $\mathfrak{p}$ be a prime ideal of A, and let $\mathfrak{P}$, $\mathfrak{Q}$ be prime ideals of B lying above $\mathfrak{p}$. Then there exists $\sigma \in \mathrm{Gal}(E/F)$ such that $\sigma(\mathfrak{P}) = \mathfrak{Q}$.*

Proof. Note that B is obviously invariant under the action of the Galois group. Let $\sigma_1, \cdots, \sigma_n$ be the distinct elements of $\mathrm{Gal}(E/F)$. By Proposition 4.4 (ii), it is sufficient to show that $\sigma_i(\mathfrak{P}) \subseteq \mathfrak{Q}$ for some i. If $\sigma_i(\mathfrak{P}) \not\subseteq \mathfrak{Q}$ for all i, then by Proposition 4.14, there exists $x \in \mathfrak{Q}$ such that $x \notin \sigma_i(\mathfrak{P})$ for any i. This implies that $\sigma_i(x) \notin \mathfrak{P}$ for any i, and so the norm $N(x)$ is a product of elements of B, none of which is in $\mathfrak{P}$. Hence $N(x) \notin \mathfrak{P}$. But $N(x) \in A$ by Proposition 6.17 (i), and obviously it is in $\mathfrak{Q}$, so $N(x) \in \mathfrak{Q} \cap A = \mathfrak{p} \subseteq \mathfrak{P}$, which is a contradiction. ■

Proposition 9.17. *Let A be a Dedekind domain, and let F be its field of fractions. Let E/F be a finite Galois extension of degree n, and let B be the integral closure of A in E. Let $\mathfrak{p}$ be a prime ideal of A, and let $\mathfrak{P}_1, \cdots, \mathfrak{P}_r$ be the primes of B lying above A. Let e_i, f_i be the ramification index and residue class degrees respectively of $\mathfrak{P}_i|\mathfrak{p}$. Then the $e_1 = \ldots = e_r$ and $f_1 = \ldots = f_r$, and if $e = e_i$ and $f = f_i$ denote the common values of the ramification index and residue class degrees, then $n = ref$.*

Proof. Since the Galois group is transitive on the primes above $\mathfrak{p}$ by Proposition 9.16, it is obvious that the ramification indices and residue class degrees of these primes must all be the same. The equality $n = ref$ follows from Theorem 9.10. ■

By contrast, we see from Example 9.15 that the ramification indices may *not* be equal for the primes above $\mathfrak{p}$ when E/F is not a normal extension.

Theorem 9.18. *Let A be a Dedekind domain, and let F be its field of fractions. Let E/F be a finite Galois extension of degree n, and let B be the integral closure of A in E. Let $\mathfrak{p}$ be a prime ideal of A, and let e, r and f be as in Proposition 9.17. Let $\mathfrak{P}$ be one of the primes of B lying above A. Assume that the residue extension of $B/\mathfrak{P}$ over $A/\mathfrak{p}$ is separable. Then there exist intermediate fields K and L such that $E \supseteq L \supseteq K \supseteq F$ where $[E : L] = e$, $[L : K] = f$ and $[K : F] = r$, having the following properties. Let C and D be the integral closures of A in K and L, respectively. Let $\mathfrak{Q} = \mathfrak{P} \cap D$, and let $\mathfrak{q} = \mathfrak{P} \cap C$. Then $\mathfrak{P}$ is the unique prime of B lying above $\mathfrak{q}$, and $e(\mathfrak{P}|\mathfrak{Q}) = e$, $f(\mathfrak{P}|\mathfrak{Q}) = 1$, $e(\mathfrak{Q}|\mathfrak{q}) = 1$, and $f(\mathfrak{Q}|\mathfrak{q}) = f$. Moreover $B/\mathfrak{P}$ is a Galois extension of degree f of $A/\mathfrak{p}$, $B/\mathfrak{P} \cong D/\mathfrak{Q}$ and $C/\mathfrak{q} \cong A/\mathfrak{p}$, and there is a surjective homomorphism from $\mathrm{Gal}(E/K)$ to the Galois group of $B/\mathfrak{P}$ over $C/\mathfrak{q}$. The kernel of this homomorphism is $\mathrm{Gal}(E/L)$.*

The assumption that the extension of $B/\mathfrak{P}$ over $A/\mathfrak{p}$ is separable is automatically satisfied if $A/\mathfrak{p}$ is perfect, which is often the case in applications. The fields K and L are commonly known as the *decomposition* and *inertia* fields of $\mathfrak{P}$ over $\mathfrak{p}$. The subgroups $\mathrm{Gal}(E/K)$ and $\mathrm{Gal}(E/L)$ of $\mathrm{Gal}(E/F)$ are called the *decomposition* and *inertia* groups, respectively.

Proof. Let $G = \mathrm{Gal}(E/F)$. We have seen in Proposition 9.16 that G acts transitively on the primes of B above $\mathfrak{p}$. Let H be the stabilizer of $\mathfrak{P}$, and let K be the fixed field of H. Since the action of G on the primes above $\mathfrak{p}$ is transitive, the number r of primes is equal to the index of the stabilizer, so the cardinality of H is ef. By the fundamental theorem of Galois theory, we therefore have $[E : K] = ef$ and $[K : F] = r$. Let $\mathfrak{q} = \mathfrak{P} \cap C$, where C is the integral closure of A in K. By Proposition 9.16, $H = \mathrm{Gal}(E/K)$ acts transitively on the primes above $\mathfrak{q}$, but by construction H does not move $\mathfrak{P}$, and therefore $\mathfrak{P}$ is the only prime of B lying above $\mathfrak{q}$. Now by Theorem 9.10 (ii) we have $e(\mathfrak{P}|\mathfrak{q}) f(\mathfrak{P}|\mathfrak{q}) = [E : K] = ef$. Furthermore by Proposition 9.12, $e(\mathfrak{P}|\mathfrak{q})$ divides $e(\mathfrak{P}|\mathfrak{p}) = e$ and $f(\mathfrak{P}|\mathfrak{q})$ divides $f(\mathfrak{P}|\mathfrak{p}) = f$, so $e(\mathfrak{P}|\mathfrak{q}) = e$ and $f(\mathfrak{P}|\mathfrak{q}) = f$. Moreover, by Proposition 9.12, $f(\mathfrak{P}|\mathfrak{q}) f(\mathfrak{q}|\mathfrak{p}) = f$, and so $f(\mathfrak{q}|\mathfrak{p}) = 1$. This means that $C/\mathfrak{q} \cong A/\mathfrak{p}$.

Let us denote $\overline{B} = B/\mathfrak{P}$ and $\overline{C} = C/\mathfrak{q}$. We have an inclusion of fields $\overline{C} \to \overline{B}$. The group $H = \mathrm{Gal}(E/K)$ acts on B stabilizing $\mathfrak{P}$, so we have a homomorphism from H to the group of field automorphisms of $\overline{B}$ fixing $\overline{C}$. We will show now that this homomorphism is surjective, and at the same time establish that $\overline{B}/\overline{C}$ is a Galois extension.

Indeed, since $\overline{C} \cong A/\mathfrak{p}$ and since $\overline{B}$ is by hypothesis a separable extension of $A/\mathfrak{p}$, $\overline{B}/\overline{C}$ is a separable extension. By the Theorem of the Primitive Element, there exists an element $\overline{x} \in \overline{B}$ which generates $\overline{B}$ over $\overline{C}$. We assume that $\overline{x}$ is the residue class of $x \in B$. Let f be the minimal polynomial satisfied by x over K. Since B is integral over C, and since C is integrally closed, f is monic and has coefficients in C by Proposition 4.7. Moreover, since E/K is Galois, f splits into linear factors over E, so we may write

$$f(t) = \prod(t - x_i),$$

where $x_i \in E$, and $x = x_1$. In fact, each $x_i \in B$, since they are integral over C, and B is the integral closure of C in E. Let $\overline{f}$ be the polynomial with coefficients in $C/\mathfrak{q}$ obtained by reducing f modulo $\mathfrak{q}$. (Of course $\overline{f}$ may not be irreducible.) Let $\overline{x}_i$ be the image of x_i in $\overline{B}$. Then

$$\overline{f}(t) = \prod(t - \overline{x}_i).$$

This shows that the generator $\overline{x}$ of $\overline{B}$ satisfies a polynomial over $\overline{C}$ which splits into linear factors in $\overline{B}$. This implies that $\overline{B}/\overline{C}$ is a normal extension, and since it is separable, it is Galois.

If $\overline{\sigma}$ is any automorphism of $B/\mathfrak{P}$ over $C/\mathfrak{q}$, then $\overline{\sigma}$ takes $\overline{x}$ to another root $\overline{x}_i$ of $\overline{f}$. Now there exists an element σ of $\mathrm{Gal}(E/K)$ such that $\sigma(x) = x_i$. Thus σ induces an automorphism of $\overline{B}$ over $\overline{C}$, which sends $\overline{x} \to \overline{x}_i$. This is therefore $\overline{\sigma}$. This proves that the induced map $\mathrm{Gal}(E/K) \to \mathrm{Gal}(\overline{B}/\overline{C})$ is surjective.

Let $I \subseteq H$ be the kernel of this homomorphism, and let L be the fixed field. Since the cardinality of $\mathrm{Gal}(\overline{B}/\overline{C})$ is equal to $[\overline{B} : \overline{C}] = f$, we have $[E : L] = [H : I] = e$. Let D be the integral closure of A in L, and let $\mathfrak{Q} = D \cap \mathfrak{P}$. Let $\overline{D} = D/\mathfrak{Q}$. We have a bijective homomorphism $\mathrm{Gal}(L/K) \to \mathrm{Gal}(\overline{B}/\overline{C})$, and so the residue class degree $f(\mathfrak{Q}|\mathfrak{q}) = f$. Therefore $e(\mathfrak{Q}|\mathfrak{q}) = 1$, and consequently $e(\mathfrak{P}|\mathfrak{Q}) = e$. ∎

Let A be a Dedekind domain, and let F be its field of fractions. By a *fractional ideal*, we mean a nonzero finitely generated A-submodule of F. Note that a fractional ideal is not necessarily an ideal in the usual sense, since it may not be contained in A. If we wish to emphasize that an ideal is contained in A, we may refer to it as an *integral ideal*. This usage is not mandatory, because the unadorned term "ideal" always means an integral ideal.

We may multiply fractional ideals, by letting $\mathfrak{ab}$ be the set of all finite sums $\sum a_i b_i$ with $a_i \in \mathfrak{a}$, $b_i \in \mathfrak{b}$. It is proved in Exercise 9.3 that the fractional ideals form a group under multiplication.

Now if E/F is a finite separable extension, and B is the integral closure of A in F, then B is also a Dedekind domain by Proposition 9.3. We will define an ideal $\mathfrak{D}_{B/A}$ of E, called the *different* of B over A. Let $\mathfrak{D}_{B/A}^{-1}$ be the set of all $x \in E$ such that $\mathrm{tr}(xB) \subseteq A$. This is a fractional ideal containing B by Proposition 6.17 (iii), and so its inverse is a fractional ideal which is contained in B, that is, an integral ideal.

In the next proposition, we will make use of the obvious fact that if B is a Dedekind domain, K its field of fractions, and C is the integral closure of B in a finite separable extension E of K, then there is a homomorphism from the group of fractional ideals of B to the group of fractional ideals of C, namely, $\mathfrak{a} \to C\mathfrak{a}$.

Proposition 9.19. *Let A be a Dedekind domain, and let F be its field of fractions. Let $E \supseteq K \supseteq F$ be finite separable extensions of F. Let B and C be the integral closures of A in K and E respectively. Then $\mathfrak{D}_{C/A} = \mathfrak{D}_{C/B}\,(C\mathfrak{D}_{B/A})$.*

Proof. Suppose that $\mu \in \mathfrak{D}_{C/B}^{-1}$ and let $\nu \in \mathfrak{D}_{B/A}^{-1}$. Then denoting by $\mathrm{tr}_{E,F}$ the trace from E to F, we have

$$
\begin{aligned}
\mathrm{tr}_{E,F}(\mu\nu C) = \mathrm{tr}_{K,F}(\mathrm{tr}_{E,K}(\mu\nu C)) &= \mathrm{tr}_{K,F}(\nu\,\mathrm{tr}_{E,K}(\mu C)) && [\text{since } \nu \in K] \\
&\subseteq \mathrm{tr}_{K,F}(\nu B) && [\text{since } \mu \in \mathfrak{D}_{C/B}^{-1}] \\
&\subseteq A && [\text{since } \nu \in \mathfrak{D}_{B/A}^{-1}].
\end{aligned}
$$

Thus $\mu\nu \in \mathfrak{D}_{C/A}^{-1}$. This shows that $\mathfrak{D}_{C/A}^{-1} \supseteq \mathfrak{D}_{C/B}^{-1}\,(C\mathfrak{D}_{B/A}^{-1}) = \mathfrak{D}_{C/B}^{-1}\mathfrak{D}_{B/A}^{-1}$. Taking inverses, it follows that $\mathfrak{D}_{C/A} = \mathfrak{D}_{C/B}\,(C\mathfrak{D}_{B/A})$.

On the other hand, let $\lambda \in \mathfrak{D}_{C/A}^{-1}$. Then

$$
\mathrm{tr}_{K,F}(\mathrm{tr}_{E,K}(\lambda C)B) = \mathrm{tr}_{K,F}(\mathrm{tr}_{E,K}(\lambda C)) \subseteq A,
$$

so $\mathrm{tr}_{E,K}(\lambda C) \subseteq \mathfrak{D}_{B/A}^{-1}$. Therefore $\mathrm{tr}_{E,K}(\lambda\mathfrak{D}_{B/A}C) = \mathfrak{D}_{B/A}\,\mathrm{tr}_{E,K}(\lambda C) \subseteq B$, and so $\lambda\mathfrak{D}_{B/A} \subseteq \mathfrak{D}_{C/B}^{-1}$. This proves that $\mathfrak{D}_{C/A}^{-1}(C\mathfrak{D}_{B/A}) \subseteq \mathfrak{D}_{C/B}^{-1}$, and therefore $\mathfrak{D}_{C/A} \supseteq \mathfrak{D}_{C/B}\,(C\mathfrak{D}_{B/A})$. ∎

Proposition 9.20. *Let A be a Dedekind domain, F its field of fractions, and let E be a finite separable extension of F. Let B be the integral closure of A in E, and let S be a multiplicative subset of A. Then $S^{-1}A$ and $S^{-1}B$ are Dedekind domains, and $S^{-1}B$ is the integral closure of $S^{-1}A$ in E. We have*

$$
S^{-1}\mathfrak{D}_{B/A} = \mathfrak{D}_{S^{-1}B/S^{-1}A}.
$$

Proof. It follows from Proposition 6.19 that $S^{-1}A$ and $S^{-1}B$ are integrally closed. It follows from Exercise 3.4 and the fact that A and B are Noetherian domains of dimension one that $S^{-1}A$ and $S^{-1}B$ are also. Moreover, it follows from Proposition 6.19 that $S^{-1}B$ is the integral closure in E of $S^{-1}A$.

To prove the assertion about the differents, one first checks that

$$
S^{-1}\mathfrak{D}_{B/A}^{-1} = \mathfrak{D}_{S^{-1}B/S^{-1}A}^{-1}.
$$

Indeed, it is clear from the definitions that

$$S^{-1}\mathfrak{D}_{B/A}^{-1} \subseteq \mathfrak{D}_{S^{-1}B/S^{-1}A}^{-1}.$$

The converse is also easy to check if one bears in mind that B is finitely generated as an A-module, by Proposition 6.17 (iii). We leave the verification to the reader. Now the map $\mathfrak{a} \to S^{-1}\mathfrak{a}$ from the group of fractional ideals of A to the group of fractional ideals of $S^{-1}A$ is a homomorphism by Exercise 9.3, and so $S^{-1}\mathfrak{D}_{B/A} = \mathfrak{D}_{S^{-1}B/S^{-1}A}$. ∎

Proposition 9.21. *Let A be a principal ideal domain, F its field of fractions, let E/F be a finite separable extension, and let B be the integral closure of A in E. Let $\mathfrak{p}$ be a prime ideal of A, and let $\mathfrak{P}$ be a prime of B lying above $\mathfrak{p}$. Assume that the extension $B/\mathfrak{P}$ over $A/\mathfrak{p}$ is separable. Then there exists an element $x \in \mathfrak{P}^{-e}$ such that $\mathrm{tr}_{E/F}(x) \notin A$, where $e = e(\mathfrak{P}|\mathfrak{p})$ is the ramification index.*

Proof. Let $\mathfrak{p}B = \prod \mathfrak{P}_i^{e_i}$ be the decomposition of $\mathfrak{p}B$ into primes, where $\mathfrak{P}_1 = \mathfrak{P}$ and $e_1 = e$. Since $B/\mathfrak{P}$ is a separable extension of $A/\mathfrak{p}$, we may find $y \in B$ such that the image $\overline{y}$ of y in $B/\mathfrak{P}$ has nonzero trace in $A/\mathfrak{p}$. Moreover, by the Chinese Remainder Theorem (Proposition 9.9), we may assume that $y \in \mathfrak{P}_i^{e_i}$ for $i \neq 1$.

We will show that $\mathrm{tr}(y) \notin \mathfrak{p}$. We may apply Theorem 9.10 (iii). In the notation there, $\mathrm{tr}_{K_1/K}(\overline{y}_1) \neq 0$ and $e_1 = 1$, while $\overline{y}_i = 0$ for $i \neq 1$. Thus $\mathrm{tr}_{E/F}(y) \neq 0$.

Since A is by assumption a principal ideal domain, let π be a generator of the ideal $\mathfrak{p}$, and consider $x = \pi^{-1}y$. We have $y \in \mathfrak{P}_i^{e_i}$ for $i \neq 1$, and $e_1 = 1$, so $x \in \mathfrak{P}^{-e}$. We have $\mathrm{tr}(x) = \pi^{-1}\mathrm{tr}(y)$, and since $\mathrm{tr}(y) \notin \mathfrak{p}$, we have $\mathrm{tr}(x) \notin A$. ∎

Theorem 9.22. *Let A be a Dedekind domain, F its field of fractions, let E/F be a finite separable extension, and let B be the algebraic closure of A in E. Let $\mathfrak{p}$ be a prime ideal of A, and let $\mathfrak{P}$ be a prime of B lying above $\mathfrak{p}$. Let $e = e(\mathfrak{P}|\mathfrak{p})$ be the ramification index. Then $\mathfrak{P}^{e-1}$ divides $\mathfrak{D}_{B/A}$. Moreover, if $B/\mathfrak{P}$ is a finite separable extension of $A/\mathfrak{p}$, then $\mathfrak{P}^e$ divides $\mathfrak{D}_{B/A}$ if and only if the characteristic of $A/\mathfrak{p}$ divides e. Only a finite number of primes of B are ramified, and all of these divide the different. Conversely, if $B/\mathfrak{P}$ is a separable extension of $A/\mathfrak{p}$, and if $\mathfrak{P}|\mathfrak{D}_{B/A}$, then $\mathfrak{P}$ is ramified.*

The hypothesis that $B/\mathfrak{P}$ is a separable extension of $A/\mathfrak{p}$ is automatic if $A/\mathfrak{p}$ is perfect. The dichotomy between the two cases where the residue characteristic does or does not divide the ramification index is important. If the characteristic of $A/\mathfrak{p}$ divides e, we say $\mathfrak{p}$ ramifies *wildly* in E, or that $\mathfrak{P}$ is *wildly ramified*. If $\mathfrak{p}$ is ramified but not wildly ramified we say it is *tamely ramified*.

Proof. Let $S = A - \mathfrak{p}$. We may replace A by $A_{\mathfrak{p}}$ and B by $S^{-1}B$, since if the Theorem is true for $A_{\mathfrak{p}}$, it follows from Proposition 9.20 and Exercise 3.4 and Exercise 9.3 that it is true for A.

Hence we may assume that $\mathfrak{p}$ is the unique prime ideal of A, and hence A is a discrete valuation ring. Let π be a generator of the maximal ideal $\mathfrak{p}$ of A, which is principal by Proposition 5.27. Let $\mathfrak{P}_i$, $i = 1, \cdots, r$ be the distinct primes of B lying above $\mathfrak{p}$. We may arrange that $\mathfrak{P} = \mathfrak{P}_1$. Let $e_i = e(\mathfrak{P}_i | \mathfrak{p})$ be the ramification indices. We will also denote $K = A/\mathfrak{p}$ and $K_i = B/\mathfrak{P}_i$.

Let

$$\mathfrak{A} = \prod_{i=1}^{r} \mathfrak{P}_i^{e_i - 1}.$$

We will show that $\mathfrak{A} | \mathfrak{D}_{B/A}$. This is equivalent to showing that $\mathfrak{A}^{-1} \subseteq \mathfrak{D}_{B/A}^{-1}$. Let $z \in \mathfrak{A}^{-1}$, and let $y \in B$. By the definition of the different, what we must show is that $\operatorname{tr}(zy) \in A$. Let $w = \pi yz$, so $\operatorname{tr}(zy) = \pi^{-1} \operatorname{tr}(w)$, and since $\pi z \in \mathfrak{p}\mathfrak{A}^{-1} = \prod \mathfrak{P}_i$, we have $w \in \mathfrak{P}_i$ for every prime i. Hence the image of w in $B/\mathfrak{P}_i$ is zero for every i, and it follows from Theorem 9.10 (iii) that $\operatorname{tr}_{E/F}(w) \in \mathfrak{p}$. Therefore $\operatorname{tr}(zy) \in A$. This proves that $\mathfrak{A}$ divides $\mathfrak{D}_{B/A}$. In particular, $\mathfrak{P}^{e-1}$ divides $\mathfrak{D}_{B/A}$.

Now assume that $B/\mathfrak{P}$ is a separable extension of $A/\mathfrak{p}$. We must show that if

$$\mathfrak{B} = \mathfrak{P}_1^{e_1} \prod_{i=2}^{r} \mathfrak{P}_i^{e_i - 1},$$

then $\mathfrak{B} | \mathfrak{D}_{B/A}$ if and only if the characteristic of K divides $e = e_1$. By the definition of the different, $\mathfrak{B}$ divides $\mathfrak{D}_{B/A}$ if and only if $\operatorname{tr}(\mathfrak{B}^{-1}) \subseteq A$. If $z \in \mathfrak{B}^{-1}$, let $w = \pi z$. The $w \in \mathfrak{P}_i$ for $i = 2, \cdots, r$, so $\pi \operatorname{tr}(z) = \operatorname{tr}(\pi z) \equiv e_1 \operatorname{tr}(\overline{w})$ modulo $\mathfrak{p}$. Thus $\operatorname{tr}(z) \in A$ if and only if $e_1 \operatorname{tr}(\overline{w}) = 0$ in $A/\mathfrak{p}$. Since E/F is separable, $\operatorname{tr} : K_i \to K$ is surjective, so it is now clear that $\operatorname{tr}(\mathfrak{B}^{-1}) \subseteq A$ if and only if the characteristic of K divides e_1. ∎

Theorem 9.23. *Let X and Y be complete nonsingular curves, and let $f : Y \to X$ be a nonconstant separable morphism of degree n. Then there exist only finitely many points P of X which are ramified in f. For all other points P, the number of elements of the fiber $f^{-1}(P)$ is exactly n. For the finite number of ramified points P, the cardinality of the fiber is strictly less than n, and there exists at least one ramified point Q in the fiber such that the ramification index $e(Q|P) > 1$. The sum of the ramification indices of all the points in the fiber is exactly n. If the morphism $f : Y \to X$ is Galois, then every point in the fiber has the same ramification index.*

Proof. Let E and F be the function fields of Y and X respectively. As in Theorem 8.7, the morphism $Y \to X$ amounts to a field homomorphism $E \to F$

over k, and we may regard F as a subfield of its finite separable extension E. We may find a finite cover $\{U_i\}$ of X by affine open sets. Let V_i be the preimage of each U_i in Y. Thus if A_i is the coordinate ring of U_i, the coordinate ring B_i of V_i is the integral closure of A_i in E by Proposition 9.4. The rings A_i and B_i are Dedekind domains, also by Proposition 9.4.

The Theorem is thus a consequence of the corresponding statement for affine curves, which is just a restatement of the results of this chapter. The finiteness of ramification is asserted in Theorem 9.22, and the remaining statements are asserted in Theorem 9.10 and Theorem 9.18. Note that since $A/\mathfrak{p} \cong k$ is algebraically closed in these Theorems, the residue class degrees $f_i = 1$, and the residue class extensions are always trivial, hence separable. ∎

Exercises

The following exercise leads to one of the most fundamental properties of Dedekind domains—the unique factorization of ideals into prime elements.

Let A be a ring. Two ideals $\mathfrak{a}$ and $\mathfrak{b}$ are called *coprime* if $\mathfrak{a} + \mathfrak{b} = A$.

Exercise 9.1 (Unique Factorization of Dedekind domains). (i) Let A be a ring, and let $\mathfrak{a}$, $\mathfrak{b}$ be coprime ideals. Prove that $\mathfrak{a} \cap \mathfrak{b} = \mathfrak{a}\mathfrak{b}$.

(ii) Show that if $\mathfrak{a}$ is coprime to $\mathfrak{b}$ and to $\mathfrak{c}$ then it is coprime to $\mathfrak{b}\mathfrak{c}$.

(iii) Show that if $r(\mathfrak{a})$ is coprime to $r(\mathfrak{b})$, then $\mathfrak{a}$ is coprime to $\mathfrak{b}$.

(iv) Show that if $\mathfrak{q}_i, \cdots, \mathfrak{q}_n$ are ideals such that their radicals $r(\mathfrak{q}_i)$ are coprime in pairs, then $\bigcap \mathfrak{q}_i = \prod \mathfrak{q}_i$.

(v) Let $\mathfrak{p}$ be a prime ideal of A. Let $\mathfrak{q}_1$ and $\mathfrak{q}_2$ be two $\mathfrak{p}$-primary ideals. Show that $\mathfrak{q}_1 \cap \mathfrak{q}_2$ is also a $\mathfrak{p}$-primary ideal.

(vi) Let A be a Dedekind domain. Let $\mathfrak{a}$ be an ideal of A. $\mathfrak{a}$ has a primary decomposition $\mathfrak{a} = \prod \mathfrak{q}_i$ where the ideals $\mathfrak{q}_i$ are primary, and mutually coprime.

Hint: Use Proposition 5.17 and parts (iv) and (v) above.

(vii) Let A be a Dedekind domain, and let $\mathfrak{q}$ be a $\mathfrak{p}$-primary ideal. Prove that $\mathfrak{q} = \mathfrak{p}^n$ for some $\mathfrak{p}$.

Hint: $\mathfrak{q}A_{\mathfrak{p}} = \mathfrak{p}^n A_{\mathfrak{p}}$ for some n by Proposition 1.1, because $A_{\mathfrak{p}}$ is a discrete valuation ring.

(viii) Let A be a Dedekind domain. Then every ideal has a factorization $\mathfrak{a} = \prod_i \mathfrak{p}_i^{e_i}$ into powers of distinct primes.

(ix) Prove that the decomposition in (viii) is unique.

Hint: Localize.

Fractional Ideals. Let A be a Dedekind domain, and let F be its field of fractions. By a *fractional ideal*, we mean a nonzero finitely generated A-submodule of F. Note that a fractional ideal is not necessarily an ideal in the usual sense, since it may not be contained in A. If we wish to emphasize that an ideal is contained in A, we may refer to it as an *integral ideal*. This usage is not mandatory, because the unadorned term "ideal" always means an integral ideal.

We may multiply fractional ideals, by letting $\mathfrak{ab}$ be the set of all finite sums $\sum a_i b_i$ with $a_i \in \mathfrak{a}$, $b_i \in \mathfrak{b}$.

Exercise 9.2. Show that the fractional ideals form a group under multiplication.

Hint: The only thing which requires checking is that fractional ideals have inverses. It is sufficient to prove this for integral ideals. Show that if $\mathfrak{a}$ is a nonzero principal ideal of A, then $\mathfrak{a}$ has an inverse; show using Exercise 9.1 that if an integral ideal $\mathfrak{a}$ has an inverse, and if $\mathfrak{b}$ is an integral ideal containing $\mathfrak{a}$, then $\mathfrak{b}$ has an inverse; and show that every nonzero ideal contains a nonzero principal ideal.

Exercise 9.3. Let A be a Dedekind domain, and let S be a multiplicative subset. Prove that $S^{-1}A$ is a Dedekind domain. Let $\mathfrak{a}$ be a fractional ideal of A. Then $S^{-1}\mathfrak{a}$ is a fractional ideal of $S^{-1}A$, and $\mathfrak{a} \to S^{-1}A$ is a homomorphism from the group of fractional ideals of A to the group of fractional ideals of $S^{-1}A$.

The next exercises concern *extension of scalars*. Let B be an A-algebra with structure map $i : A \to B$, and let M be an A-module. Then B becomes an A-module, with the multiplication $A \times B \to B$ given by $(a, b) \to i(a)\, b$. The module $B \otimes_A M$ is naturally a B-module, with multiplication $(b, b' \otimes m) \to bb' \otimes m$. $B \otimes M$ is called the B-module obtained from M by *extension of scalars*.

Exercise 9.4. Basic Properties of Extension of Scalars. (i) Let A be a ring and M an A-module. Show that $A \otimes_A M \cong M$ as A-modules.

(ii) Let A be a ring. Let B be an A-algebra, and let C be a B-algebra. Then C is naturally also an A-algebra. Show that

$$C \otimes_B (B \otimes_A M) \cong C \otimes_A M.$$

(iii) **Extension and Restriction of Scalars are Adjoint Functors.** Let B be an A-algebra. Let M be an A-module, and let N be a B-module. Then N may be regarded as an A-module by "restriction of scalars." That is, if $i : A \to B$ is the structure map, then the multiplication $A \times N \to N$ is given by $(a, n) \to i(a)\, N$. Prove that

$$\mathrm{Hom}_B(B \otimes_A M, N) \cong \mathrm{Hom}_A(M, N).$$

Exercise 9.5. Let A be an integral domain, and let F be its field of fractions. Let V be a finite-dimensional vector space over F, and let M be an A-submodule of V such that the F-span of M is V. Let $m_1, \cdots, m_n$ be an F-basis of V which is contained in M. Show that every element of $F \otimes M$ may be written in the form $\sum_{i=1}^{n} c_i \otimes m_i$ with $c_i \in F$.

Exercise 9.6. In the setting of the previous exercise, prove that $F \otimes_A M \cong V$ as F-vector spaces, where $F \otimes M$ is an F-module by extension of scalars.

Hint: Multiplication $F \times M \to V$ is an A-bilinear map, hence induces an A-module homomorphism $F \otimes M \to V$, which is clearly also F-linear. Show that this is an isomorphism, using Exercise 9.5 to establish injectivity.

Exercise 9.7. Let A be a ring and M an A-module. Let $\mathfrak{a}$ be an ideal of A. Show that $(A/\mathfrak{a}) \otimes_A M \cong M/\mathfrak{a}M$ as $A/\mathfrak{a}$-modules, where $(A/\mathfrak{a}) \otimes_A M$ is an $A/\mathfrak{a}$-module by extension of scalars.

Hint: Construct inverse isomorphisms $(A/\mathfrak{a}) \otimes_A M \to M/\mathfrak{a}M$ and $M/\mathfrak{a}M \to (A/\mathfrak{a}) \otimes_A M$.

Free Modules. Let A be a ring. An A-module M is called *free* if it is isomorphic to a cartesian product of copies of A. Thus there exists an indexing set X such that M is isomorphic to the set $A^{(X)}$ of all mappings from $X \to A$, are zero for all but finitely many elements of X. $A^{(X)}$ is an A-module with componentwise operations: if $f, g \in A^{(X)}$, $a, b \in A$, then for $x \in X$, the A-module structure on $A^{(X)}$ is given by $(af + bg)(x) = af(x) + bg(x)$. Thus $A^{(X)}$ is the direct sum of copies of A, indexed by X. If X is a finite set $\{1, \cdots, n\}$, then we may interpret $A^{(X)}$ as the set of ordered n-tuples of elements of A, and denote it as usual by A^n.

Exercise 9.8. Suppose that $A^{(X)} \cong A^{(Y)}$. Prove that the cardinalities of X and Y are equal.

Hint: Let m be any maximal ideal of A. Show that the cardinality of X is equal to the dimension of the vector space $(A/\mathfrak{m}) \otimes_A A^{(X)}$.

The *rank* of a free A-module M is the cardinality of X where $M \cong A^{(X)}$. This is well-defined by Exercise 9.8.

Exercise 9.9. Let M and N be A-modules. Let $m_1, \cdots, m_r \in M$ and $n_1, \cdots, n_r \in N$ such that $\sum m_i \otimes n_i = 0$ in $M \otimes_A N$. Prove that there exist finitely generated submodules M' and N' of M and N respectively, such that $\sum m_i \otimes n_i = 0$ in $M' \otimes_A N'$.

Hint: Recall the definition of $M \otimes N$ as the free A-module on the symbols $m \otimes n$, modulo a submodule of relations.

Flatness. Let A be a ring, and N an A-module. N is called *flat* if given any injective homomorphism $i : M \to M'$ of A-modules, the induced map $N \otimes_A M \to N \otimes_A M'$ is also injective.

Exercise 9.10. (i) Prove that a free module is flat.

Hint: Use the distributive property of tensor product over direct sums.

(ii) Prove that if M is an A-module such that every finitely generated submodule of M is contained in a free A-submodule of M, then M is flat.

Hint: Use Exercise 9.9.

Exercise 9.11. Prove that if A is an integral domain, and $S \subseteq A$ a multiplicative subset, then $S^{-1}A$ is a flat A-module.

Hint: Use Exercise 3.4 (vi).

Exercise 9.12. Prove the following generalization of Exercise 2.6. Let A be an integral domain, and let F be its field of fractions. Let B be a ring such that $A \subseteq B \subseteq F$, and suppose that B is flat as an A-module. Let V be a finite-dimensional vector space over F, and let M be an A-submodule such that the F-span of M is V. Prove that $B \otimes_A M \cong BM \subseteq V$.

Hint: Multiplication $B \times M \to BM$ is an A-bilinear map, hence by the universal property of the tensor product induces an A-homomorphism $B \otimes M \to BM$, which is clearly also a B-module homomorphism, and evidently surjective. To prove that it is injective, consider the commutative diagram

$$
\begin{array}{ccc}
B \otimes M & \longrightarrow & BM \\
\downarrow & & \downarrow \\
B \otimes V & \longrightarrow & V
\end{array}
$$

The left arrow is injective by flatness, and the bottom arrow is an isomorphism, with inverse map $V \to B \otimes V$ given by $v \to 1 \otimes v$.

Exercise 9.13. In Example 9.15, determine which of the two points in the fiber above ± 1 has ramification index two, and which has ramification index one.

Exercise 9.14. Let X and Y be affine varieties of equal dimension, and let $\phi : Y \to X$ be a dominant separable morphism. Let $n = [F(Y) : F(X)]$. Prove that there is an open set $U \subseteq X$ such that if $x \in U$, then the cardinality of the fiber $\phi^{-1}(x)$ is precisely n.

Hint: Unless $\dim(X) = \dim(Y)$, the results of this chapter are not applicable. Instead, show that there exists a principal open set U' such that if $V' = \phi^{-1}U'$, and if the coordinate ring A, B are the coordinate rings of U' and V', respectively, then $B = A[\phi]$ for ϕ a separable polynomial with coefficients in A. Let $U = U'_g$, where g is the discriminant of ϕ.

Although in the case of curves over an algebraically closed field, the residue class degrees in Theorem 9.9 are always one, nevertheless, there are geometric examples where the residue class degree is strictly greater than one, namely, for prime ideals corresponding to points on curves over a field which is not algebraically closed, or for prime ideals corresponding to subvarieties of codimension one for a higher dimensional variety. These examples are studied in the next two exercises.

Exercise 9.15 Curves over a Field which is not Algebraically Closed. Let $k = \mathbb{R}$, and let $A = k[x]$ be a polynomial ring in one indeterminate. Let $B = k[x, y]$, where x and y satisfy $x^2 + y^2 = 1$. Show that A is a Dedekind domain, and B is its integral closure in the field of fractions of B. Let $\mathfrak{p}$ be the maximal ideal of all polynomials in A vanishing at $x = a$. Compute r and the e_i, f_i in Theorem 9.10 for different values of a.

Exercise 9.16 Subvarieties of Codimension One. Let X and Y be normal affine varieties of dimension r, and let $\phi : Y \to X$ be a finite dominant separable morphism. Let

$A \subseteq B$ be the coordinate rings of X and Y respectively, and let F and E be the function fields. Let $W \subseteq X$ be a closed subvariety of codimension one, and let $\mathfrak{p}$ be its prime ideal. Then A is not a Dedekind domain, but $A_{\mathfrak{p}}$ is a discrete valuation ring, so we may apply Theorem 9.10. Let $S = A - \mathfrak{p}$, and let $B' = S^{-1}B$. Show that B' is the integral closure of $A_{\mathfrak{p}}$, hence is a Dedekind domain. Now specialize to the following case. Let $X = \mathbf{A}^2$, and let $A = k[x, y]$ be its coordinate ring. Let $B = A[z]$, where z satisfies the equation $z^2 = y$.

Hint: Assume $\mathrm{char}(k) \neq 2$, so this equation is separable. Let W be a line $ax + b = 0$. Compute r and the e_i, f_i in Theorem 9.10 for different values of a and b.

10. Completions

In this chapter, we will introduce the powerful technique of *completion*. In complex analysis, one makes frequent use of the Taylor expansion of a function. Similarly, in algebraic geometry, one may consider expansions in formal power series. In contradistinction to complex analysis, these may not be regarded as giving analytic continuation to a function. Nevertheless, they are a potent tool. The formation of a ring or field of formal power series is an example of the process of completion.

Like those of the previous chapter, the techniques of this chapter are of importance in algebraic number theory, where p-adic completions are a pervasive tool.

Recall that in a topological abelian group A, a *fundamental system of neighborhoods of the identity* is a collection $\mathcal{V}$ of open sets containing the identity, such that any open set containing the identity contains an element of $\mathcal{V}$. The topology may be reconstructed from $\mathcal{V}$, since a subset U of A is open if and only if it satisfies the requirement that for all $x \in U$ there exists a $V \in \mathcal{V}$ such that $x + V \subseteq U$. Conversely, suppose that we are given an abelian group A, and a set $\mathcal{V}$ of subsets containing the identity. We may try to make A into a topological abelian group, by specifying that a set U is open if it satisfies this last description. This will make A into a topological abelian group if the following two axioms are satisfied: (1) If $V, V' \in \mathcal{V}$ then there exists $V'' \in \mathcal{V}$ such that $V'' \subseteq V \cap V'$; and (2) If $V \in \mathcal{V}$, then there exists $V' \in \mathcal{V}$ such that $V' + V' \subseteq V$ and $-V' \subseteq V$. The first axiom guarantees that the open sets form a topology, and the second axiom guarantees that the group laws are continuous. Note that the second axiom is automatic if the elements of $\mathcal{V}$ are subgroups of A. The group will be Hausdorff if and only if $\bigcap \mathcal{V} = \{0\}$.

Let A be a Hausdorff topological abelian group, written additively. A sequence $\{a_n\}$ of elements of A is called a *Cauchy sequence* if for any neighborhood U of the identity, there exists an N such that if $n, m \geq N$, then $a_n - a_m \in U$. If every Cauchy sequence converges, we say that A is *complete*. Every Hausdorff abelian group may be embedded in a complete group as follows. We define an equivalence relation on the set of Cauchy sequences of A, in which two Cauchy sequences $\{a_n\}$ and $\{b_n\}$ are equivalent if for any neighborhood U of the identity, there exists an M such that if $n \geq M$, then $a_n - b_n \in U$. We denote $\{a_n\} \sim \{b_n\}$. Let $\widehat{A}$ be the set of equivalence classes of Cauchy sequences. It is easily checked that if $\{a_n\}$ and $\{b_n\}$ are Cauchy sequences, then so is $\{a_n + b_n\}$. Moreover, if $\{a_n\} \sim \{a_n'\}$ and $\{b_n\} \sim \{b_n'\}$, then $\{a_n + b_n\} \sim \{a_n' + b_n'\}$. Hence the set $\widehat{A}$ of equivalence classes of Cauchy sequences forms a group, called the *completion* of A. We have a canonical homomorphism $A \to \widehat{A}$, where $a \in A$ is represented in $\widehat{A}$ by the constant Cauchy sequence $\{a\}$. The Hausdorff axiom implies that this map is injective, so we may regard A as a subgroup of H.

To specify the topology on $\widehat{A}$, we will describe a fundamental system of neighborhoods of the identity. If $V \in \mathcal{V}$, let $\widehat{V}$ be the set of elements of $\widehat{A}$ represented by a Cauchy sequence $\{a_n\}$ such that for some $V' \in \mathcal{V}$ we have $a_n + V' \subseteq V$ for all sufficiently large n. It is easy to see that if $\{b_n\} \sim \{a_n\}$, then $\{b_n\}$ also has this property. Evidently $\widehat{V} \cap A = V$. Let $\widehat{\mathcal{V}} = \{\widehat{V} | V \in \mathcal{V}\}$. We take $\widehat{\mathcal{V}}$ to be a fundamental system of neighborhoods of the identity in $\widehat{A}$.

Proposition 10.1. *Let A be a Hausdorff topological abelian group, and let $\widehat{A}$ be its completion. Then $\widehat{A}$ is complete. Moreover, A has the subspace topology with respect to $\widehat{A}$. If A is complete, then the inclusion $A \to \widehat{A}$ is an isomorphism.*

Proof omitted. See Exercise 10.1. ∎

Example 10.2. Let A be a ring, and let $\mathfrak{a} \subseteq A$ be any ideal such that $\bigcap_{n=0}^{\infty} \mathfrak{a}^n = 0$. Topologize the additive group of A by taking as a system of neighborhoods of the identity the powers $\mathfrak{a}^n$. (These sets are closed as well as open, since the complement of $\mathfrak{a}^n$ is a union of cosets of $\mathfrak{a}^n$, which are open.) This is called the $\mathfrak{a}$-*adic topology* on A. The Hausdorff axiom is easily seen to follow from the hypothesis that $\bigcap \mathfrak{a}^n = 0$. Moreover, if $\{a_n\}$ and $\{b_n\}$ are Cauchy sequences in this topology, then so is $\{a_n b_n\}$. Thus $\widehat{A}$ inherits a ring structure from A. (See Exercise 10.2.)

Proposition 10.3. *Let A and $\mathfrak{a}$ be as in* Example 10.2. *Let $\widehat{\mathfrak{a}}$ be the closure of $\mathfrak{a}$ in $\widehat{A}$. Then $\widehat{\mathfrak{a}}$ is an ideal of $\widehat{A}$, and the inclusion of A in $\widehat{A}$ induces an isomorphism $A/\mathfrak{a} \cong \widehat{A}/\widehat{\mathfrak{a}}$.*

Proof. The closure $\widehat{\mathfrak{a}}$ of $\mathfrak{a}$ in $\widehat{A}$ consists of the limits of Cauchy sequences which are eventually in $\mathfrak{a}$, and is clearly an ideal. It is evident that $\widehat{\mathfrak{a}} \cap A = \mathfrak{a}$, and so the canonical map $A/\mathfrak{a} \to \widehat{A}/\widehat{\mathfrak{a}}$ is injective. It is necessary to check that it is surjective, which will follow if we show that $A + \widehat{\mathfrak{a}} = \widehat{A}$. Let $0 \neq a \in \widehat{A}$, and let a_n be a Cauchy sequence in A converging to a. Let N be sufficiently large that if $n, m \geq N$, then $a_n - a_m \in \mathfrak{a}$. Then $a - a_N \in \widehat{\mathfrak{a}}$, and so $a = a_N + (a - a_N) \in A + \widehat{\mathfrak{a}}$. ■

Example 10.4. As a special case of Example 10.2, let K be a field, $A = K[x]$ be the polynomial ring, and let $\mathfrak{m}$ be the principal ideal generated by x. No nonzero polynomial is divisible by x^n for all n, so $\bigcap \mathfrak{m}^n = 0$. We will show now that $\widehat{A}$ may be identified with the *ring* $K[[x]]$ *of formal power series*, whose elements are formal expressions $\sum_{n=0}^{\infty} a_n x^n$ with $x \in K$. Indeed, we may associate with a formal series $\sum a_n x^n$ its sequence of partial sums. This is a Cauchy sequence in the $\mathfrak{m}$-adic topology, conversely any Cauchy sequence is equivalent to such a sequence of partial sums. The field of fractions of $K[[x]]$ is the *field $K((x))$ of formal power series*, whose elements are formal expressions $\sum_{n=-N}^{\infty} a_n x^n$, where now the summation is allowed to start at a negative value of n.

Example 10.5. As another special case of Example 10.2, let $A = \mathbb{Z}$, and let $\mathfrak{a}$ be the principal ideal generated by a rational prime p. Then the completion $\widehat{A}$ is just the ring $\mathbb{Z}_p$ of p-adic integers.

Example 10.6. Let F be a field with a discrete valuation v. Define a fundamental system of neighborhoods of the identity to be the "fractional ideals" $U_n = \{x \in F | v(x) \geq n\}$. It is again true that if $\{a_n\}$ and $\{b_n\}$ are Cauchy sequences in this topology, then so is $\{a_n b_n\}$. Therefore the completion $\widehat{F}$ has the structure of a topological ring. In fact, it is a field, since if a_n is a Cauchy sequence which converges to $a \neq 0$ in $\widehat{A}$, then a_n^{-1} is a Cauchy sequence, and its limit is the inverse to a, showing that nonzero elements are invertible. Let $\mathfrak{o} = \{x \in F | v(x) \geq 0\}$ be the ring of integers in F, and let $\mathfrak{m}$ be its maximal ideal. Then if $n \geq 0$, U_n is just the ideal $\mathfrak{m}^n$, so we see that the $\mathfrak{a}$-adic topology on $\mathfrak{o}$ agrees with the subspace topology inherited from F, so the completion $\widehat{\mathfrak{o}}$ is embedded in the completion $\widehat{F}$.

In connection with this last example, we have

Proposition 10.7. *Let F be a field with a discrete valuation v, and let $\widehat{F}$ be its completion with respect to v. Then v extends to a discrete valuation of $\widehat{F}$.*

Proof. Let $0 \neq a \in \widehat{F}$. Let a_n be a Cauchy sequence converging to a. It is easily checked that $v(a_n)$ converges to a limit which is independent of the choice of the Cauchy sequence a_n, and that if we define $v(a)$ to be this limit, then v is a discrete valuation of $\widehat{F}$ extending v. (See Exercise 10.4.) ■

If F is a field with a discrete valuation v, and if F is complete with respect to the topology of Example 10.6, we say that F is *complete with respect to the discrete valuation v*.

Proposition 10.8. *Let F be a field complete with respect to a discrete valuation v. Let $\mathfrak{o}$ be the discrete valuation ring associated with v, and let $\mathfrak{m}$ be its maximal ideal. Suppose that $\mathfrak{o}$ contains a subfield K such that the canonical map $K \to \mathfrak{o} \to \mathfrak{o}/\mathfrak{m}$ is an isomorphism. Then $\mathfrak{o}$ and F are respectively isomorphic to the ring and field of formal power series $K[[t]]$ and $K((t))$. More precisely, if x is any generator of $\mathfrak{m}$, then the canonical map ϕ from the polynomial ring $K[t]$ to $\mathfrak{o}$ such that $\phi(t) = x$ may be extended to an isomorphism $K[[t]] \to \mathfrak{o}$.*

A necessary and sufficient condition for the existence of a subfield K is that F and $\mathfrak{o}/\mathfrak{m}$ have equal characteristic. If $\mathfrak{o}/\mathfrak{m}$ is perfect, this is proved in Serre [**25**], II.4. However, the subfield K is only unique in special cases. In the applications to algebraic geometry, a canonically determined subfield K such that $K \to \mathfrak{o} \to \mathfrak{o}/\mathfrak{m}$ is automatically supplied.

Proof. Extend the map $\phi : K[t] \to \mathfrak{o}$ to a map $K[[t]] \to \mathfrak{o}$ as follows. Let $\sum a_n t^n \in K[[t]]$. Let $\alpha_n = \sum_{i=0}^{n} a_i x^i \in \mathfrak{o}$. This is a Cauchy sequence, since if $n, m \geq N$, then $\alpha_n - \alpha_m \in \mathfrak{m}^N$. We define $\phi(\sum a_n t^n)$ to be the limit of this Cauchy sequence. It is easily checked that ϕ thus extended is an injective homomorphism $K[[t]] \to \mathfrak{o}$. To verify that ϕ is surjective, let $a \in \mathfrak{o}$. We will recursively define a sequence a_n of elements of K such that $a - \sum_{i=0}^{n} a_i x^i \in \mathfrak{m}^{n+1}$. Indeed, since $K \to \mathfrak{o} \to \mathfrak{o}/\mathfrak{m}$ is an isomorphism, there exists a unique $a_0 \in K$ such that $a - a_0 \in \mathfrak{m}$. Then $a - a_0$ is a multiple of x, since by Proposition 5.27, $\mathfrak{m}$ is the principal ideal generated by x. Thus we may find a unique a_1 such that $a_1 - (a - a_0)/x \in \mathfrak{m}$, or $a - (a_0 + a_1 x) \in \mathfrak{m}^2$. Continuing in this way, if $a_0, \cdots, a_{n-1}$ have been defined such that $a - \sum_{i=0}^{n-1} a_i x^i \in \mathfrak{m}^n$, we define a_n by the requirement that $a_n - (a - \sum_{i=0}^{n-1} a_i x^i)/x^n \in \mathfrak{m}$, which implies that $a - \sum_{i=0}^{n} a_i x^i \in \mathfrak{m}^{n+1}$. Now it is clear that $\phi(\sum a_n t^n) = a$. Hence $\phi : K[[t]] \to \mathfrak{o}$ is surjective, and so $K[[t]] \cong \mathfrak{o}$. Now since $K((t))$ is the field of fractions of $K[[t]]$, and F is the field of fractions of $\mathfrak{o}$, the isomorphism extends to an isomorphism of $K((t)) \to F$. ∎

Proposition 10.9. *Let X be an affine curve, let F be its function field, and let P be a smooth point on X. The local ring $\mathcal{O}_P(X)$ is a discrete valuation ring, and if t is any generator of its maximal ideal, then the completion of $\mathcal{O}_P(X)$ may be identified with $k[[t]]$, and its field of fractions, which is the completion of F with respect to the discrete valuation associated with P, may be identified with $k((t))$.*

Proof. The fact that $\mathcal{O}_P(X)$ is a discrete valuation ring follows from Proposition 5.28. By Example 10.6 and Proposition 10.7, its completion is a complete discrete valuation ring, and t is a generator of its maximal ideal. The field k of constants is of course embedded in $\mathcal{O}_P(X)$, and if $\mathfrak{m}$ is the maximal ideal then the canonical map $k \to \mathcal{O}_P(X) \to \mathcal{O}_P(X)/\mathfrak{m}$ is an isomorphism. By Proposition 10.3, this remains true if $\mathcal{O}_P(X)$ is replaced by its completion, and $\mathfrak{m}$ by the maximal ideal in the latter discrete valuation ring. Now Proposition 10.9 follows from Proposition 10.8. ■

In this context, we will refer to the completion of F at the discrete valuation associated with P as *the completion of F at P*.

Proposition 10.10. *Let F be a field, and let v_1 and v_2 be two discrete valuations on F. If v_1 and v_2 induce the same topology on F, they are equal.*

Proof. It follows from the remark to Proposition 5.27 that if the valuation rings of F corresponding to the valuations v_1 and v_2 are equal, then v_1 and v_2 are equal. Now the valuation ring corresponding to the discrete valuation v_1 is

$$F - \{x \in F | x^{-n} \to 0 \text{ as } n \to \infty\}.$$

This topological characterization shows that the valuation ring, and hence the valuation, is determined by the topology. ■

Proposition 10.11. *Let F be a field complete with respect to a discrete valuation v, and let E/F be a finite extension. Let w be a discrete valuation of E such that $w|F = ev$ for some integer e. Then E is a topological field, and F has the subspace topology. Let $x_1, \cdots, x_n$ be a basis of E over F, and let a_i be a sequence of elements of E. Write $a_i = \sum_{j=1}^{n} a_{ij} x_j$. Then $a_i \to 0$ if and only if $a_{ij} \to 0$ is for each j. The field E is complete with respect to the valuation w.*

Proof. Note that since the restriction of w to F is a multiple of v, F has the topology induced as a subset of E.

Clearly if $a_{ij} \to 0$ for each j then $a_i \to 0$. We must prove the converse. We will prove by induction on r that if $\sum_{j=1}^{r} a_{ij} x_j \to 0$, then $a_{ij} \to 0$ for $i = 1, \cdots, r$. This is obvious if $r = 1$. Assume that it is true for $r - 1$, and that $\sum_{j=1}^{r} a_{ij} x_j \to 0$. We will assume that a_{ir} does not tend to zero, and derive a contradiction. If a_{ir} does not tend to zero, then there exists a constant N and a subsequence such that $v(a_{ir}) < N$ for all a_{ir} in the subsequence. Passing to the subsequence, we may therefore assume that $v(a_{ir}) < N$ for all i. Now $a_{ir}^{-1}(\sum_{j=1}^{r} a_{ij} x_j) \to 0$, and thus multiplying the given sequence by a_{ir}^{-1}, we may assume that $a_{ir} = 1$ for all i. Now $\sum_{j=1}^{r-1} a_{ij} x_j = \sum j = 1^r a_{ij} x_j - x_r \to -x_r$ as $i \to \infty$. This means that $\sum_{j=1}^{r-1} (a_{ij} - a_{kj}) x_j \to 0$ as $i, k \to \infty$. By the induction hypothesis it follows that $a_{ij} - a_{kj} \to 0$ as $i, k \to \infty$. Hence for fixed j, a_{ij} is a Cauchy sequence, and since F is complete, it tends to a limit, so that $a_{ij} \to a_j$ for some $a_j \in F$. Now $\sum_{j=1}^{r-1} a_{ij} x_j \to \sum_{j=1}^{r-1} a_j x_j$. On the other hand, $\sum_{j=1}^{r-1} a_{ij}x_j = -x_r$, and so $-x_r = \sum_{j=1}^{r-1} a_j x_j$. This is a contradiction since the x_j are linearly independent.

This proves that if $\sum_{j=1}^{r} a_{ij}x_j \to 0$, then $a_{ir} \to 0$. Now $\sum_{j=1}^{r-1} a_{ij}x_j = \sum_{j=1}^{r} a_{ij}x_j - a_{ir}x_r \to 0$. By induction we get $a_{ij} \to 0$ for $j < r$, also, as required. This completes the induction, and proves the first assertion.

To prove the second, suppose that a_i is a Cauchy sequence in E. Write $a_i = \sum_{j=1}^{n} a_{ij}x_j$. Then $\sum_{j=1}^{n}(a_{ij} - a_{kj})x_j \to 0$ as i, $k \to \infty$. By the first assertion, this implies that $a_{ij} - a_{kj} \to 0$ as i, $k \to \infty$ for each j. Now since F is complete, a_{ij} tends to a limit a_j for each j, and so $a_i = \sum_{j=1}^{n} a_{ij}x_j \to \sum_{j=1}^{n} a_j x_j$ also tends to a limit. This proves that E is complete. ∎

Proposition 10.12. *Let F be a field complete with respect to a discrete valuation v, and let E be a finite separable extension of F. Then there is a unique discrete valuation w of E such that $w|F$ is a multiple of v.*

The hypothesis that E/F be separable is not necessary. However, this is all we will require. By abuse of language, the valuation w is said to *extend* v.

Proof. First we prove existence. Let A be the valuation ring of F corresponding to the discrete valuation v, let $\mathfrak{m}$ be its maximal ideal, and let B be the integral closure of A in E. By Proposition 9.3, B is a Dedekind domain. Let $\mathfrak{M}$ be any prime of B lying above $\mathfrak{m}$. Then $B_{\mathfrak{M}}$ is a discrete valuation ring of E by Proposition 9.5. Let w be the corresponding discrete valuation of E. It follows from Proposition 9.11 that w restricted to F is e times v, where e is the ramification index of $\mathfrak{M}$ over $\mathfrak{m}$. Thus we have existence.

To prove uniqueness, we note that the topology of E induced by w is determined by the topology of F, by Proposition 10.11, since a topology is determined by knowledge of its convergent sequences. Now by Proposition 10.10, this means that the valuation w is determined. This proves uniqueness. ∎

Theorem 10.13 (Hensel's Lemma). *Let F be a field complete with respect to a discrete valuation v. Let A be the corresponding valuation ring, and let $\mathfrak{m}$ be its maximal ideal. Let $K = A/\mathfrak{m}$ be the residue field. Let $f(x) \in A[x]$ be a monic polynomial with coefficients in A, and let $\overline{f}(x)$ be the image of f in the polynomial ring $K[x]$ over K. Assume that $f(x)$ has a simple zero $\overline{\alpha} \in K$. Then there exists a zero α of f in A whose image in K is $\overline{\alpha}$.*

For a stronger statement, see Exercise 10.8.

Proof. We recall "Taylor's Formula without denominators." If $f(x) = \sum_{n=0}^{N} a_n x^n$ is a polynomial with coefficients in any ring R, define

$$f^{[r]}(x) = \sum_{n=r}^{N} a_n \binom{n}{r} x^{n-r}.$$

If R is a field of characteristic zero, we have $f^{[r]} = \frac{1}{r!} f^{(r)}$, where $f^{(r)}$ is the r-th derivative of f. However if R has positive characteristic, this representation is not valid since $r!$ may not be invertible. We have $f^{[0]} = f$ and $f^{[1]} = f'$. Taylor's Formula without denominators is then the identity

$$(10.1) \qquad f(a + h) = f(a) + h\,f'(a) + h^2\,f^{[2]}(a) + h^3\,f^{[3]}(a) + \dots .$$

To prove this, note that if $f(x) = x^n$, this is the binomial theorem, and the general case follows from this particular one. We also have

$$(10.2) \qquad f'(a + h) = f'(a) + 2h\,f^{[2]}(a) + 3h^2\,f^{[3]}(a) + \dots .$$

Let α_0 be an arbitrary element of A whose image in K is $\overline{\alpha}$. Since $\overline{\alpha}$ is a simple zero of $\overline{f}$, we have $v\big(f(\alpha_0)\big) > 0$ while $v\big(f'(\alpha_0)\big) = 0$. We define recursively

$$\alpha_{i+1} = \alpha_i - f(\alpha_i)/f'(\alpha_i).$$

We will show that $v\big(f(\alpha_i)\big) \geq 2^i$, while $v\big(f'(\alpha_i)\big) = 0$. Note that this implies that $f'(\alpha_i)$ is a unit in A, so that α_{i+1} is defined. If $i = 0$, our induction hypothesis is given. Assume that it is true for i. We will establish that it is true for $i + 1$. Indeed, we have by Eq. (10.1) with $a = \alpha_i$, $h = -f(\alpha_i)/f'(\alpha_i)$, that h^2 divides $f(\alpha_{i+1})$, and since $v(h) \geq 2^i$, we have $v\big(f(\alpha_{i+1})\big) \geq 2^{i+1}$. Similarly, by Eq. (10.2), $f'(\alpha_{i+1}) - f'(\alpha_i)$ is divisible by h, and hence is in $\mathfrak{m}$. Since $f'(\alpha_i) \notin \mathfrak{m}$, therefore $f'(\alpha_{i+1}) \notin \mathfrak{m}$.

Now since $\alpha_{i+1} - \alpha_i \to 0$, it follows from completeness that α_i converges to some $\alpha \in A$. Since f is continuous, and since $f(\alpha_i) \to 0$, we have $f(\alpha) = 0$. It is easy to see that α and α_0 have the same image $\overline{\alpha}$ in $A/\mathfrak{m}$, as required. ■

As a first application, we show that the situation of Proposition 10.8 is preserved under field extensions, at least when the residue field is perfect.

Proposition 10.14. *Let F be a field complete with respect to a discrete valuation v. Let A be the discrete valuation ring associated with v, and let $\mathfrak{m}$ be its maximal ideal. Suppose that A contains a field K such that the canonical map $K \to A \to A/\mathfrak{m}$ is an isomorphism. Assume that K is perfect. Let E/F be a finite separable extension. By Proposition 10.12, the discrete valuation v may be extended to a unique discrete valuation w of E, and E is complete with respect to w. Let B be the corresponding discrete valuation ring, and let $\mathfrak{M}$ be its maximal ideal. Let L be the algebraic closure of K in E. Then L is contained in B, and the canonical map $L \to B \to B/\mathfrak{M}$ is an isomorphism.*

Proof. Since K is a field, the algebraic closure L of K in E is the same as the integral closure of K in E. The latter is contained in every valuation ring of E which contains K by Exercise 2.5, and in particular, $L \subseteq B$. Since L is a field, the canonical map $L \to B \to B/\mathfrak{M}$ is clearly injective. We must show that it is surjective. If $\overline{\alpha} \in B/\mathfrak{M}$, then $\overline{\alpha}$ is the root of a monic polynomial $\overline{f}$ with coefficients in $A/\mathfrak{m}$. Moreover, since $A/\mathfrak{m}$ is perfect, $B/\mathfrak{M}$ is a separable extension, and we may therefore choose $\overline{f}$ to be a separable polynomial. Since $K \to A/\mathfrak{m}$ is an isomorphism, we may represent $\overline{f}$ by a polynomial $f(x)$ with coefficients in K. We will show that f has a root α in L, and that the image of α in $B/\mathfrak{M}$ is $\overline{\alpha}$. Indeed, this follows from Hensel's Lemma: $\overline{\alpha}$ is a simple root of $\overline{f}$ by separability, and so Theorem 10.13 is applicable. The root α constructed there is algebraic over K, hence lies in L by the the definition of L. ∎

Let E and K be two different fields containing a field F. Let us consider how many different ways there are of combining E and K in a common field. By a *compositum* of E and K over F we mean a field L containing F together with injections $i : E \to L$ and $j : K \to L$ which are the identity on F, and such that L is generated as a field by the images of i and j. If L' is another compositum, with injections $i' : E \to L'$ and $j' : K' \to L'$, we say that L and L' are *isomorphic* if there exists a field isomorphism $\phi : L \to L'$ such that $\phi \circ i = i'$ and $\phi \circ j = j'$.

Proposition 10.15. *Let E/F be a finite separable extension, and let K be any field containing F. There are only finitely many isomorphism classes of composita of E and K over F. Let K_i, $i = 1, \cdots, r$ be a set of representatives of these isomorphism classes. Then K_i/K is a finite separable extension for each i, and $K \otimes_F E \cong \bigoplus K_i$ as K-algebras. If $x \in E$ then $\mathrm{tr}_{E/F}(x) = \sum \mathrm{tr}_{K_i/K}(x)$ and $N_{E/F}(x) = \prod N_{K_i/K}(x)$.*

The ring structure of $K \otimes E$ is described in Exercise 4.1. It is also a K-module by extension of scalars (Exercise 9.4), and hence a K-algebra.

Proof. By the Theorem of the Primitive Element (Lang [18], Theorem V.4.6 on p. 243), $E = F(\xi)$, where $\xi \in E$ is the root of an irreducible separable polynomial $f(x) \in F[x]$. In $K[x]$, f may no longer be irreducible. Let $f = \prod_{i=1}^{r} f_i$ be its factorization. Let ξ_i be a root of f_i in an algebraic closure of K. Then $K(\xi_i)$ is a compositum of K and E. The injection $E \to K_i$ is given by $\phi(\xi) \mapsto \phi(\xi_i)$ for $\phi(x)$ in the polynomial ring $F[x]$, and the injection $K \to K_i$ is the identity. We will show that the K_i are a set of representatives for the isomorphism classes of composita. This means that if L is any compositum of E and K, then there exists a unique compositum K_i which is isomorphic to L. To see this, if $i : E \to L$ is the injection, let $\xi' = i(\xi) \in L$. Then $L = K(\xi')$. Now ξ' is a root of $f(x)$, hence a root of a unique $f_i(x)$. We see that L is isomorphic to K_i under the map $\phi(\xi') \mapsto \phi(\xi_i)$ for $\phi \in K[x]$. It is also clear that L cannot be isomorphic to more than one of the K_i, for the polynomials f_i are coprime in pairs by the separability of $f(x)$. This shows that $K_1, \cdots, K_r$ are a set of representatives of the classes of composita of E and K over F.

Now let us show that $K \otimes E \cong \bigoplus K_i$. Indeed, for each i, the map $K \times E \to K_i$ given by multiplication $(x, y) \mapsto xy$ is F-bilinear, and hence induces a map $K \otimes E \to K_i$ such that $x \otimes y \mapsto xy$, and this is a K-algebra homomorphism. Putting these homomorphisms $K \otimes E \to K_i$ together, we get a homomorphism $K \otimes E \to \bigoplus K_i$. We will show that this is an isomorphism.

First let us show that it is injective. An F-basis of E is $1, \xi, \xi^2, \cdots, \xi^{n-1}$, where $n = \deg(f)$. Consequently a K-basis of $K \otimes E$ is $1 \otimes 1, 1 \otimes \xi, \cdots, 1 \otimes \xi^{n-1}$. This means that an element u of $K \otimes E$ has a unique representation in the form

$$u = a_0 \otimes 1 + a_1 \otimes \xi + \ldots + a_{n-1} \otimes \xi^{n-1}, \qquad a_j \in K.$$

We will show that if the image of this element u in K_i is zero for all i, then the a_j are all zero. Consider the polynomial $\phi(x) = \sum a_j x^j$ in $K[x]$. The image of u in K_i is $\phi(\xi_i)$. If this is zero then since f_i is the minimal polynomial of ξ_i, we have $f_i | \phi$. Since the f_i are coprime, it follows that $f = \prod f_i | \phi$. But the degree of ϕ is $n - 1 < n = \deg(f)$, and so this implies that $\phi = 0$. Thus all a_j vanish and so $u = 0$. This shows that the map $K \otimes E \to \bigoplus K_i$ is injective. To show that it is an isomorphism, note these two K-algebras have the same dimension, since $\dim_K(K \otimes E) = \dim_F(E) = \deg(f) = n$, while $\dim_K \bigoplus K_i = \sum \dim_K(K_i) = \sum \deg(f_i) = n$.

To prove the identities for the trace and norm, let $x \in E$. Consider the endomorphism $y \to xy$ of E over F. The trace of this endomorphism is $\mathrm{tr}_{E/F}(x)$, and the determinant is $N_{E/F}(x)$. These are equal, therefore, to the trace and determinant of the endomorphism $\xi \to (1 \otimes x)\xi$ of $K \otimes E$. To compute this, we consider the induced endomorphism on the isomorphic K-algebra $\bigoplus K_i$. The endomorphism $y \to xy$ of $\bigoplus K_i$ leaves each summand K_i invariant, and induces multiplication by x on that summand, so the trace and determinant of the restriction to K_i are $\mathrm{tr}_{K_i/K}(x)$ and $N_{K_i/K}(x)$, respectively. Hence $\mathrm{tr}_{E/F}(x) = \sum \mathrm{tr}_{K_i/K}(x)$ and $N_{E/F}(x) = \prod N_{K_i/K}(x)$. ■

Proposition 10.16. *Let F be a field with a discrete valuation v, and let E/F be a finite separable extension. Let w_i be the distinct extensions of v to valuations of E/F. Let $\overline{F}$ be the completion of F with respect to v, and let $\overline{E}_i$ be the completions of E with respect to the w_i. Then the E_i are a set of representatives of the composita of $\overline{F}$ with E over F. We have an isomorphism of $\overline{F}$-algebras, $\overline{F} \otimes E \cong \bigoplus \overline{E}_i$. If $x \in E$, then $\mathrm{tr}_{E/F} = \sum \mathrm{tr}_{\overline{E}_i/\overline{F}}(x)$, and $N_{E/F} = \prod N_{\overline{E}_i/\overline{F}}(x)$.*

Proof. Firstly, let L be any compositum of E and $\overline{F}$ over F. Then L is a finite separable extension of $\overline{F}$, and by Proposition 10.12, there is a unique extension of v to a valuation w of L. The restriction of w to E is an extension of v, hence one of the w_i. Now L is a complete field containing E whose valuation induces w_i on E, and hence L contains a copy of $\overline{E}_i$. However by definition, L is generated by $\overline{F}$ and by E, which are both contained in $\overline{E}_i$, and so L is isomorphic to $\overline{E}_i$. This shows that each compositum of E and $\overline{F}$ is isomorphic to one of the $\overline{E}_i$. Conversely, we must show that each $\overline{E}_i$ is a compositum of E with $\overline{F}$. Indeed, $\overline{E}_i$ contains copies of E and $\overline{F}$, and it is only necessary to show that it is generated by the images of these. Let L be the subfield generated by the images of E and $\overline{F}$. By Proposition 10.12, there admits a unique extension of v to a valuation of L, which must obviously agree with w. Also by Proposition 10.12, L is complete, and therefore closed in $\overline{E}_i$. Since E is dense in $\overline{E}_i$, this implies that $L = \overline{E}_i$.

We have established the first part of the Proposition, that the composita of E with $\overline{F}$ are precisely the $\overline{E}_i$. The remaining parts of the Proposition now follow from Proposition 10.15. ■

Proposition 10.17. *Let F be field complete with respect to a discrete valuation v, and let E/F be a finite Galois extension. Let w be the valuation of E extending F, unique by* Proposition 10.12. *Let $\sigma \in \mathrm{Gal}(E/F)$, and $x \in E$. Then $w\big(\sigma(x)\big) = w(x)$.*

Proof. Indeed, $x \mapsto w\big(\sigma(x)\big)$ defines another discrete valuation of E extending v. By Proposition 10.12, there is a unique such valuation, and consequently $w(x) = w\big(\sigma(x)\big)$. ■

Proposition 10.18. *Let E be a field complete with respect to a discrete valuation v, and let E/F be a finite separable extension. Let f be the residue class degree of E/F, and let w be the valuation of E extending v, unique by* Proposition 10.12. *Then for $x \in E$, we have $v\big(N_{E/F}(x)\big) = f\,w(x)$.*

Proof. Let E'/F be the normal closure of E/F, and let w' be the extension of v to a valuation of E'. Let e and e' be the ramification indices of E/F and E'/E, respectively. Let $\sigma_1, \cdots, \sigma_m$ be the distinct embeddings of E into E' over F, so that $m = [E : F] = ef$. We have $w(x) = \frac{1}{e'}w'(x)$, and by Proposition 10.17, we have $w(x) = \frac{1}{e'}w'(\sigma_i x)$ for each i. Multiplying these identities together gives $ef\,w(x) = \frac{1}{e'}\sum w'(\sigma_i x) = \frac{1}{e'}w'(N_{E/F}x)$. Hence $f\,w(x) = \frac{1}{ee'}w'\big(N_{E/F}(x)\big) = v\big(N_{E/F}(x)\big)$, as required. ■

Exercises

Exercise 10.1. Supply proofs for Proposition 10.1.

Exercise 10.2. Verify the claim in Example 10.2 that if $\{a_n\}$ and $\{b_n\}$ are Cauchy sequences in the topological ring A, then so is $\{a_n b_n\}$, and show that this makes $\hat{A}$ into a topological ring.

Exercise 10.3. Show that the ring $K((t))$ described in Example 10.4 is a field, and that it is the field of fractions of $K[[t]]$.

Exercise 10.4. Verify the assertion in the proof of Proposition 10.7 that $v(a_n)$ converges to a limit which is independent of the choice of Cauchy sequence a_n, and that this limit extends v to a discrete valuation of $\hat{A}$.

Exercise 10.5. Let X be the singular affine curve $y^2 = x^2(x+1)$. Assume that the ground field k has characteristic not equal to 2. Let P be the singular point at the origin, and let $A = \mathcal{O}_P(X)$ be the corresponding local ring. We consider the completion $\hat{A}$ of A with respect to the maximal ideal $\mathfrak{m}$ of A, as in Example 10.2. Show that $(y - \xi)(y + \xi) = 0$ in $\hat{A}$, where

$$\xi = x + \tfrac{1}{2}\, x^2 - \tfrac{1}{8}\, x^3 + \tfrac{1}{16}\, x^4 - \tfrac{5}{128}\, x^5 + \dots,$$

the series being convergent in $\hat{A}$. Thus although A is an integral domain, $\hat{A}$ is not.

Exercise 10.6. Let A a Dedekind domain, F its field of fractions. Let E be a finite separable extension of degree n of F, and let B be the integral closure of A in E. Let $\mathfrak{p}$ be a prime ideal of A. Let $\mathfrak{p}B = \prod \mathfrak{P}_i^{e_i}$ be the decomposition of $\mathfrak{p}$ in B. Let e_i and f_i be the ramification indices and residue class degrees of $\mathfrak{P}_i$ over $\mathfrak{p}$, so that $n = \sum e_i f_i$ by Theorem 9.10. Let v be the discrete valuation of F corresponding to the discrete valuation ring $A_\mathfrak{p}$, and let w_i be the discrete valuation of E corresponding to the discrete valuation ring $B_{\mathfrak{P}_i}$. By Proposition 9.11, the restriction of w_i to F is $e_i\, v$. Let $\overline{F}$ and $\overline{E}_i$ be the completions of F and E with respect to the valuations v and w, respectively. Show that $\overline{E}_i$ is an extension of degree $e_i f_i$ of $\overline{F}$.

Exercise 10.7. Hilbert's Ramification Groups. A reference for this exercise is Serre [25], IV.1. Let F be a field complete with respect to a discrete valuation v. Let E/F be a finite Galois extension, with Galois group $G = \mathrm{Gal}(E/F)$. Our aim is to show that the inertia subgroup G_0 of E is solvable. Let L be the inertia field, that is, the fixed field of G_0. Then the extension E/L is totally ramified, with Galois group G_0, and the inertia group in G_0 is G_0 itself. Therefore we may replace F with L, and *assume that E/F is totally ramified*.

By Proposition 10.12, there exists a unique valuation w of E extending v. This means that $v(x) = e\, w(x)$ for $x \in F$, where $e = [E : F]$ is the ramification index. Let A and B be the valuation rings of F and E respectively corresponding to these valuations, and let $\mathfrak{p} \subseteq A$ and $\mathfrak{P} \subseteq B$ be their maximal ideals. Let $K = B/\mathfrak{P}$ be the residue field. Let π be a generator of $\mathfrak{P}$. Let $U = \{x \in B \,|\, w(x) = 0\}$ be the group of units in B. For $i > 0$, let $U^i = 1 + \mathfrak{P}^i$. This is a subgroup of U.

(i) Prove that $A/\mathfrak{p} \cong B/\mathfrak{P}$, and that $E = F(\pi)$.

(ii) Show that if $\sigma \in G$, then $\sigma \circ w = w \circ \sigma$. Moreover, $\sigma(\mathfrak{P}^i) = \sigma(\mathfrak{P}^i)$ for all $i > 0$, so σ induces an automorphism of the ring $B/\mathfrak{P}^i$.

(iii) Define a homomorphism $\theta_0 : G_0 \to K^\times$ as follows. If $\sigma \in G_0$, let $\theta_0(\sigma)$ be the residue class of $\sigma(\pi)/\pi$ in K. Show that this does not depend on the choice of the local parameter π, and that θ_0 is a homomorphism.

(iv) If $i > 0$, let G_i be the subgroup of G acting trivially on $B/\mathfrak{P}^{i+1}$ (cf. part (ii)). Show that $G_0 \supseteq G_1 \supseteq G_2 \supseteq \cdots$, and that for sufficiently large i, $G_i = 0$.

(v) Show that the kernel of θ_0 is G_1.

(vi) Let $\theta_i : G_i \to U^i/U^{i+1}$ be defined as follows. If $\sigma \in G_i$, show that $\sigma(\pi)/\pi \in U^i$. Let $\theta_i(\sigma)$ be the residue class of $\sigma(\pi)/\pi$ in U^i/U^{i+1}. Show that this does not depend on the choice of the local parameter π, and that θ_i is a homomorphism.

(vii) Show that the kernel of θ_i is G_{i+1}.

(viii) Show that the inertia group G_0 is solvable. (We have shown that every factor G_i/G_{i+1} is isomorphic to a subgroup of an abelian group $K^\times$ or U^i/U^{i+1}.)

(ix) If the characteristic of K is zero, show that U^i/U^{i+1} is torsion free, while every finite subgroup of $K^\times$ is cyclic. Conclude that $G_1 = \{1\}$, and that the inertia group G_0 is cyclic.

(x) If the characteristic of K is $p > 0$, show that every element of U^i/U^{i+1} has order p, while every finite subgroup of $K^\times$ is cyclic of order prime to p. Conclude that G_1 is the unique p-Sylow subgroup of G_0.

Exercise 10.8. (Hensel's Lemma). Prove the following refinement of Theorem 10.13. Let F be a field complete with respect to a discrete valuation v. Let A be the corresponding valuation ring, and let m be its maximal ideal. Let $K = A/\text{m}$ be the residue field. Let $f(x) \in A[x]$ be a monic polynomial with coefficients in A, and let $\overline{f}(x)$ be the image of f in the polynomial ring $K[x]$ over K. Assume that $\alpha_0 \in A$ such that $v(f(\alpha_0)) > 2v(f'(\alpha_0))$. Then there exists a zero α of f in A whose image in K is $\overline{\alpha}$.

Note that this contains Theorem 10.13 as a special case, where $v(f(\alpha_0)) > 0$ and $v(f'(\alpha_0)) = 0$, since $\overline{f}$ has a simple zero in K.

Hint: define the sequence α_i as in the proof of Theorem 10.13. Show recursively that $v(f(\alpha_i)) - 2v(f'(\alpha_i)) \geq 2^i$. Note that $v(f'(\alpha_{i+1})) = v(f'(\alpha_i))$.

Exercise 10.9. Show by example that the assumption of separability in Proposition 10.16 is necessary. That is, show that if E/F is inseparable, it is possible to find K such that $K \otimes E$ is not a direct sum of fields.

Hint: Let $x \in E$ such that $x^p \in F$ but $x \notin F$. Consider $x \otimes 1 - 1 \otimes x \in E \otimes E$. Show that this is a nonzero nilpotent element.

11. Differentials and Residues

We recall the classical theorem that if M is a compact Riemann surface, and if ω is any meromorphic differential 1-form on M, then the sum of the residues of ω is equal to zero. This may be proved as follows. We start with a *canonical dissection* of M. For example, if the genus of M is two, we cut it as follows:

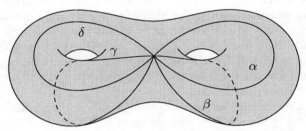

The complement of the cut lines is simply connected, and we may now lay the surface out "flat," on a table of constant negative curvature. It looks like this:

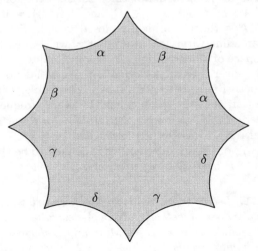

According to Cauchy's integral formula, the sum of the residues of ω is equal to the integral of $(2\pi i)^{-1}\omega$ around the perimeter. Observe that each of the cut lines α, β, γ and δ is traversed twice, in opposite directions, so that the integrals cancel. Hence the sum of the residues is zero.

Let X be a complete nonsingular curve over the complex numbers. Then X may be given the structure of a compact Riemann surface, so the theorem just proved is valid. The problem of this chapter is to reformulate the result so as to be valid for a complete nonsingular curve over an arbitrary field. This was first accomplished by Hasse. Our treatment differs from other published accounts in making use of the Cartier operators in characteristic p. Although we state and prove the global results over an algebraically closed field k, it is our concession to greater generality that we state and prove the key local result, Theorem 11.19, over a perfect field. It is possible to finesse some of the difficulties in characteristic p by showing that the characteristic zero case implies certain algebraic identities which can then be employed in the case of characteristic p. This type of argument is used in the treatments of Lang [17] and Serre [26]. The proof we give, based on Whaples [34] and Cartier [7] is more complicated, but has the advantage of revealing a beautiful structure underlying the local residue theorem in characteristic p. The Cartier operators themselves are rather fascinating.

The first step is to define the module of *Kähler differentials*, which are an algebraic analog of the meromorphic differential 1-forms.

Let A be a ring, B an A-algebra with structure map $i : A \to B$, and M a B-module. Then M is also an A-module. By a *derivation* $D : B \to M$ of B over A, we mean an A-linear map which satisfies the following properties: $D(bb') = D(b)\,b' + b\,D(b')$ for b, $b' \in B$, and $D\big(i(a)\big) = 0$ for $a \in A$. For example, if $A = k$ is the algebraically closed ground field, $B = \mathcal{O}_x(X)$ is the local ring of an affine variety X at a point x, we defined the tangent space $T_x(X)$ to be the space of all derivations $B \to k$ in Chapter 5. We will also refer to a derivation of B over A as an *A-derivation*. We will denote by $\mathrm{Der}_A(B, M)$ the set of all A-derivations $B \to M$. It is a B-module, with multiplication $(bD)(b') = b\,D(b')$. If A is the ground field k, as will frequently be the case, we will denote $\mathrm{Der}_A(B, M)$ as simply $\mathrm{Der}(B, M)$. Moreover, $\mathrm{Der}_A(B, B)$ will be denoted as $\mathrm{Der}_A(B)$, or as $\mathrm{Der}(B)$ in the case $A = k$.

The module $\Omega_{B/A}$ of Kähler differentials is defined to be the *universal* module of derivations of B over A. More precisely,

Proposition 11.1. *Let B be an A-algebra. There exists a B-module $\Omega_{B/A}$, unique up to isomorphism, together with an A-derivation $d : B \to \Omega_{B/A}$, such that if M is a B-module and $D : B \to M$ is any A-derivation, there exists a unique B-module homomorphism $\tau : \Omega_{B/A} \to M$ such that $D = \tau \circ d$. $\Omega_{B/A}$ is generated by the image of d.*

The universal property thus means that for any A-module M,

$$\operatorname{Der}_A(B, M) \cong \operatorname{Hom}_B(\Omega_{B/A}, M).$$

Proof. As usual, $\Omega_{B/A}$ is characterized up to isomorphism by the universal property. To show that such a module exists, begin with the free B-module on the symbols $d(b)$ for $b \in B$, and divide by the submodule of relations, generated by the elements of the form $d(bb') - b\, d(b') - b'\, d(b)$ for $b\, b' \in B$, and $d(i(a))$ for $a \in A$. Note that the resulting map $d : B \to \Omega_{B/A}$ is A-linear, since if $a \in A$ then $d(ab) - a\, d(b)$ is in the submodule of relations. This construction shows that B is generated by the image of d. ■

In the rational function field $k(x_1, \cdots, x_n)$, which is the field of fractions of the polynomial ring $k[x_1, \cdots, x_n]$, we have n canonical derivations $\partial/\partial x_i$.

Proposition 11.2. *Let $F = k(x_1, \cdots, x_d)$ be the field of rational functions of transcendence degree d over k. The F-vector space $\operatorname{Der}(F)$ of k-derivations $F \to F$ has dimension d, and a basis is given by $D_i = \partial/\partial x_i$ for $i = 1, \cdots, d$.*

Proof. Clearly D_i are k-derivations. They are linearly independent over F since if $\sum f_i D_i = 0$, with $f_i \in F$, then $f_j = (\sum f_i D_i)(x_j) = 0$ for each j. Let D be an arbitrary k-derivation, and let $f_i = Dx_i$. We will show that $D = \sum f_i D_i$. Indeed, let $D_0 = D - \sum f_i D_i$. Then D_0 is a k-derivation which annihilates each x_i. Since by the chain rule

$$D_0(f) = \sum_{i=1}^{n} \frac{\partial f}{\partial x_i} D_0(x_i),$$

this implies that $D_0 = 0$. ■

Proposition 11.3. *Let F be a field containing a field K, and let E be a finite separable extension of F.*

(i) Let $D : F \to F$ be an K-derivation. Then there exists a unique extension of D to an K-derivation $E \to E$.

(ii) The E-vector spaces $\operatorname{Der}_K(E)$ and $E \otimes \operatorname{Der}_K(F)$ are isomorphic, and moreover, $\dim_F \operatorname{Der}_K(F) = \dim_E \operatorname{Der}_K(E)$.

Proof. Let us prove (i). By the Theorem of the Primitive Element, (Lang [18], Theorem V.4.6 on p. 243) $E = F(\xi)$, where the minimal polynomial $f(x) \in F[x]$ satisfied by ξ is separable. Let $\eta = f'(\xi)$. Since f is separable, we have $\eta \neq 0$.

If $\phi(x) = \sum a_i x^i \in F[x]$, let ϕ^D be the polynomial $\phi^D(x) = \sum D(a_i)\, x^i$. Since D is a derivation, it is easily checked that $(\phi_1 \phi_2)^D = \phi_1^D\, \phi_2 + \phi_1\, \phi_2^D$.

First we will prove uniqueness. Note that given any extension of D to a K-derivation of E, then for $\phi \in F[x]$, $D\phi(\xi) = \phi^D(\xi) + \phi'(\xi) D\xi$, and so it is sufficient to show that $D\xi$ is uniquely determined. We have $f(\xi) = 0$, and so $0 = Df(\xi) = f^D(\xi) + \eta D\xi$, and so necessarily $D\xi = -f^D(\xi)/\eta$. This shows that the extension of D to K is uniquely determined.

To prove existence, let $\zeta = -f^D(\xi)/\eta$, and define

$$D\phi(\xi) = \phi^D(\xi) + \phi'(\xi)\, \zeta.$$

We leave it to the reader to check that this is well-defined, and a derivation $E \to E$. This completes the proof of (i).

To prove (ii), let D_i, $i \in I$ be an F-basis of $\mathrm{Der}_K(F)$. (We do not assert that the indexing set I is necessarily finite.) By part (i), they extend to derivations $E \to E$. We must show that the extended derivations are then an E-basis of $\mathrm{Der}_K(E)$. This will show that $\mathrm{Der}_K(E)$ has an E-basis which is contained in $\mathrm{Der}_K(F)$, and which is an F-basis of $\mathrm{Der}_K(F)$. This implies that $\mathrm{Der}_K(E) \cong E \otimes \mathrm{Der}_K(F)$, and that $\dim_E \mathrm{Der}_K(E) = \dim_F \mathrm{Der}_K(F)$.

Let us show that the extended D_i span $\mathrm{Der}_K(E)$. Let $D : E \to E$ be an arbitrary K-derivation. We will express D as a linear combination of the D_i. Choose a basis $\xi_1, \cdots, \xi_d$ of E over F. Let $\lambda_1, \cdots, \lambda_d$ be the dual basis of F-linear functionals $E \to F$. We have $\sum_i \xi_i \lambda_i(y) = y$ for any $y \in E$; indeed, it is sufficient to check this for $y = \xi_j$, in which case it follows from the fact that $\lambda_i(\xi_j) = \delta_{ij}$ (Kronecker's δ).

Now $\lambda_j \circ D|F$ is a K-derivation of F, and hence $\lambda_j \circ D|F = \sum_{i \in I} \alpha_{ij} D_i$ for some $\alpha_{ij} \in F$. For each j, only finitely many α_{ij} are nonzero. We show that $D = \sum_{i,j} \xi_j \alpha_{ij} D_i$. By the uniqueness assertion of part (i), it is sufficient to show that $D(x) = \sum_{i,j} \xi_j \alpha_{ij} D_i(x)$ for $x \in F$. The right hand side equals $\sum_j \xi_j \lambda_j D(x)$, which equals $D(x)$, as required. This shows that the D_i span the $\mathrm{Der}_K(E)$. We leave it to the reader to show that they remain linearly independent over E. ∎

Recall from Chapter 3 that an *abstract function field* over k is a field containing k, and finitely generated over k.

Proposition 11.4. *Let F be an abstract function field over k, and let d be the transcendence degree of F. Then $\mathrm{Der}_k(F)$ is a d-dimensional vector space over F.*

Proof. Let $x_1, \cdots, x_d$ be a separating transcendence basis of F over k. Let $K = k(x_1, \cdots, x_d)$. By Proposition 11.3 (ii), the dimension over F of the space $\mathrm{Der}_k(F)$ of k-derivations $F \to F$ is the same as the dimension over K of $\mathrm{Der}_k(K)$. By Proposition 11.2, the latter dimension is d. ∎

Proposition 11.5. *Let K be a field and let F be a field containing K. There exists a natural dual pairing $\mathrm{Der}_K(F) \times \Omega_{F/K} \to F$ of F-vector spaces. In particular, $\Omega_{F/K}$ and $\mathrm{Der}_K(F)$ have the same dimension.*

Proof. To construct the dual pairing, let $D \in \mathrm{Der}_K(F)$ and $\omega \in \Omega_{F/K}$. By the universal property of $\Omega_{F/K}$, there is a unique F-linear map $\tau_D : \Omega_{F/K} \to F$ such that $D = \tau_D \circ d$. We define $\langle D, \omega \rangle = \tau_D(\omega)$. To show that this is a dual pairing, we must verify that (1) if $\tau_D(\omega) = 0$ for all ω, then $D = 0$; and (2) if $\tau_D(\omega) = 0$ for all D, then $\omega = 0$. We have, for $x \in F$, $Dx = \tau_D(dx)$, and so (1) is obvious. To prove (2), suppose that $\omega \neq 0$. Then there exists an F-linear functional λ on $\Omega_{F/K}$ such that $\lambda(\omega) \neq 0$. Now $D = \lambda \circ d \in \mathrm{Der}_K(F)$. Thus $\lambda = \tau_D$, and therefore $\tau_D(\omega) \neq 0$. ∎

Suppose that A is a ring, and B and C are A-algebras, with an A-algebra homomorphism $j : B \to C$. If $d : C \to \Omega_{C/A}$ is the canonical derivation, then $d \circ j : B \to \Omega_{C/A}$ is a derivation, and by the defining universal property of $\Omega_{B/A}$ (Proposition 11.1), there is induced a B-module homomorphism $\Omega_{B/A} \to \Omega_{B/C}$.

Proposition 11.6. *Let F be a field containing a field K, and let E be a finite separable extension of F. Then the natural map $\Omega_{F/K} \to \Omega_{E/K}$ is an injection, and $\Omega_{E/K} \cong E \otimes \Omega_{F/K}$ as E-vector spaces.*

Proof. The modules $\Omega_{E/K}$ and $\Omega_{F/K}$ are, by Proposition 11.5, the dual spaces of $\mathrm{Der}_K(E)$ and $\mathrm{Der}_K(F)$ as vector spaces over E and F respectively, and by Proposition 11.3 (ii), $\mathrm{Der}_K(E) \cong E \otimes \mathrm{Der}_K(F)$. The assertion follows by dualizing. ∎

Proposition 11.7. *Let $F = k(x_1, \cdots, x_d)$ be the field of rational functions of transcendence degree d over k. Then $dx_1, \cdots, dx_d$ are a basis for $\Omega_{F/k}$.*

Proof. Indeed, the $\partial/\partial x_i$ are a basis of $\mathrm{Der}(F)$ by Proposition 11.2. It follows from the definitions that the elements of the dual basis of $\Omega_{F/k}$ under the dual pairing of Proposition 11.5 are the dx_i. ∎

We recall *MacLane's Criterion* for separability. As in Chapter 6, a field extension E/F is called *separably generated* if E admits a separating transcendence basis over F. The extension E/F is called *separable* if every subfield which is finitely generated over F is separably generated. (If E/F is algebraic, this coincides with the usual notion.) Extensions E/F and L/F, both contained within a larger field Ω, are called *linearly disjoint* if every subset of E which is linearly independent over F remains linearly independent over L. (This relation is symmetric in E and L.) Linearly disjoint extensions are discussed in Lang [18], Section VIII.3, and separable extensions are discussed in [18], Section VIII.4.

Proposition 11.8 (MacLane's Criterion). *Let E/F be an extension of fields of characteristic $p > 0$. The following are equivalent:*

(i) E is linearly disjoint from $F^{1/p}$;

(ii) E is linearly disjoint from F^{1/p^m} for all $m \geq 1$;

(iii) E/F is separable.

Proof. See Lang [18], Proposition VIII.4.1 on p. 362. ∎

Proposition 11.9. *Let K be a perfect field, and L an extension. Let $a \in L$ such that $a^{1/p} \notin L$. Then L is separably generated over $K(a)$.*

Proof. By MacLane's Criterion (Proposition 11.8), it is sufficient to show that L and $K(a^{1/p})$ are linearly disjoint over $K(a)$. The irreducible polynomial satisfied by $a^{1/p}$ over K is $X^p - a = 0$, and since $a^{1/p} \notin L$, this is also the irreducible polynomial satisfied by $a^{1/p}$ over L. Hence $[L(a^{1/p}) : L] = p = [K(a^{1/p}) : K]$, proving that L and $K(a^{1/p})$ are linearly disjoint. ∎

Proposition 11.10. *Let F be an abstract function field of dimension d.*

(i) $\Omega_{F/k}$ is a d-dimensional F-vector space.

(ii) Let $x_1, \cdots, x_d$ be a transcendence basis of F over k. Then they are a separating transcendence basis if and only if $dx_1, \cdots, dx_d$ are a basis of $\Omega_{F/k}$.

Proof. Part (i) follows from Proposition 11.4 and Proposition 11.5.

As for part (ii), first assume that $x_1, \cdots, x_d$ are a separating transcendence basis. Let $K = k(x_1, \cdots, x_d)$. We will identify $\Omega_{K/k}$ with its image in $\Omega_{F/k}$ by Proposition 11.6. The dx_i are a K-basis of $\Omega_{K/k}$ by Proposition 11.7, and so they are an F-basis of $\Omega_{F/k}$ by Proposition 11.6.

To prove the converse, assume that the dx_i are a basis of $\Omega_{F/k}$. Let $K = k(x_1, \cdots, x_d)$ as before. Let L be the separable closure of K in F. We wish to prove that $L = F$. Suppose not. F/L is purely inseparable, so there exists an $x \in F$ such that $x \notin L$ but $x^p = a \in L$. Since $a^{1/p} \notin L$, it follows from Proposition 11.9 that L is separably generated over $k(a)$. Note that a is transcendental over k, since k is perfect (in fact algebraically closed), and $a \in k$ would imply $x \in k \subseteq L$. Hence we may find a separating transcendence basis of L over k which contains a as its first element. Let $y_1 = a, y_2, \cdots, y_d$ be this separating transcendence basis. We have just proven that the dy_i are a basis $\Omega_{L/k}$. In particular, $da = dy_1 \neq 0$ in $\Omega_{L/k}$. Since L/K is separable, $x_1, \cdots, x_d$ are also a separating transcendence basis, and so $dx_1, \cdots, dx_d$ are also a basis of $\Omega_{L/k}$. It follows that there exist constants $c_i \in L$ such that $da = \sum c_i \, dx_i$ in $\Omega_{L/k}$, and since $da \neq 0$, the c_i are not all zero. The inclusion $L \to F$ induces an L-linear map $\Omega_{L/k} \to \Omega_{F/k}$. We consider the image of the relation $da = \sum c_i\, dx_i$ in $\Omega_{F/k}$. In the latter space, we have $0 = d(x^p) = da$, and so $0 = \sum c_i dx_i$ in $\Omega_{F/k}$. Since the c_i are not all zero, this contradicts the assumption that the dx_i are a basis of $\Omega_{F/k}$. ■

Now let X be a complete nonsingular curve, and let F be its function field. By a *local uniformizing parameter* (or simply a *local parameter*) at P, we mean an element t of F which generates the maximal ideal in the discrete valuation ring $\mathcal{O}_P(X)$. Thus if $v = v_P$ is the discrete valuation of F associated with this discrete valuation ring, t is a local parameter if and only if $v(t) = 1$.

Proposition 11.11. *Let X be a complete nonsingular curve, $P \in X$, and let t be a local parameter at P. Then the function field F of X is separable over $k(t)$, and $dt \neq 0$ in $\Omega_{F/k}$. Moreover, if F_P is the completion of F at P, then $dt \neq 0$ in $\Omega_{F_P/k}$.*

Proof. By Proposition 11.10 (ii), the separability assertion will follow when we establish that $dt \neq 0$ in $\Omega_{F/k}$. Moreover, since image of dt under the canonical map $\Omega_{F/k} \to \Omega_{F_P/k}$ is dt, it is really only necessary to check that $dt \neq 0$ in $\Omega_{F_P/k}$. By Proposition 10.9, we have $F_P \cong k((t))$. There is an obvious derivation $D : F_P \to F_P$, namely differentiation with respect to t in the field of formal power series. By the universal property of $\Omega_{F_P/k}$, there is induced a F_P-linear map $\tau : \Omega_{F_P/k} \to F_P$ such that $\tau(dt) = Dt = 1$. Therefore $dt \neq 0$ in $\Omega_{F_P/k}$. ■

Let X be a complete nonsingular curve, and let F be its function field. We will denote $\Omega = \Omega_{F/k}$ in the sequel. We see that $\dim_F \Omega = 1$. The one dimensional F-vector space Ω is a substitute for the space of meromorphic differentials in the theory of a compact Riemann surface. Our objective is to define the residue of $\omega \in \Omega$ at $P \in X$, to show that this residue $\mathrm{res}_P\, \omega$ is zero for all but finitely many P, and to prove that $\sum_P \mathrm{res}_P\, \omega = 0$.

We would like to imitate the construction in the complex case as follows. Let $P \in X$. By Proposition 10.9, if t is a local parameter, we may identify the completion of F with respect to v with the field $k((t))$ of formal power series in t. Now by Proposition 11.10 and Proposition 11.11, $\omega = f\, dt$ for some $f \in F$. Write f as a formal power series $f(t) = \sum a_n t^n \in k((t))$. By analogy with the complex case, we would like to define the residue to be "the -1 coefficient in the Taylor expansion of ω/dt at P," that is, $\mathrm{res}_P\, \omega = a_{-1}$. The problem with this definition is that it is not clear that it is independent of the choice of t. This is particularly hard to establish in the case where $\mathrm{char}(k) = p > 0$.

There are various ways of handling this point. We will base our discussion on the use of *Cartier operators*, which were introduced in Cartier [7]. Eventually we will obtain Theorem 11.19, whose proof is more difficult when the characteristic is $p > 0$. An alternative approach, due to Hasse, is followed by Serre [26] and Lang [17]; in the alternative approach, the case of positive characteristic is deduced from the case of characteristic zero.

Proposition 11.12. *Let K be a perfect field of characteristic $p > 0$, A a K-algebra. For $i = 0, 1, 2, \cdots$ let $\delta_i : A \to \Omega = \Omega_{A/K}$ be the map $\delta_i(x) = x^{p^i - 1} \, dx$. Define $\Sigma^i = \Sigma^i_{A/K} \subseteq \Omega$ recursively by $\Sigma^0 = 0$, and*

$$\Sigma^{i+1} = \{\omega \in \Omega | \omega - \delta_i(x) \in \Sigma^i \text{ for some } x \in A\}.$$

(i) If $x, y \in A$, then $\delta_i(xy) = x^{p^i} \delta_i(y) + y^{p^i} \delta_i(x)$. If u is in the prime field, then $\delta_i(ux) = u \, \delta_i(x)$.

(ii) If $x, y \in A$, then $\delta_i(x + y) - \delta_i(x) - \delta_i(y) \in \Sigma^i$.

(iii) Σ^i is a K-vector subspace of Ω.

(vi) Σ^i is the abelian group generated by the $\delta_j(A)$ for $j < i$.

(v) Let $\Omega^i = \Omega^i_{A/K} = \Omega/\Sigma^i$. There exists a well-defined A-module structure on Ω^i with the multiplication $A \times \Omega^i \to \Omega^i$ given by $x[\omega] = [x^{p^i} \omega]$, where $[\omega]$ denotes the class of $\omega \in \Omega$ in Ω^i. Let $d_i : A \to \Omega^i$ be $d_i(x) = [\delta_i(x)]$. Then d_i is a derivation.

Proof. For (i), it is straightforward to show that $\delta_i(xy) = x^{p^i} \delta_i(y) + y^{p^i} \delta_i(x)$. If u is in the prime field, then $\delta_i(u) = 0$ and $u^{p^i} = u$, so this implies that $\delta_i(ux) = u \, \delta_i(x)$.

By (i), we have $\delta_i(xy) = x^{p^i} \delta_i(y)$ if $x \in K$. Since K is perfect, any element of K has the form x^{p^i}. Therefore $\delta_i(A)$, and hence all the sets Σ^i are closed under multiplication by elements of K. Similarly since $\delta_i(-x) = -\delta_i(x)$, the Σ^i are closed under $x \to -x$. The only thing to be established in (iii) is that Σ^i is closed under addition.

With this in mind, we will prove (ii) and (iii) simultaneously by induction on i. Both parts are clear if $i = 0$, and (iii) is clear if $i = 1$. Assume that (iii) is true for i. We will prove that (ii) is true for i, and that (iii) is true for $i + 1$.

It follows from the congruence $\binom{p^i - 1}{r} \equiv (-1)^r \bmod p$, valid if $0 \leq r \leq p^i - 1$ and the binomial theorem that $\delta_i(x + y) - \delta_i(x) - \delta_i(y)$ equals

$$\left[\sum_{r=0}^{p^i - 2} (-1)^r \, x^r \, y^{p^i - 1 - r} \, dx \right] + \left[\sum_{r=1}^{p^i - 1} (-1)^r \, x^r \, y^{p^i - 1 - r} \, dy \right]$$

$$= \sum_{r=1}^{p^i - 1} (-1)^{r-1} \, y^{p^i - 1 - r} \, x^{r-1} \, (y \, dx - x \, dy).$$

By construction, Σ^i contains $\delta_j(A)$ for $j < i$. We will prove that each term in the preceding formula is in $\delta_j(A)$ for some $j < i$. Specifically, if $1 \leq r \leq p^i - 1$, let $r = p^j u$, where $p \nmid u$. Then u is invertible in the prime field, and we find that

$$\delta_j(u^{-1}\, y^{p^{i-j}-u}\, x^u) = u^{-1}\delta_j(y^{p^{i-j}-u}\, x^u) = y^{p^i-r-1}\, x^{r-1}\, (y\, dx - x\, dy).$$

This proves that each term in our formula for $\delta_i(x+y) - \delta_i(x) - \delta_i(y)$ is in Σ^i. Our induction hypothesis is that Σ^i is closed under addition, and this proves (ii). Now to prove (iii) for Σ^{i+1}, we have already pointed out that the only nontrivial point is to establish closure under addition. Suppose that ω and $\omega' \in \Sigma^{i+1}$. Then $\omega - \delta_i(x)$ and $\omega' - \delta_i(x') \in \Sigma^i$ for some x, $x' \in A$. Now

$$\omega + \omega' - \delta_i(x + x') = \big(\omega - \delta_i(x)\big) + \big(\omega' - \delta_i(x')\big) + \big(\delta_i(x) + \delta_i(x') - \delta_i(x + x')\big).$$

By (ii), just established, the last term is in Σ^i. Since Σ^i is closed under addition, this proves that $\omega + \omega' \in \Sigma^{i+1}$. Therefore Σ^{i+1} is closed under multiplication. This concludes the proof by induction of (ii) and (iii).

Given (ii) and (iii), part (iv) is easily established, and we leave this to the reader.

In (v), we must establish that the multiplication described is well-defined. We must show that $x^{p^i} \Sigma^i \subseteq \Sigma^i$ for $x \in A$, and by (iv), it is enough to show that $x^{p^i} \delta_j(A) \subseteq \delta_j(A)$. This follows from the relation $\delta_j(x^{p^{i-j}} y) = x^{p^i} \delta_j(y)$, which is valid if $i > j$. The assertion that d_i is a derivation is now a simple restatement of (i). ∎

Now let K be a perfect field, A a K-algebra. If the characteristic of K is p, let $\Sigma_{A/K}$ be the union of the $\Sigma^i_{A/K}$ defined in Proposition 11.12. On the other hand, if the characteristic of K is zero, let $\Sigma_{A/K} = d(A) \subseteq \Omega_{A/K}$. In either case, $\Sigma_{A/K}$ is a K-vector subspace of $\Omega_{A/K}$. Suppose that $f : A \to B$ is a homomorphism of K-algebras. There is induced a map $\Omega_{A/K} \to \Omega_{B/K}$, and the spaces $\Sigma^i_{A/K}$ (resp. $\Sigma_{A/K}$) are mapped into $\Sigma^i_{B/K}$ (resp. $\Sigma_{B/K}$). Hence are induced maps $\Omega^i_{A/K} \to \Omega^i_{B/K}$.

Let F be a field which is complete with respect to a discrete valuation v. Let $\mathfrak{o}$ be the valuation ring of F corresponding to this discrete valuation, and let $\mathfrak{m}$ be its maximal ideal. Suppose that $\mathfrak{o}$ contains a subfield K such that the canonical map $K \to \mathfrak{o} \to \mathfrak{o}/\mathfrak{m}$ is an isomorphism. By Proposition 10.8, $F \cong K((x))$. We may consider $\Omega_{F/K}$. However, this module is really too large for our purposes. We would like to consider a modified module $\Omega^{(c)}_{F/K}$ which is a universal module for *continuous* derivations of the topological field F. This may be defined as follows. By Proposition 11.5, $\Omega_{F/K}$ and $\mathrm{Der}_K(F)$ are naturally dual spaces. let $\mathrm{Der}^{(c)}_K(F)$ be the subspace of continuous derivations of F. A continuous derivation of F satisfies

$$(11.1) \qquad D\left(\sum a_n x^n\right) = \left(\sum n a_n x^{n-1}\right) Dx,$$

and the space of such derivations is one-dimensional over F (Exercise 11.6). Let $\Omega_{F/K}^{(c)}$ be the dual space, which is therefore a one-dimensional subspace over F. It is naturally a quotient space of $\Omega_{F/K}$, and comes equipped with a derivation $d : F \to \Omega_{F/K}^{(c)}$. It follows from Eq. (11.1) by duality that

$$(11.2) \qquad d\left(\sum a_n x^n\right) = \left(\sum n a_n x^{n-1}\right) dx,$$

valid in $\Omega_{F/K}^{(c)}$. Since $\Omega_{F/K}^{(c)}$ is a one-dimensional vector space over the topological field F, it has a natural topology. Let $\Sigma_{F/K}^{(c)}$ be the closure of the image of $\Sigma_{F/K}$ in $\Omega_{F/K}$. (Note: If the characteristic of K is zero, then the image of $\Sigma_{F/K}$ is itself closed. If the characteristic of K is not zero, then the image of each space $\Sigma_{F/K}^i$ is closed, but it may be seen that their union is not. Hence it is necessary to take the closure.)

Proposition 11.13. *Let F be a field complete with respect to a discrete valuation v. Let $\mathfrak{o}$ be the valuation ring of F corresponding to v, and let $\mathfrak{m}$ be its maximal ideal. Suppose that $\mathfrak{o}$ contains a subfield perfect K such that the canonical map $K \to \mathfrak{o} \to \mathfrak{o}/\mathfrak{m}$ is an isomorphism. Then $\Sigma_{F/K}^{(c)}$ has codimension one as a K-vector subspace of $\Omega_{F/K}^{(c)}$. Let $x \in F^\times$. If the characteristic of K is zero, then $x^{-1}\, dx \in \Sigma_{F/K}^{(c)}$ if and only if $v(x) = 0$. If the characteristic of K is $p > 0$, then $x^{-1}\, dx \in \Sigma_{F/K}^{(c)}$ if and only if $p|v(x)$. In either case, $\Sigma_{F/K}^{(c)}$ contains the image of $\Omega_{\mathfrak{o}/K}$ in $\Omega_{F/K}^{(c)}$.*

Although we will use the realization of F as a field of formal power series in our verification that $\Sigma_{F/K}^{(c)}$ has the indicated properties, it is useful that at least the definition of $\Sigma_{F/K}^{(c)}$ does not depend on this realization, since we are spared the necessity of verifying that the definition is independent of the choice of parameter.

Proof. First we treat the case where the characteristic of K is zero. Let t be an element of F with $v(t) = 1$. We identify $\mathfrak{o}$ and F respectively with $K[[t]]$ and $K((t))$ by Proposition 10.8. Then every element ω of $\Omega_{F/K}^{(c)}$ may be uniquely written as $\left(\sum a_n t^n\right) dt$. If $a_{-1} = 0$, then $\omega = d\left(\sum n^{-1} a_{n-1} t^n\right)$, and so $\omega \in \Sigma_{F/K}^{(c)}$. On the other hand, if $a_{-1} \neq 0$, clearly $\omega \notin \Sigma_{F/K}^{(c)}$. Therefore $\Sigma_{F/K}^{(c)}$ is characterized by the relation $a_{-1} = 0$. This proves that $\Sigma_{F/K}^{(c)}$ is of codimension one. Note that if $v(u) = 0$, then $u^{-1}\, du$ has the form $f(t)\, dt$ with $f(t) \in k[[t]]$, and any element of this form is in $\Sigma_{F/K}^{(c)}$. Now if $x \in F^\times$, write

$x = u\,t^r$ where $v(u) = 0$ and $r = v(x)$. Then $x^{-1}\,dx = u^{-1}\,du + r\,t^{-1}\,dt$. Here $u^{-1}\,du \in \Sigma_{F/K}^{(c)}$, and so $x^{-1}\,dx = (\sum a_n t^n)\,dt$ with $a_{-1} = r = v(x)$. This shows that $x^{-1}\,dx \in \Sigma_{F/K}^{(c)}$ if and only if $v(x) = 0$. It is also clear that if $x \in \mathfrak{o}$, then $dx \in \Sigma_{F/K}^{(c)}$, since if $x = \sum_{n=0}^{\infty} b_n t^n$, then $dx = \sum_{n=0}^{\infty}(n+1)b_{n+1}t^n$, and the coefficient of t^{-1} is zero.

Now let us consider the case where the characteristic of K is $p > 0$. In this case, it may be checked that the image of $\Sigma_{F/K}^i$ in $\Omega_{F/K}^{(c)}$ consists of the elements of the form $(\sum a_n t^n)\,dt$ where $a_{n-1} = 0$ if $p^{i+1}|n$. Therefore the closure of the union $\Sigma_{F/K}^{(c)}$ of the images of the $\Sigma_{F/K}^i$ is again characterized by the relation $a_{-1} = 0$. The rest of the proof is the same as in the characteristic zero case. ■

We may now define the residue of a differential. In the setting of the last Proposition, let $R_{F/K} = \Omega_{F/K}^{(c)}/\Sigma_{F/K}^{(c)}$. Thus $R_{F/K}$ is a one-dimensional vector space over K, and if $v(x) = 1$, the image of dx/x in $R_{F/K}$ is nonzero. Moreover, the image of dx/x in $R_{F/K}$ does not depend on the choice of the local parameter x, since if u is a unit, $d(ux)/ux = (dx/x) + (du/u)$, and $du/u \in \Sigma_{F/K}^{(c)}$. Hence there exists a unique K-linear map $\rho = \rho_{F/K} : R_{F/K} \to K$ such that $\rho(dx/x) = 1$ if $v(x) = 1$. By abuse of notation, we use the same letter ρ to denote the composition of ρ with the canonical map $\Omega_{F/K} \to R_{F/K}$.

Proposition 11.14. *In the setting of* Proposition 11.13, *assume that the characteristic* $p > 0$. *Let* $\Omega_{F/K}^{i\,(c)}$ *be the quotient of* $\Omega_{F/K}^{(c)}$ *by the image of* $\Sigma_{F/K}^i$, *defined in* Proposition 11.12. *Then* $R_{F/K}$ *is naturally isomorphic to the direct limit*

$$\varinjlim \Omega_{F/K}^{i\,(c)}$$

in the category of topological groups.

Proof. As in the proof of Proposition 11.13, let t be an element of F with $v(t) = 1$. We will show that the image of $\lambda = (\sum a_n t^n)\,dt \in \Omega^{(c)}$ is zero in this direct limit if and only if $a_{-1} = 0$, i.e. if and only if $\lambda \in \Sigma^{(c)}$. Let

$$\lambda_i = \sum_{n \equiv -1 (\mathrm{mod}\ p^i)} a_n t^n\,dt \in \Omega^{(c)}.$$

Then λ and λ_i have the same image in $\Omega^{i\,(c)}$, hence also in $\varinjlim \Omega_{F/K}^{i\,(c)}$. On the other hand, the sequence λ_i converges in $\Omega^{(c)}$ to $a_{-1}\,t^{-1}\,dt$. Thus the image of λ in the direct limit is zero if and only if $a_{-1} = 0$. Thus

$$\varinjlim \Omega_{F/K}^{i\,(c)} \cong \Omega^{(c)}/\Sigma^{(c)} = R_{E/K}.$$

This completes the proof. ■

Proposition 11.15. *The residue, defined in the context of* Proposition 11.13, *has the following characteristics.*

(i) If ω is in the image of the natural map $\Omega_{o/K} \to \Omega_{F/K}$, then $\rho(\omega) = 0$.

(ii) If $f \in F$, then $\rho(f^{n-1}\, df) = 0$ unless $n = 0$.

(iii) If $f \in F$, then $\rho(f^{-1}\, df) = v(f) \cdot 1$, the image of $v(f)$ in K.

Proof. Since by Proposition 11.13, the image of $\Omega_{o/K}$ in $\Omega_{F/K}^{(c)}$ is contained in the kernel $\Sigma_{F/K}^{(c)}$ of ρ, part (i) is clear. As for (ii), if the characteristic of K is zero, $f^{n-1}\, df = d(n^{-1}\, f^n)$ is in $\Sigma^{(c)}$ by construction provided $n \neq 0$. Hence $\rho(f^{n-1}\, df) = 0$. If the characteristic of K is $p > 0$, then write $n = mp^i$ where $p \nmid m$. Then, provided $m \neq 0$, $f^{n-1}\, df = \delta_i(m^{-1}\, f^m)$ is in $\Sigma_{F/K}^{(c)}$, and hence we have $\rho(f) \neq 0$. This proves (ii). Finally, to prove (iii), write $f = u\, x^r$ where u is a unit, $v(x) = 1$ and $r = v(f)$. Then $f^{-1}\, df = u^{-1}\, du + r\, x^{-1}\, dx$. Now $u^{-1}\, du$ is in the image of $\Omega_{o/K}$, hence $\rho(u^{-1}\, du) = 0$ by (i). On the other hand, $\rho(x^{-1}\, dx) = 1$ by construction. Hence $\rho(f^{-1}\, df) = r \cdot 1$. ∎

Proposition 11.16. *In the context of* Proposition 11.13, *let t be an element of F with $v(t) = 1$. Identify F with $K((t))$ by* Proposition 10.8. *Then every element ω of $\Omega_{F/K}^{(c)}$ may be written as $(\sum a_n t^n)\, dt$, and $\rho(\omega) = a_{-1}$.*

Proof. Since $\Omega_{F/K}^{(c)}$ is a quotient of $\Omega_{F/K}$, it is generated as an F-module by elements of the form df with $f \in F$, and it is sufficient to show that such an element has the form $(\sum a_n t^n)\, dt$. However by Proposition 10.8, we may write $f = \sum b_n t^n$, and now this follows from Eq. (11.2). Now if $\omega \in \Omega_{F/K}^{(c)}$, we may write

$$\omega = \left(\sum_{n=-N}^{-1} a_n t^n + f_0 \right) dt,$$

where $f_0\, dt = \left(\sum_{n=0}^{\infty} a_n t^n \right) dt$ is in the image of $\Omega_{o/K}$. Thus by Proposition 11.15 (i), $\rho(f_0\, dt) = 0$. By Proposition 11.15 (ii), $\rho(a_n t^n\, dt) = 0$ if $n \neq -1$, while by Proposition 11.15 (iii), $\rho(a_{-1} t^{-1}\, dt) = a_{-1}$. ∎

Proposition 11.17 (Artin-Schreier). *Let F be a field of characteristic p, and let E/F be a cyclic extension of degree p. Then $E = F(x)$ where x is the solution to an irreducible equation of the form $X^p - X + \alpha = 0$, with $\alpha \in F$. Moreover, if $F = K((t))$, where K is a perfect field, and if E/F is totally ramified, then we may assume that $v(\alpha) = -m$, where m is a positive integer prime to p.*

Proof. Let σ be a generator of $\mathrm{Gal}(E/F)$. Since E/F is separable, there exists an element $\theta \in E$ such that $\mathrm{tr}_{E/F}(\theta) = 1$. Let

$$x = \sigma(\theta) + 2\sigma^2(\theta) + 3\sigma^3(\theta) + \ldots + (p-1)\,\sigma^{p-1}(\theta).$$

We have $x+1 = x+\mathrm{tr}(\theta) = \sum(i+1)\sigma^i(\theta)$, and so $\sigma(x+1) = x$, or $\sigma(x) = x-1$. Now we have $\sigma(x^p - x) = (x-1)^p - (x-1) = x^p - x$, so α, defined to be $-(x^p - x)$ is in F. On the other hand, $\sigma(x) \neq x$, so $x \notin F$, and therefore the polynomial $x^p - x + \alpha$ is irreducible.

Now assume that $F = K((t))$, where K is perfect, and that E/F is totally ramified. We claim that α cannot be an element of $K[[t]]$. Indeed, it is easily seen that if $\mathfrak{p}$ is the maximal ideal of $K[[t]]$, then the map $X \to X^p - X$ of $\mathfrak{p}$ to itself is surjective, since if $\xi \in \mathfrak{p}$, the coefficients in a power series solution to $X^p - X = \xi$ may be constructed recursively. (Or use Hensel's Lemma.) Now suppose that $\alpha \in K[[t]]$, and let a_0 be the constant term of α. Let $u \in \mathfrak{p}$ be a solution to $X^p - X = a_0 - \alpha$. Consider $(x+u)^p - (x+u) = a_0$. Since E/F is totally ramified, the residue class degree is one, and K is algebraically closed in E. Since $x+u$ satisfies a polynomial equation over K, it therefore lies in K. As $u \in F$, this implies that $x \in F$, which is a contradiction. Therefore $\alpha \notin K[[t]]$, i.e. if v is the valuation of F, we have $v(\alpha) = -m$, with $m > 0$. Assume that x is chosen with m minimal. We will prove that $p \nmid m$. Indeed, if $\alpha = \sum_{n=-m}^{\infty} a(n)t^n$, with $p|m$, since K is perfect, we may find $b \in K$ such that $b^p = a(-m)$. Now we may replace x by $x - b\,t^{-m/p}$ and α by $\alpha + b^p\,t^{-m} - b\,t^{m/p}$ and reduce the value of m, contradicting its assumed minimality. ∎

Proposition 11.18. *Let F be a field complete with respect to a discrete valuation v. Suppose that E/F is a finite separable extension of degree m which is totally ramified. Thus E and F have the same residue class field K. Assume that an image of K is contained as a subfield of F, so that* Proposition 10.8 *is applicable. Assume that the characteristic of K does not divide m. Let w be the valuation of E extending v, so that $w(x) = m\,v(x)$ for $x \in F$. Then there exists an element $u \in E$ such that $w(u) = 1$, $E = F(u)$, and $u^m \in F$.*

Proof. Let $u \in E$ and $t \in F$ be given such that $w(u) = v(t) = 1$. Identifying E with $K((u))$ by Proposition 10.8, $w(t) = m$ and so $t = \sum_{n=m}^{\infty} a_n u^n$. Dividing t by a_m if necessary, we have $t = u^m c$, where $c = 1 + a_{m+1}u + \ldots$. We have $c \equiv 1 \bmod \mathfrak{P}$, where $\mathfrak{P}$ is the maximal ideal in the valuation ring B of E. We now apply Hensel's Lemma (Theorem 10.13) to the polynomial $f(x) = x^m - c$. Modulo $\mathfrak{P}$, this equation is $x^m - 1$, which has 1 as a simple zero, and therefore $x^m = c$ has a zero in E, which is congruent to one modulo $\mathfrak{P}$. If ξ is such a root, then $(u\,\xi)^m = t$. Thus replacing u by $u\xi$, we are done. ∎

Theorem 11.19. *Let F be a field complete with respect to a discrete valuation v, and let E/F be a finite separable extension. Thus by Proposition 10.12, there is a unique discrete valuation w of E such that $w(x) = e\,v(x)$, where e is the ramification index. Let A and B be the valuation rings of F and E respectively associated with the valuations v and w, and let $\mathfrak{p}$ and $\mathfrak{P}$ be their respective maximal ideals. Assume that A contains a perfect field K such that the composition $K \to A \to A/\mathfrak{p}$ is an isomorphism. Let L be the algebraic closure of K in E. Then the composition $L \to B \to B/\mathfrak{P}$ is also an isomorphism. Hence residue maps $\Omega_{F/K} \to K$ and $\Omega_{E/L} \to L$ are defined. Let $x \in F$ and let $y \in E$. Then we have*

$$(11.3) \qquad \operatorname{tr}_{L/K}\big(\rho_{E/L}(y\,dx)\big) = \rho_{F/K}\big(\operatorname{tr}_{E/F}(y)\,dx\big).$$

This result was first established by Hasse. Our proof is based on (though different from) Whaples [**34**]. For further historical discussion see Serre [**26**], p. 25.

Proof. The first assertion, that the composition $L \to B \to B/\mathfrak{P}$ is an isomorphism, is asserted by Proposition 10.14.

Lemma 11.20. *Let $F \subseteq E \subseteq E'$ be a tower of separable extensions. If Eq. (11.3) is valid for E'/E and E'/F, then it is valid for E/F.*

Indeed, let L' be the algebraic closure of K in E'. Since E'/E is separable, if $y \in E$, we may find $z \in E'$ such that $\operatorname{tr}_{E'/E}(z) = y$. Now if $x \in F$, we have $\rho_{E/L}(y\,dx) = \operatorname{tr}_{L'/L} \rho_{E'/L'}(z\,dx)$ because Eq. (11.3) is assumed for E'/E. Applying $\operatorname{tr}_{L/K}$ to this relation we get $\operatorname{tr}_{L/K} \rho_{E/L}(y\,dx) = \operatorname{tr}_{L'/K} \rho_{E'/L'}(z\,dx)$. Moreover $\rho_{F/K}\big(\operatorname{tr}_{E/F}(y)\,dx\big) = \rho_{F/K}\big(\operatorname{tr}_{E'/F}(z)\,dx\big) = \operatorname{tr}_{L'/K} \rho_{E'/L'}(z\,dx)$ because Eq. (11.3) is assumed for E'/F. Combining these, we see that it is true for E/F. $\quad\square$

As a particular case, we may take E' to be the normal closure of E/F. We see therefore that it is sufficient to establish Eq. (11.3) in the special case where E/F is Galois, which we now assume. By Theorem 9.18, there exists an intermediate field $E \supseteq E_0 \supseteq F$ such that E/E_0 is Galois and totally ramified, and E_0/F is Galois and nonramified. Moreover, the extension E/E_0 is solvable (Exercise 10.7).

Lemma 11.21. *If $F \subseteq E \subseteq E'$ is a tower of separable extensions, and if Eq. (11.3) is valid for E/F and E'/E, then it is valid for E'/F.*

Proof. This is similar to that of Lemma 11.20. $\quad\square$

We are therefore reduced to proving Eq. (11.3) in the special cases where E/F is Galois and nonramified; and where E/F is totally ramified, and cyclic of prime degree.

Lemma 11.22. *Let $t \in F$ such that $dt \neq 0$ in $\Omega^{(c)}_{E/K}$. Assume that* Eq. (11.3) *is valid when $x = t$ and $y \in E$ is arbitrary. Then* Eq. (11.3) *is valid for all $x \in F$ and $y \in E$.*

Proof. Our hypothesis implies that $dt \neq 0$ in $\Omega^{(c)}_{F/K}$ as well, since if $dt = 0$ in $\Omega^{(c)}_{F/K}$ its image under the canonical map to $\Omega^{(c)}_{E/K}$ would also vanish. In particular, since by Proposition 11.16, $\Omega^{(c)}_{F/K}$ is one-dimensional over F, if $x \in F$, $dx = f\,dt$ for some $f \in F$. This relation is valid in $\Omega^{(c)}_{F/K}$, and also in $\Omega^{(c)}_{E/L}$. Now let $y_1 = f\,y$. Then $\mathrm{tr}_{E/F}(y_1) = f\,\mathrm{tr}_{E/F}(y)$. Using the assumption that Eq. (11.3) is true when $x = t$, we have

$$
\begin{aligned}
\mathrm{tr}_{L/K}\big(\rho_{E/L}(y\,dx)\big) &= \mathrm{tr}_{L/K}\big(\rho_{E/L}(y_1\,dt)\big) \\
&= \rho_{F/K}\big(\mathrm{tr}_{E/F}(y_1)\,dt\big) \\
&= \rho_{F/K}\big(\mathrm{tr}_{E/F}(y)\,dx\big),
\end{aligned}
$$

as required. $\square$

Now let us treat the case where E/F is Galois and unramified. We have $\mathrm{Gal}(E/F) \cong \mathrm{Gal}(L/K)$. If $t \in F$ such that $v(t) = 1$, then by Proposition 10.8, E and F are identified respectively with $L((t))$ and $K((t))$. Then the isomorphism of $\mathrm{Gal}(E/F)$ with $\mathrm{Gal}(L/K)$ may be made very explicit as follows. If $\sigma \in \mathrm{Gal}(L/K)$, then the action of σ may be extended to E/F by $\sigma(\sum a_n t^n) = \sum \sigma(a_n)\,t^n$ for $\sigma \in \mathrm{Gal}(L/K)$. It is therefore clear that if $y = \sum a_n t^n \in E$ then $\mathrm{tr}_{E/F}(y) = \sum \mathrm{tr}_{L/K}(a_n)\,t^n$. Now

$$
\mathrm{tr}_{L/K}\big(\rho_{E/L}(y\,dt)\big) = \mathrm{tr}_{L/K}(a_{-1}) = \rho_{F/K}\big(\mathrm{tr}_{E/F}(y)\,dt\big).
$$

Thus Eq. (11.3) is valid when $x = t$, which by Lemma 11.22 is sufficient.

Lemma 11.23. *Suppose that E/F is Galois and totally ramified. Thus $L = K$. Then there exists a constant $c \in K$ such that*

(11.4)
$$
\rho_{F/K}\big(\mathrm{tr}_{E/F}(y)\,dx\big) = c\,\rho_{E/K}(y\,dx)
$$

for all $y \in E$, $x \in F$.

Proof. The Galois group $G = \text{Gal}(E/F)$ acts on $\Omega^{(c)}_{E/K}$ in an evident way. Thus if $\sigma \in G$, there is induced a map $\sigma : \Omega_{E/K} \to \Omega_{E/K}$, and this induces a map on $\Omega^{(c)}_{E/K}$ such that the diagram:

$$
\begin{array}{ccc}
E & \xrightarrow{\;d\;} & \Omega^{(c)}_{E/K} \\
\Big\downarrow{\sigma} & & \Big\downarrow{\sigma} \\
E & \xrightarrow{\;d\;} & \Omega^{(c)}_{E/K}
\end{array}
$$

commutes for every y. Clearly if $x \in F$ and $y \in E$, we have $\sigma(y\,dx) = \sigma(y)\,dx$. Summing over $\sigma \in G$ gives us a *trace map* $\text{tr} : \Omega^{(c)}_{E/K} \to \Omega^{(c)}_{E/K}$, and the image is contained in $\Omega^{(c)\,G}_{E/K} = \Omega^{(c)}_{F/K}$. We have a commutative diagram:

(11.5)
$$
\begin{array}{ccc}
E & \xrightarrow{\;d\;} & \Omega^{(c)}_{E/K} \\
\Big\downarrow{\text{tr}} & & \Big\downarrow{\text{tr}} \\
F & \xrightarrow{\;d\;} & \Omega^{(c)}_{F/K}
\end{array}
$$

If $x \in F$ and $y \in E$, then

(11.6)
$$
\text{tr}(y\,dx) = \text{tr}(y)\,dx.
$$

If the characteristic of E is $p > 0$, we also have commutative diagrams:

$$
\begin{array}{ccc}
E & \xrightarrow{\;\delta_i\;} & \Omega^{(c)}_{E/K} \\
\Big\downarrow{\sigma} & & \Big\downarrow{\sigma} \\
E & \xrightarrow{\;\delta_i\;} & \Omega^{(c)}_{E/K}
\end{array}
$$

If $t \in F$ such that $dt \neq 0$ in $\Omega^{(c)}_{F/K}$, then since E/F is separable, $dt \neq 0$ in $\Omega^{(c)}_{E/K}$ also, by Proposition 11.6. Every element ω of $\Omega^{(c)}_{E/K}$ may be written in the form $f\,dt$ with $f \in E$, since $\Omega^{(c)}_{E/K}$ is a one-dimensional vector space over E. Then $f\,dt$ is in the image of $\Omega^{(c)}_{F/K}$ if and only if $f \in F$, that is, if ω is invariant under G. We see that the natural map $\Omega^{(c)}_{F/K} \to \Omega^{(c)}_{E/K}$ is injective, and that the image is precisely the subspace $\Omega^{(c)\,G}_{E/K}$ of G-fixed differentials in $\Omega^{(c)}_{E/K}$. We therefore identify this space with $\Omega^{(c)}_{F/K}$.

In characteristic zero, the cokernels of the horizontal arrows are $R_{E/K}$ and $R_{F/K}$ and we have immediately that there exists a commutative diagram:

(11.7)
$$
\begin{array}{ccc}
\Omega^{(c)}_{E/K} & \longrightarrow & R_{E/K} \\
\downarrow{\scriptstyle \mathrm{tr}} & & \downarrow{\scriptstyle \mathrm{tr}} \\
\Omega^{(c)}_{F/K} & \longrightarrow & R_{F/K}
\end{array}
$$

The commutativity of this diagram immediately implies that if $\rho_{E/K}(\omega) = 0$, then $\rho_{F/K}(\mathrm{tr}\ \omega) = 0$. Thus we may define a map $\lambda : K \to K$ by the requirement that the diagram:

(11.8)
$$
\begin{array}{ccccc}
\Omega^{(c)}_{E/K} & \longrightarrow & R_{E/K} & \xrightarrow{\rho_{E/K}} & K \\
\downarrow{\scriptstyle \mathrm{tr}} & & \downarrow{\scriptstyle \mathrm{tr}} & & \downarrow{\scriptstyle \lambda} \\
\Omega^{(c)}_{F/K} & \longrightarrow & R_{F/K} & \xrightarrow{\rho_{F/K}} & K
\end{array}
$$

is commutative. Since λ is K-linear, it has the form $\lambda(x) = cx$ for some $x \in K$. The existence of this commutative diagram and Eq. (11.6) imply Eq. (11.4).

In characteristic p, however, we must work harder to obtain Eq. (11.8). In this case, the cokernels in Eq. (11.5) are the groups $\Omega^{1\,(c)}$. We have then a commutative diagram:

(11.9)
$$
\begin{array}{ccc}
\Omega^{(c)}_{E/K} & \longrightarrow & \Omega^{1\,(c)}_{E/F} \\
\downarrow{\scriptstyle \mathrm{tr}} & & \downarrow{\scriptstyle \mathrm{tr}} \\
\Omega^{(c)}_{F/K} & \longrightarrow & \Omega^{1\,(c)}_{F/K}
\end{array}
$$

Now let $d_1 : E \to \Omega^{1\,(c)}_{E/K}$ be the map defined in Proposition 11.12, or rather the composition of that map $E \to \Omega^1_{E/K}$ with the projection to $\Omega^{1\,(c)}_{E/K}$. For each $\sigma \in G$, we have a commutative diagram:

$$
\begin{array}{ccc}
E & \xrightarrow{d_1} & \Omega^{1\,(c)}_{E/K} \\
\downarrow{\scriptstyle \sigma} & & \downarrow{\scriptstyle \sigma} \\
E & \xrightarrow{d_1} & \Omega^{1\,(c)}_{E/K}
\end{array}
$$

The sum over G of the maps $\sigma : \Omega^{1\,(c)}_{E/K} \to \Omega^{1\,(c)}_{E/K}$ is the rightmost vertical arrow in Eq. (11.9), and has its image in $\Omega_{F/K}^{1\ (c)}$. Thus we get a commutative diagram:

$$
\begin{array}{ccc}
E & \xrightarrow{\ d_1\ } & \Omega_{E/K}^{1\ (c)} \\
\downarrow{\scriptstyle\mathrm{tr}} & & \downarrow{\scriptstyle\mathrm{tr}} \\
F & \xrightarrow{\ d_1\ } & \Omega_{F/K}^{1\ (c)}
\end{array}
$$

Passing to the cokernel, we get a diagram:

$$
\begin{array}{ccc}
\Omega_{E/K}^{(c)} & \longrightarrow & \Omega_{E/F}^{2\ (c)} \\
\downarrow{\scriptstyle\mathrm{tr}} & & \downarrow{\scriptstyle\mathrm{tr}} \\
\Omega_{F/K}^{(c)} & \longrightarrow & \Omega_{F/K}^{2\ (c)}
\end{array}
$$

Continuing in this fashion, and passing to the direct limit, and identifying $R_{E/K} = \varinjlim \Omega_{E/K}^{i\ (c)}$ by Proposition 11.14, we get the diagram Eq. (11.4) in this case also, and so in this case also we obtain Eq. (11.8). From this, we proceed as before. $\square$

This means that in the last remaining case where E/F is totally ramified and cyclic of prime degree, to verify Eq. (11.3) we really have only to evaluate the constant c in Lemma 11.23. We may do this by verifying Eq. (11.3) in one particular case, provided that the left and right sides are not both zero.

Let us consider the case where E/F is totally ramified, of degree $[E : F] = m$ not divisible by the characteristic. In this case by Proposition 11.18, $E = F(u)$ for some u with $w(u) = 1$, where w is the valuation on E, and $u^m \in F$. Now let $x = u^m$, so $v(x) = 1$, and choose $y = x^{-1}$, in Lemma 11.23. We have $x^{-1}\,dx = m\,u^{-1}\,du$, so $\rho_{E/K}(y\,dx) = \rho_{E/K}(x^{-1}\,dx) = m\,\rho_{E/K}(u^{-1}\,du) = m$. On the other hand, $\mathrm{tr}_{E/F}(y) = m\,x^{-1}$ since $y = x^{-1} \in F$ and the degree of E/F is m. Thus $\rho_{F/K}\big(\mathrm{tr}(y)\,dt\big) = \rho_{F/K}(m\,x^{-1}\,dx) = m$ by Proposition 11.15 (iii). Hence the constant c in Lemma 11.23 equals 1, whence Eq. (11.3).

Finally, there remains the case where E/F is totally ramified and cyclic of degree p, equal to the characteristic. In this case, by Proposition 11.17, Then $E = F(z)$ where z is the solution to an irreducible equation of the form $X^p - X + t = 0$, with $t \in F$ and $v(t) = -m$, where m is a positive integer prime to p. Let $y = z^{-1}$ in Lemma 11.23, and $x = t$. We have $y^p - t^{-1}\,y^{p-1} + t^{-1} = 0$. The coefficient of y^{p-1} in this polynomial is the trace of y, so $\mathrm{tr}_{E/F}(y) = t^{-1}$. Also in $\Omega_{E/K}^{(c)}$ we have $dz = dt$, which follows from applying d to the identity $z^p - z + t = 0$. Note that $w(z) = v(t) = -m$ by Exercise 5.11. Now $\rho_{E/K}(y\,dx) = \rho_{E/K}(z^{-1}\,dz) = -m$ by Proposition 11.15 (iii). On the other hand $\rho_{F/K}\big(\mathrm{tr}(y)\,dx\big) = \rho_{F/K}(t^{-1}\,dt) = -m$. We see that the constant c in Lemma 11.23 equals 1, so Eq. (11.3) is also valid in this case. $\blacksquare$

Now let X be a complete nonsingular curve over an algebraically closed field k. Let F be the function field of X. Let $P \in X$, and let $\omega \in \Omega_{F/k}$. Let v_P be the discrete valuation of F corresponding to the prime P, and let F_P be the corresponding completion. We define the residue of ω at P to be $\rho_{F_P/k}$ applied to the image of ω in $\Omega_{F_P/k}^{(c)}$. We denote this residue by $\operatorname{res}_P(\omega)$.

Theorem 11.24. *The residue* $\operatorname{res}_P(\omega)$ *is nonzero for all but finitely many* P, *and*

$$\sum_P \operatorname{res}_P(\omega) = 0.$$

Proof. First let us establish this $X = \mathbf{P}^1$, so that $F = k(x)$ is a rational function field. In this case, any element of $\Omega_{F/k}$ has the form $f(x)\,dx$ where $f(x)$ is a rational function. $f(x)$ has a partial fractions decomposition as the sum of a polynomial, and terms of the form $c\,(x-a)^{-n}$. It is therefore sufficient to prove the statement for polynomials and for rational functions of the form $(x-a)^{-n}$. Moreover, any latter such rational function may, by means of the transformation $x \to x + a$, which does not change dx, be put in the form x^{-n}. We see therefore, that it is sufficient to establish the Theorem for ω of the form $x^n\,dx$, where n may be positive or negative. It is easily checked that the theorem is valid for differentials of this type (Exercise 11.7).

Now let Y be a general curve with function field E. Let $x \in E$ such that E is separable over $F = k(x)$. As in Theorem 8.7, the inclusion $F \to E$ induces a separable morphism $\phi : Y \to X$, where $X = \mathbf{P}^1$ is the rational curve just considered. We will show that the validity of the Theorem for X implies it for Y. By Proposition 11.11, any differential $\omega \in \Omega_{E/k}$ may be written as $y\,dx$ for some $y \in E$.

To show that this ω has only finitely many nonzero residues, we may find an affine open set $U \subseteq Y$ such that y and x are both regular on U. We will show that if $P \in U$, then $\operatorname{res}_P(\omega) = 0$. Let A be the coordinate ring of U. Then $x, y \in A$. Now let $P \in U$, and let v be the corresponding discrete valuation of F. Let $\mathfrak{o}$ be the valuation ring in the completion F_P of F with respect to v. Then $A \subseteq \mathfrak{o}$, and so $x, y \in \mathfrak{o}$. Thus $y\,dx$ may be realized as an element of $\Omega_{\mathfrak{o}/k}$. Now by Proposition 11.15 (i), ω has vanishing residue at P. This proves that there are only finitely many points where ω has a nonzero residue.

Let $\operatorname{tr}(\omega)$ denote the differential $\operatorname{tr}_{E/F}(y)\,dx \in \Omega_{F/k}$. We will prove that for $P \in X$,

$$(11.10) \qquad \sum_{Q \in \phi^{-1}(P)} \operatorname{res}_Q(\omega) = \operatorname{res}_P(\operatorname{tr} \omega).$$

Since we have already shown that $\sum_P \operatorname{res}_P(\operatorname{tr} \omega) = 0$, this implies that $\sum_Q \operatorname{res}_Q(\omega) = 0$.

So consider $P \in X$, and let v be the corresponding discrete valuation of F. Let $Q_1, \cdots, Q_r$ be the points in the fiber $\phi^{-1}(P)$. Let $w_1, \cdots, w_r$ be the corresponding discrete valuations of E extending v. Let $\overline{F}$ be the completion of F with respect to v, and let $\overline{E}_i$ be the completions of E with respect to the E_i. By Proposition 10.16, for $y \in E$, we have $\mathrm{tr}_{E/F}(y) \cong \sum \mathrm{tr}_{\overline{E}_i/\overline{F}}(y)$. Thus if $\omega = y\,dx$, we have $\mathrm{tr}(\omega) = \sum \mathrm{tr}_{\overline{E}_i/\overline{F}}(y)\,dx$. Using Theorem 11.19, we have then

$$
\begin{aligned}
\mathrm{res}_P(\mathrm{tr}\ \omega) &= \rho_{\overline{F}/k}\big(\mathrm{tr}(y)\,dx\big) \\
&= \sum \rho_{\overline{F}/k}\big(\mathrm{tr}_{\overline{E}_i/\overline{F}}(y)\,dx\big) \\
&= \sum \rho_{\overline{E}_i/k}(y\,dx) \\
&= \sum \mathrm{res}_{Q_i}(\omega),
\end{aligned}
$$

whence Eq. (11.10). This completes the proof. ∎

Exercises

Exercise 11.1. The purpose of this exercise is to give an alternative description of $\Omega_{B/A}$, where B is an A-algebra. In this exercise, $\otimes$ will always mean tensor product over A. $B \otimes B$ has the structure of a B-algebra, with structure map $b \to b \otimes 1$. We have a B-algebra homomorphism $B \otimes B \to B$ such that $b_1 \otimes b_2 \to b_1 b_2$. Let I be the kernel of this homomorphism, which is an ideal in $B \otimes B$. Let $\Omega = I/I^2$, which inherits a B-module structure from $B \otimes B$ via the structure map. Let $d : B \to \Omega$ be the A-linear map such that $d(b) = 1 \otimes b - b \otimes 1$ modulo I^2. Check that d is a derivation. To show that $\Omega \cong \Omega_{B/A}$, it is sufficient to show that Ω has the universal property of Proposition 11.1. Let M be any B-module, and $D : B \to M$ an A-derivation. By the universal property of the tensor product, there is a map $B \otimes B \to M$ such that $b \otimes b' \to b\,Db'$. Show that I^2 is in the kernel of this map, and let $\rho : \Omega \to M$ be the induced map. Show that this ρ is the unique B-module homomorphism such that $D = \rho \circ d$.

Exercise 11.2. (i) Let R be a ring, and let $M' \xrightarrow{i} M \xrightarrow{p} M'' \to 0$ be a sequence of R-modules. Show that this sequence is exact if and only if the induce sequence

$$
0 \to \mathrm{Hom}_R(M'', N) \xrightarrow{p^*} \mathrm{Hom}_R(M, N) \xrightarrow{i^*} \mathrm{Hom}_R(M', N)
$$

is exact for every R-module N. Moreover, if i^* is surjective, then i is injective.

Hints: Assuming the exactness of the Hom sequence, to show that p is surjective, take $N = \mathrm{coker}(p)$; to show that $p \circ i = 0$, take $N = M''$; to show that $\ker(p) \subseteq \mathrm{im}(i)$, take $N = \mathrm{coker}(i)$. To prove the final assertion, take $N = M'$.

(ii) Let R, A and B be rings, with homomorphisms $R \to A \to B$. Let N be a B-module, which then inherits the structures of, respectively, an A-module and an R-module. Show that the sequence

$$0 \to \mathrm{Der}_A(B, N) \to \mathrm{Der}_R(B, N) \to \mathrm{Der}_R(A, N)$$

is exact.

(iii) Let R, A and B be rings, with homomorphisms $R \to A \to B$. Show that there is an exact sequence

$$B \otimes_A \Omega_{A/R} \to \Omega_{B/R} \to \Omega_{B/A} \to 0.$$

Hints: $d : B \to \Omega_{B/A}$ is an R-derivation, hence there is induced a B-module homomorphism $\Omega_{B/R} \to \Omega_{B/A}$. The composite $A \to B \to \Omega_{B/R}$ is an R-derivation, and hence there is induced an A-module homomorphism $\Omega_{A/R} \to \Omega_{B/R}$. Now by Exercise 2.4 (ii), there is induced a B-module homomorphism $B \otimes_A \Omega_{A/R} \to \Omega_{B/R}$. This gives us the maps in question. To prove exactness, use the criterion of (i). The exactness is given by (ii) and Exercise 2.4.

Exercise 11.3. Let R be a ring, A an R-algebra, which you may assume to be an integral domain. Let $S \subseteq A$ be a multiplicative set.

(i) Let N be an $S^{-1}A$-module. Show that any R-derivation $D : A \to N$ may be extended uniquely to a derivation $S^{-1}A \to N$ by means of the "quotient rule" $D(s^{-1}a) = s^{-1}D(a) - s^{-2}a\,D(s)$.

(ii) Prove that $\Omega_{S^{-1}A/R} \cong S^{-1}\Omega_{A/R}$ as $S^{-1}A$-modules.

Hints: Use Exercise 11.2 (iii). Show that $\Omega_{S^{-1}A/A} = 0$, and deduce that the canonical map $S^{-1}\Omega_{A/R} \cong S^{-1}A \otimes_A \Omega_{A/R} \to \Omega_{S^{-1}A/R}$ is surjective. To prove that it is injective, return to the criterion of Exercise 11.2 (i). The surjectivity of i^* follows from part (i) of this exercise.

Exercise 11.4. Show that if F is a perfect field of prime characteristic, then any derivation $F \to F$ is zero. If K is the prime subfield of F, show that if $\Omega_{F/K} = 0$.

Exercise 11.5. Let R be a ring, A an R-algebra, and let I be an ideal of A. Let $B = A/I$. Let N be a B-module.

(i) Show that there is an exact sequence of B-modules

$$0 \to \mathrm{Der}_R(B, N) \to \mathrm{Der}_R(A, N) \to \mathrm{Hom}_B(I/I^2, N).$$

Furthermore, if the composite map $R \to A \to B$ is an isomorphism, then the final arrow is surjective.

Hints: The map $\mathrm{Der}_R(A, N) \to \mathrm{Hom}_B(I/I^2, N)$ is described as follows. If $D : A \to N$ is an R-derivation, then the restriction of D to I is an A-homomorphism, which is trivial on I^2. To prove the final surjectivity assumption, generalize the proof of Proposition 5.1.

(ii) Show that there is an exact sequence

$$I/I^2 \to B \otimes_A \Omega_{A/R} \to \Omega_{B/R} \to 0.$$

Moreover, if the composition $R \to A \to B$ is an isomorphism, the first arrow is injective.

Hints: Use Proposition 11.3 (i).

(iii) Deduce Proposition 5.1 from Exercise 11.6 (ii).

Exercise 11.6. Let D be a continuous derivation of the topological field $F = K((x))$ of formal power series. Show that $D(f(x)) = f'(x)\,Dx$, where if $f(x)$ is a formal power series $f'(x)$ is its derivative. Conclude that the module $\mathrm{Der}_K(F)$ of continuous derivations is a one-dimensional vector space over F.

Exercise 11.7. Let $X = \mathbf{P}^1$ be the rational curve whose function field is the field of rational functions $k(x)$. Let $\omega = x^n\,dx$. Verify that

$$\mathrm{res}_P(\omega) = \begin{cases} 1 & \text{if } n = -1 \text{ and } P = 0; \\ -1 & \text{if } n = -1 \text{ and } P = \infty; \\ 0 & \text{otherwise.} \end{cases}$$

12. The Riemann-Roch Theorem

The proof of the Riemann-Roch theorem which we will present is Weil's adelic proof. This proof was first sketched in 1938 in Weil [29]. Proofs of the Riemann-Roch theorem based on Weil's adelic approach may be found in Lang [17], Serre [26] and in Weil [33]. (See also [31].) In [33], Weil proved the Riemann-Roch theorem for function fields over a finite field. By contrast, we will prove the Riemann-Roch theorem over an algebraically closed field, and when we need the finite-field version in Chapter 14, we will deduce it from the algebraically closed case.

Let X be a complete nonsingular curve, and let F be its function field. By a *divisor* D on X, we mean an element of the free abelian group on X, that is, a finite formal sum with multiplicities of points of X. If $D = \sum n_P P$ is a divisor, then by definition $n_P = 0$ for all but finitely many points P of X, and we define the *degree* of D to be $\deg(D) = \sum n_P$. We call the finite set of P with $n_P \neq 0$ the *support* of the divisor D.

For example, if $f \in F$ is a nonzero function, we may associate with f a divisor, called *the divisor of f* and denoted (f), to be described as follows. If $P \in X$, let v_P denote the corresponding valuation of F. The coefficient of P in the divisor (f) is $v_P(f)$. It is easy to show that this is nonzero for only finitely many P.

Proposition 12.1. *(i) If D_1 and D_2 are any divisors, then $\deg(D_1 + D_2) = \deg(D_1) + \deg(D_2)$.*

(ii) If f_1, $f_2 \in F$, then $(f_1 f_2) = (f_1) + (f_2)$.

(iii) Let $f \in F$. Then the degree $\deg\big((f)\big) = 0$.

Proof. Parts (i) and (ii) are obvious. Let us establish (iii) first for the rational curve $X = \mathbf{P}^1$ whose function field is the field of rational functions $k(x)$. Then f is a product of terms of the form $(x - a)^n$, where n may be positive or negative. By (i) and (ii), it is sufficient to establish (iii) for $f = (x - a)^n$. It is easy to see then that

$$v_P(f) = \begin{cases} n & \text{if } x = a; \\ -n & \text{if } x = \infty; \\ 0 & \text{otherwise.} \end{cases}$$

Hence (iii) is true in the case of a rational function field.

Let Y be a more general curve with function field E. Choose $x \in E$ such that E is separable over $F = k(x)$. As in Theorem 8.7, there is then a separable morphism $\phi : Y \to X$, where X is the rational curve just considered. Let $m = [E : F]$. We will show that if $f \in E$, then

$$(12.1) \qquad v_P(N_{E/F}\, f) = \sum_{Q \in \phi^{-1}(P)} w_Q(f),$$

where v_P is the valuation of F corresponding to the point P, and w_Q are the valuations of E extending v_P corresponding to the points Q in the fiber. Thus we get $\sum w_Q(f) = \sum v_P(N_{E/F}(f))$, and the latter is zero by the case just considered of the rational function field.

Let $v = v_P$. Let $Q_1, \cdots, Q_r$ be the points in $\phi^{-1}(P)$. let $w_i = w_{Q_i}$ be the valuation of E corresponding to each Q_i. Let $\overline{F}$ be the completion of F with respect to v, and let $\overline{E}_i$ be the completion of E with respect to w_i. Let e_i be the ramification degree of $\overline{E}_i$ over $\overline{F}$. Note that $\overline{E}_i$ is totally ramified over $\overline{F}$ since the residue field of F is algebraically closed, so $f_i = 0$. Thus $[\overline{E}_i : \overline{F}] = e_i$, and $\sum e_i = m$. By Proposition 10.16, we have $N_{E/F}(f) = \prod N_{\overline{E}_i/\overline{F}}(f)$. Therefore $v(N_{E/F}(f)) = \sum v(N_{\overline{E}_i/\overline{F}}(f))$, and in order to prove Eq. (12.1) it is sufficient to show that $v(N_{\overline{E}_i/\overline{F}}(f)) = w_i(f)$. Indeed, this follows from Proposition 10.18. ■

We say that two divisors D_1 and D_2 are *linearly equivalent* if their difference $D_1 - D_2$ is the divisor of a function. We will use the notation $D_1 \sim D_2$ to denote this equivalence relation. We will call the set of divisors linearly equivalent to D the *linear equivalence class* of D.

A divisor $D = \sum n_P . P$ is called *positive* or *effective* if all the coefficients $n_P \geq 0$. We define a partial ordering $\leq$ on the divisors, in which $D \leq E$ if $E - D$ is effective. (We also write $E \geq D$.)

We may write an arbitrary divisor $D = \sum n_P . P$ as the difference between two effective divisors, $D = D_+ - D_-$, where

$$D_+ = \sum_{n_P > 0} n_P . P, \qquad D_- = - \sum_{n_P < 0} n_P . P \,.$$

In particular if $D = (f)$ is the divisor of a function, we denote $D_+ = (f)_0$ and $D_- = (f)_\infty$. These are called the *divisor of zeros* and the *divisor of poles* of

f, respectively. This terminology is consistent with the usage which we expect from the case of a Riemann surface: if $v_P(f) > 0$, then f has a zero at P, and $v_P(f)$ is its order; similarly if $v_P(f) < 0$, then f has a pole at P, and $-v_P(f)$ is its order.

Proposition 12.2. *Let f be a nonconstant element of F, and assume that $F/k(f)$ is separable.*

(i) The inclusion $k(f) \to F$ induces a ramified covering $\phi : X \to \mathbf{P}^1$. Let $\phi^{-1}(\infty) = \{Q_1, \cdots, Q_r\}$. Let e_i be the ramification index of Q_i over ∞. Then $(f)_\infty = \sum e_i Q_i$.

(ii) We have $\deg((f)_0) = \deg((f)_\infty) = [F : k(f)]$.

Proof. Let us establish (i). Let v be the valuation of $k(f)$ corresponding to the point $\infty \in \mathbf{P}^1$, and let $w_1, \cdots, w_r$ be the valuations of F extending v. These correspond to the points $Q_1, \cdots, Q_r$ which map to ∞ under ϕ. Let e_i be the corresponding ramification indices. If $Q \in X$, and if w is the corresponding valuation, then Q occurs with nonzero coefficient in $(f)_\infty$ if and only if $w(f) < 0$. Since v is the unique valuation of $k(f)$ with $v(f) < 0$, a necessary and sufficient condition for $w(f) < 0$ is that w extends v, i.e. $\phi(Q) = \infty$. Consequently, the points which occur with nonzero coefficient in $(f)_\infty$ are precisely $Q_1, \cdots, Q_r$. Moreover, the coefficient $n_{Q_i} = -w_i(f) = -e_i v(f) = e_i$. Thus (i) is clear.

As for (ii), it follows from Theorem 9.10 (ii) that $\deg((f)_\infty) = \sum n_{Q_i} = \sum e_i = n$. Furthermore $\deg((f)_0) = \deg((f)_\infty)$ by Proposition 12.1 (iii). ∎

We will need to know that the conclusion of Proposition 12.2 (ii) is valid without the hypothesis that $F/k(f)$ is separable. Let us therefore take a moment to discuss the structure of separable morphisms. Let K be any field. If F is a finitely generated extension of K such that K is algebraically closed in K, then we call F a *function field* over K. If the transcendence degree of F/K is d, we say that d is the *dimension* of the function field F. If $K = k$ is our fixed algebraically closed field, this agrees with the definition of an abstract function field in Chapter 3. Actually much of the theory of function fields is the same over a perfect field as over an algebraically closed field, and we prove the next Proposition for function fields over a perfect field. If F is a field of characteristic p, and if q is a power of p, recall that the *Frobenius map* $\mathcal{F}_q : F \to F$ defined by $\mathcal{F}_q(x) = x^q$ is a field automorphism of $F \to \mathcal{F}(F)$.

Proposition 12.3. *Let K be a perfect field of characteristic p, and let $E \supset F$ be function fields of dimension one over K. Suppose that E/F is purely inseparable of degree q. Then $\mathcal{F}_q$ is an isomorphism of E onto F.*

Proof. We may build E up from F by stages, adjoining one p-th root at a time. If the Proposition is true for each stage, it is true for E/F. Therefore, we may assume that $q = p$. Let $\xi \in E$ such that $\xi \notin F$. Thus $E = F(\xi)$. Since the extension E/F is purely inseparable of degree p, $\xi^p = x \in F$. Since K is perfect, we have $\mathcal{F}_p^{-1} K(x) = K(\xi)$. Now $[K(\xi) : K(x)] = [F(\xi) : F] = p$. This implies that the field extension $\mathcal{F}_p^{-1} K(x) = K(\xi)$ and $F(\xi)$ are linearly disjoint over F, and so by Mac Lane's criterion (Proposition 11.8) $F/K(x)$ is separable. By the Theorem of the Primitive Element (Lang [18], Theorem V.4.6 on p. 243), we may write $F = K(x, y)$ where y is the root of a separable polynomial over $K(x)$. Let $\eta = \mathcal{F}_p^{-1}(y)$ in an algebraic closure of E. Obviously $K(\xi, \eta)/K(\xi)$ is separable, and $\mathcal{F}_p$ is an isomorphism of $K(\xi, \eta)$ onto $K(x, y)$. Thus what we must show is that $E = K(\xi, \eta)$. Obviously $K(\xi, \eta) \supseteq K(x, y, \xi) = F(\xi) = E$. Thus it is sufficient to show that $[E : K(\xi)] \geq [K(\xi, \eta), K(\xi)]$. Since $\mathcal{F}_p$ maps $K(\xi, \eta)$ isomorphically onto $K(x, y) = F$, while mapping the subfield $K(\xi)$ onto $K(x)$, we have $[K(\xi, \eta) : K(\xi)] = [F : K(x)]$. Now the separable degree $[E : K(x)]_s$ of $E/K(x)$ equals $[F : K(x)]_s = [F : K(x)]$ since E/F is purely inseparable, and $F/K(x)$ is separable. Also $K(\xi) : K(x)$ is purely inseparable, so $[E : K(\xi)]_s = [E : K(x)]_s$. Thus $[E : K(\xi)] \geq [E : K(\xi)]_s = [K(\xi, \eta) : K(\xi)]$, as required.   ∎

We now return to the case where F is the function field of a curve X over an algebraically closed field k. We may now remove the hypothesis that $F/k(f)$ is separable from Proposition 12.2 (ii).

Proposition 12.4. *Let f be a nonzero element of F. Then $[F : k(f)] = \deg\big((f)_\infty\big) = \deg\big((f)_0\big)$.*

Proof. Let us consider the separable closure of $k(f)$ in F. By the Theorem of the Primitive Element, this has the form $k(f, g)$, where $g \in F$. Let q be the degree of the purely inseparable extension $F/k(f, g)$. By Proposition 12.3, $F = k(f_1, g_1)$, where $f_1^q = f$ and $g_1^q = g$. Then $F/k(f_1)$ is separable because it is mapped isomorphically onto the separable extension $k(f, g)/k(f)$ by the Frobenius map, and so by Proposition 12.2 (ii), we have $[F : k(f_1)] = \deg\big((f_1)_\infty\big) = \deg\big((f_1)_0\big)$. Since $f = f_1^q$ we have $\deg\big((f)_\infty\big) = \deg\big((f)_0\big) = q[F : k(f_1)]$. This equals $[F : k(f_1)][k(f_1) : k(f)] = [F : k(f)] = q[F : k(f_1)]$, as required.   ∎

Let us mention that the concepts of divisors and linear equivalence is not special to curves. Let X be a variety which is nonsingular in codimension one, and let F be its function field. A *Weil divisor* on X is defined to be an element of the free abelian group generated by the irreducible subvarieties of codimension one. By Proposition 6.23, the local ring of any subvariety E of codimension one is regular, and so by Proposition 5.28, it is a discrete valuation ring. Let v_E be the corresponding valuation. If f is any function, we may associate with f and E the integer $v_E(f)$, and the divisor of f is then defined to be $\sum_E v_E(f) E$. Again, two divisors are linearly equivalent if their difference is the divisor of a function. It turns out that this is a useful concept only if X is *complete*. For a complete variety X, nonsingular in codimension one, the concepts of divisors and linear equivalence are valid and very important. However we will concentrate on curves.

We now describe the important notion of a *linear system* of divisors, which arises in connection with the problem of embedding X into a projective space. Suppose that D is a divisor. The *complete linear system* is the set $|D|$ of effective divisors linearly equivalent to D. The set $|D|$ may be empty, but if $|D|$ is chosen appropriately, it will not be. Every element of $|D|$ may be written as $(f) + D$ for some $f \in F$. Thus we are led to define $L(D)$ as the set of functions $\{f \in F | (f) + D \geq 0 \text{ or } f = 0\}$. Then $L(D)$ is a vector space over k, and $f \to (f) + D$ gives a mapping of $L(D)$ onto $|D|$. Two functions $f_1, f_2 \in L(D)$ have the same image if and only if they have the same divisors, so that f_1/f_2 is a function without zeros or poles. This means that $f_1/f_2 \in k$ by Exercise 12.1. We see that there is a surjective map from the $L(D) - \{0\} \to |D|$ in which two elements have the same image if and only if they are proportional. Therefore $|D|$ may be identified with the projective space $\mathbf{P}(L(D))$.

We may now sketch the connection between linear systems and projective embeddings. An example will be worked out in detail in Chapter 13, where we will show that every curve of genus one may be embedded into $\mathbf{P}^2$ by studying the linear system $L(D)$, where D is a divisor of degree three.

Suppose that we have a rational map $X \to \mathbf{P}^n$. For simplicity assume that this is an embedding, so we may regard X as a subvariety of $\mathbf{P}^n$. We obtain a linear system as follows. Let $H \subset \mathbf{P}^n$ be any hyperplane, i.e. a linear subspace of codimension one. Let $Y_1, \cdots, Y_r$ be the irreducible components of $H \cap X$. These are subvarieties of X of codimension one. It is necessary to count each with a certain *intersection multiplicity* $n_i = i(H, X; Y_i)$ which is defined in the intersection theory. (See Hartshorne [12], Section I.7.) We have $n_i > 0$, and we have $n_i > 1$ if H is tangent to X along one of the components Y_i. Let $D_H = \sum n_i Y_i$. Then D is an effective divisor on X. We claim that the D_H, as H runs through the hyperplanes of $\mathbf{P}^n$, form a linear system. In particular, if H' is another hyperplane, we assert that $D_H \sim D_{H'}$. To see this, let $\mathbf{P}^n = \mathbf{P}(V)$, where $\dim V = n+1$. Let U and U' be the vector subspaces of V of codimension one corresponding to the hyperplanes H and H'. Let T and T' be linear functionals on V which vanish on U and U', respectively. Consider the function $f : V - U' \to k$ defined by $f(v) = T(v)/T'(v)$. This function is constant on one-dimensional subspaces of V, hence induces a function f on X, which is regular except on $H \cap U'$. This f thus represents an element of the function field F of X, and its divisor is $D_H - D_{H'}$. This shows that $D_H \sim D_{H'}$. It may be seen that the set of divisors D_H forms a linear system. We call this the *linear system cut out by hyperplanes*, with respect to the given projective embedding of X.

Conversely, given a linear system L, one may try to construct a projective embedding of X (or at least a rational map into projective space) which realizes L as the linear system cut out by hyperplanes. Sometimes this cannot be done, since a linear system may be empty. However, if L is "large enough," it may always be done—see Proposition 13.8 for an illustration of how this may be accomplished. Hence divisors and linear systems are at the heart of the problem of constructing embeddings of X into projective space. It becomes essential to understand the projective space $|D|$ more precisely, or equivalently, the vector space $L(D) = \{f \in F | (f) \geq -D\}$. When X is a curve, the Riemann-Roch Theorem is a formula for its dimension.

For further examples of linear systems in action, see Zariski's book [35], where they are used systematically in the theory of surfaces.

We now specialize to the case where X is a complete nonsingular curve. We will introduce a certain ring A called the *adele ring* or ring of *repartitions* over X. This is a subring of the Cartesian product $\prod_{P \in X} F_P$ of all the completions of F. The latter Cartesian product is too large to be of any use. If $x = (x_P) \in \prod F_P$, we say that x is an *adele* if $v_P(x_P) \geq 0$ for all but finitely many P. Of course some components x_P may be zero, and our convention is that $v_P(x_P) = \infty \geq 0$. The adeles clearly form a ring.

The ring A of adeles of X contains F as a subring, injected along the diagonal: if $f \in F$, we identify f with the adele whose P-th component is f for all P. Now if $D = \sum n_P.P$ is a divisor, let

$$\Lambda(D) = \{x \in A | v_P(x_P) \geq -n_P \text{ for all } P \in X\}.$$

Evidently $L(D) = \Lambda(D) \cap F$.

Proposition 12.5. *Let R be a ring, M an R-module, and let V, V' and U be R-submodules of M. Suppose that $V \supseteq V'$. Then there exists an exact sequence*

$$0 \to (V \cap U)/(V' \cap U) \to V/V' \to (V + U)/(V' + U) \to 0.$$

Proof. Consider the composite map $V \to V + U \to (V + U)/(V' + U)$. Clearly V' is in the kernel, and so there is an induced homomorphism $V/V' \to (V + U)/(V' + U)$. It is easy to see that this homomorphism is surjective. Similarly, $V' \cap U$ is the kernel of the composition $V \cap U \to V \to V/V'$, and so there is induced an injective map $(V \cap U)/(V' \cap U) \to V/V'$. It remains to be shown that the sequence

$$0 \to (V \cap U)/(V' \cap U) \to V/V' \to (V + U)/(V' + U) \to 0$$

is exact at V/V', which we leave to the reader. ∎

Now suppose that $D \geq E$, so that $L(D) \supseteq L(E)$ and $\Lambda(D) \supseteq \Lambda(E)$. We apply Lemma 5.5 with $R = k$, $V = \Lambda(D)$, $V' = \Lambda(E)$ and $U = F$. We see that

$$(12.2) \quad \dim\big(L(D)/L(E)\big) + \dim\big((\Lambda(D)+F)/(\Lambda(E)+F)\big) = \dim\big(\Lambda(D)/\Lambda(E)\big).$$

We have not yet proved that these dimensions are finite. Let us now remedy that defect.

Proposition 12.6. *We have $L(0) = k$. If $\deg(D) < 0$, then $L(D) = 0$.*

Proof. Clearly if $f \in k$, then $(f) = 0$, so $k \subseteq L(0)$. On the other hand, suppose that $f \in L(0)$. This means that f has no poles. By Proposition 12.4, the number of poles of f is equal to the number of zeros (counting with multiplicity) and so f also has no zeros, and $(f) = 0$. Now by Exercise 12.1, $f \in k$.

If $\deg(D) < 0$, and f is a nonzero element of $L(D)$, then $(f) \geq -D$, so $\deg(f) > 0$. This contradicts Proposition 12.1 (iii). ■

Proposition 12.7. *Let D and E be divisors on the complete nonsingular curve X such that $D \geq E$.*
(i) We have $\dim\big(\Lambda(D)/\Lambda(E)\big) = \deg(D) - \deg(E)$.
(ii) We have $\dim\big(L(D)\big) < \infty$.
(iii) We have $\dim\big(\Lambda(D) + F\big)/\big(\Lambda(E) + F\big) < \infty$.

Proof. Let us establish (i). It is clearly sufficient to prove this when $D = E + P$, with $P \in X$. Let $E = \sum n_Q . Q$. Clearly we have $\Lambda(D) \neq \Lambda(E)$ since we may construct an adele x such that $v_P(x_P) = -n_P - 1$, and $v_Q(x_Q) \geq -n_Q$ for $Q \neq P$. Then $x \in \Lambda(D)$ but $x \notin \Lambda(E)$. On the other hand, if y is any other element of $\Lambda(D)$, we may find a constant $c \in k$ such that $v_P(y - cx) \geq n_P$, and so $y - cx \in \Lambda(E)$. This shows that $\Lambda(D)/\Lambda(E)$ is precisely one-dimensional if $D = E + P$. This proves (i).

It follows immediately from Eq. (12.2) that $L(D)/L(E)$ and $\big(\Lambda(D) + F\big)/\big(\Lambda(E) + F\big)$ are finite-dimensional. Thus (iii) follows right away. To prove (ii), we only have to prove (given D) that $L(E)$ is finite-dimensional for some $E \leq D$. By Proposition 12.6 we have $L(0) = k$; in particular $L(0)$ is one-dimensional. Then E may be chosen so that $E \leq D$ and $E \leq 0$. Thus $L(E) \subseteq L(0) = k$, so $L(E)$ is finite-dimensional. ■

In view of Proposition 12.7 (ii), we define $l(D) = \dim L(D)$. If U and U' are subspaces of a vector space V such that $U \cap U'$ is of finite index in V, we define the generalized codimension $(U : U')$ to be

$$\dim\big(U/(U \cap U')\big) - \dim\big(U'/(U \cap U')\big).$$

This has the usual properties of codimension, namely $(U : U') + (U' : U'') = (U : U'')$ when both sides are defined. Unlike the usual codimension, it is defined whether or not $U \supseteq U'$.

We may rewrite Eq. (12.2) as

$$l(D) - l(E) + \big(\Lambda(D) + F : \Lambda(E) + F\big) = \deg(D) - \deg(E).$$

Let $r(D) = \deg(D) - l(D)$. Then we have

(12.3) $$r(D) - r(E) = \big(\Lambda(D) + F : \Lambda(E) + F\big).$$

In proving this, we have assumed that $D \geq E$. However, it is true without this hypothesis, since we may always find a divisor E' such that $D \geq E'$ and $E \geq E'$. Then

$$r(D) - r(E') = \big(\Lambda(D) + F : \Lambda(E') + F\big)$$

and

$$r(E) - r(E') = \big(\Lambda(E) + F : \Lambda(E') + F\big),$$

by the special case of Eq. (12.3) with $D \geq E$, and subtracting these relations we obtain Eq. (12.3) in general.

Proposition 12.8. *(i) If $D_1 \sim D_2$ then $l(D_1) = l(D_2)$ and $r(D_1) = r(D_2)$. (ii) if $D \leq E$ then $r(D) \leq r(E)$.*

Proof. We have $D_1 = (g) + D_2$ for some $g \in F$. Now $(f) \geq -D_1$ if and only if $(fg) = (f) + (g) \geq -D_2$, so $f \to fg$ is an isomorphism of $L(D_1)$ onto $L(D_2)$. Hence $l(D_1) = l(D_2)$. It follows from Proposition 12.1 that $\deg(D_1) = \deg(D_2)$, and so $r(D_1) = r(D_2)$. This proves (i). As for (ii), this is a consequence of Eq. (12.3). ■

Proposition 12.9. *There exists a constant c depending only on X such that $r(D) = \deg(D) - l(D) \leq c$ for all divisors D.*

Proof. Let us fix $x \in F$ such that $F/k(x)$ is separable. There is a morphism $\phi : X \to \mathbf{P}^1$ associated with the inclusion $k(x) \to F$ by Theorem 8.7. Let $\mathbf{P}^1$ be decomposed as $\mathbf{A}^1 \cup \{\infty\}$, and let $\phi^{-1}(\infty) = \{Q_1, \cdots, Q_r\}$. Let $D_0 = (x)_\infty$ be the divisor of poles x. Then $D_0 = \sum e_i Q_i$, where e_i are the ramification indices, and $n = [F : k(x)] = \deg(D_0)$ by Proposition 12.4. Let $y_1, \cdots, y_n$ be a basis of F over $k(x)$. We may multiply these by polynomials in x so that they are integral over $k[x]$. This implies that every pole of y_i is among the poles Q_i of x. For if P is a point of X where x does not have a pole, then $k[x]$

is contained in the local ring $\mathcal{O}_P$, and hence the integral closure of $k[x]$ in F is contained in this valuation ring also by Exercise 2.5. Thus $y_i \in \mathcal{O}_P$, so P is not a pole of y_i. Therefore $y_1, \cdots, y_n \in L(m_0 D_0)$ for some sufficiently large positive integer m_0. Now if $m \geq m_0$, consider the $n(m - m_0 + 1)$ functions $x^s y_i$ for $0 \leq s \leq m - m_0$, $1 \leq i \leq n$. These are clearly linearly independent, and are clearly in $L(mD_0)$. Hence $l(mD_0) \geq n(m - m_0 + 1)$. On the other hand, $\deg(mD_0) = mn$. Thus $r(mD_0) \leq c$ where $c = (m_0 - 1)n$, for all m.

Now let D be an arbitrary divisor. Let $D = D_1 + D_2$, where D_1 is a linear combination of the Q_i, and the Q_i appear with coefficient zero in D_2. We can find a polynomial $f \in k[x]$ having zeros of high order at the points in the support of D_2. Thus we may arrange that $D - (f)$ has nonpositive coefficients at every point except possibly the Q_i. Then $D - (f) \leq mD_0$ for some D_0. Now using Proposition 12.8 (i), $r(D) = r\big(D - (f)\big)$, and by Proposition 12.8 (ii) $r\big(D - (f)\big) \leq r(mD_0) \leq c$. ■

Proposition 12.10. *There exists a divisor such that* $A = \Lambda(D) + F$.

Proof. Indeed, by Proposition 12.9, choose E in Eq. (12.3) so that $r(E)$ is maximal. Suppose that $\Lambda(E) + F \neq A$. Let $x \in A$ such that $x \notin \Lambda(E) + F$. We may find $D \geq E$ such that $x \in \Lambda(D)$, and then $\big(\Lambda(D) + F : \Lambda(E) + F\big) > 0$. By Eq. (12.3), $r(D) > r(E)$, contradiction the maximality of $r(E)$. ■

Proposition 12.11 (Riemann). *There exists a function* $\delta(D)$ *on divisors, taking values in the non-negative integers, such that if* $D \geq E$ *then* $\delta(D) \leq \delta(E)$, *and such that* $\delta(D) = 0$ *for some divisor* D, *and a non-negative integer* g *such that*

$$(12.4) \qquad l(D) = \deg(D) + 1 - g + \delta(D)$$

for all divisors D.

This inequality is already a usable form of the Riemann-Roch theorem. The non-negative integer g is called the *genus* of X.

Proof. By Proposition 12.10, let D_0 be chosen so that $\Lambda(D_0) + F = A$. Then by Eq. (12.3), we have $r(D_0) - r(D) = \big(A : \Lambda(D) + F\big)$. This proves that $\big(A : \Lambda(D) + F\big) < \infty$. We therefore define $\delta(D) = \big(A : \Lambda(D) + F\big)$. The asserted properties of δ are obvious. We may now rewrite Eq. (12.3) as

$$r(D) - r(E) = \delta(E) - \delta(D).$$

This shows that $r(D) + \delta(D) = \deg(D) - l(D) + \delta(D)$ is constant, and we define g so as to make Eq. (12.4) true. It remains only to be shown that $g \geq 0$. To see this, take $D = 0$ in Eq. (12.4). By Proposition 12.6, we have $l(0) = 1$, and therefore $g = \delta(0) \geq 0$. ■

The Riemann-Roch Theorem refines Eq. (12.4) by identifying $\delta(D)$ as the dimension of a space of differentials.

If $\omega \in \Omega_{F/k}$ is a nonzero differential, we may also associate with ω a divisor, called *the divisor of* ω, to be denoted (ω). If $P \in X$, we define the coefficient n_P of P in (ω) as follows. Let $t \in F$ be a local parameter at P. Then by Proposition 11.11, $dt \neq 0 \in \Omega_{F/k}$, and $\omega = f\,dt$ for some $f \in F$. We define $n_P = v_P(f)$. Note that this does not depend on the choice of t. It is necessary to check that $n_P \neq 0$ for all but finitely many P, so that $(\omega) = \sum n_P P$ is a divisor. This may be seen as follows. Let $x \in F$ such that $dx \neq 0$, and let y be chosen so that $\omega = y\,dx$. There exists an affine open set U such that x, y and y^{-1} are all regular on U. Since there are only a finite number of ramified points in U for the ramified cover $\phi : X \to \mathbf{P}^1$ induced by the inclusion of fields $k(x) \to F$, we may exclude these from U. That is, we may assume that if $P \in U$, and if $Q = \phi(P) \in \mathbf{P}^1$, then $e(P|Q) = 1$. We will show that $n_P = 0$ for $P \in U$. Indeed, regarding x as a function on U, let a be its value at P. Then $x - a$ vanishes at P, so $v_P(x - a) > 1$. Now $x - a$ is a local parameter at $\phi(P)$ in $k(x)$, and since P is nonramified, $x - a$ is also a local parameter at P in F. We have $\omega = y\,dx = y\,d(x - a)$, and so by definition $n_P = v_P(y)$. However y and y^{-1} are both in the local ring at P, and therefore $v(y) = 0$. This proves that $n_P \neq 0$ for all but finitely many P, and hence $(\omega) = \sum n_P P$ is a divisor.

If f is a function, then it is clear that $(f\omega) = (f) + (\omega)$. Since $\Omega_{F/k}$ is one-dimensional by Proposition 11.10, it follows that the linear equivalence class of (ω) is uniquely determined. We call this class the *canonical class*. We will call any divisor in the canonical class a *canonical divisor*.

Example 12.12. Let us compute the canonical class of $\mathbf{P}^1$. Let $F = k(x)$ be the function field. We consider the divisor of the differential form dx. We write $\mathbf{P}^1 = \mathbf{A}^1 \cup \{\infty\}$. If $P = a \in \mathbf{A}^1$, then a local parameter at a is $t = x - a$, and $dx = 1.dt$. Since $v_P(1) = 0$, we see that for $P \in \mathbf{A}^1$, the coefficient n_P in (dx) is zero. On the other hand, if $P = \infty$, a local parameter at P is $t = x^{-1}$. Differentiating the relation $x = t^{-1}$, we have $dx = -t^{-2}\,dt$, and $v_P(t^{-2}) = -2$. Therefore we calculate $(dx) = -2.\infty$. We see that the degree of the canonical class is -2.

Let Δ be the set of all k-linear functionals on A which are trivial on $\Lambda(D) + F$ for some divisor D. Δ has the structure of a vector space over F, since if $T \in \Delta$, $f \in F$, we define $(fT)(a) = T(fa)$ for $a \in A$. If T is trivial on $\Lambda(D) + F$, then fT is trivial on $\Lambda\big(D + (f)\big)$.

Proposition 12.13. Δ *is at most one-dimensional over* F.

Proof. Let T_1, T_2 be elements of Δ. We will show that they are linearly dependent. Suppose on the contrary that they are not. For some D, both T_i vanish on $\Lambda(D)+F$. Let E be any divisor, and let $y \in L(E)$. Then yT_i vanishes on $\Lambda(D-E)$. If $y_1, \cdots, y_r$ is a k-basis of $L(E)$, then the linear functionals $y_j T_i$ are linearly independent over k, for if not, we have coefficients $a_{ij} \in k$ not all zero such that $\sum a_{ij} y_j T_i = 0$. Then if $a_i = \sum_j a_{ij} y_j$, by the linear independence of the y_j, a_1 and a_2 are not both zero, yet $a_1 T_1 + a_2 T_2 = 0$, contradicting the assumed independence of T_1 and T_2. Moreover, the $y_j T_i$ are clearly trivial on $\Lambda(D-E) + F$. Thus there exist $2r$ independent linear functionals on A which are trivial on $\Lambda(D) + F$, where $r = 2L(E)$, and so $\delta(D-E) \geq 2l(E)$. Now using Eq. (12.4) twice, we have

$$
\begin{aligned}
l(D-E) &= \deg(D) - \deg(E) + 1 - g + \delta(D-E) \\
&\geq \deg(D) - \deg(E) + 1 - g + 2l(E) \\
&\geq \deg(D) - \deg(E) + 1 - g + 2\big(\deg(E) + 1 - g\big) \\
&= \deg(D) + \deg(E) + 3 - 3g.
\end{aligned}
$$

Yet if E is chosen sufficiently large, we may make $l(D-E) = 0$ and $\deg(E)$ as large as we wish, so this is a contradiction. ∎

Now let us show how we can construct elements of Δ. Let $\omega \in \Omega_{F/k}$. If $a = (a_P) \in A$, define

$$
T_\omega(a) = \sum_{P \in X} \mathrm{res}_P(a_P \omega).
$$

By Theorem 11.24, T_ω is trivial on F. It is clear from the definitions that if $D = (\omega)$, then T_ω is trivial on $\Lambda(D)$.

Proposition 12.14. *Every element of Δ has the form T_ω for some $\omega \in \Omega_{F/k}$.*

Proof. It is clear that $\omega \to T_\omega$ is a nonzero F-linear map between the F-vector spaces $\Omega_{F/k}$ and Δ. The former is one-dimensional over F by Proposition 11.4, and the latter is one-dimensional by Proposition 12.13. Hence this map is an isomorphism. ∎

Proposition 12.15. *The number $\delta(D)$ is equal to the dimension of the vector space of $\omega \in \Omega_{F/k}$ such that $(\omega) \geq D$.*

Proof. Indeed, it is clear that T_ω is trivial on $\Lambda(D) + F$ if and only if $(\omega) \geq D$. ∎

Theorem 12.16 (Riemann-Roch). *Let K be a divisor in the canonical class. Then for any divisor D, we have $\delta(D) = l(K - D)$, and*

$$(12.5) \qquad\qquad l(D) = \deg(D) + 1 - g + l(K - D).$$

Proof. Assume that $K = (\omega)$. Now $f \to f\omega$ is an isomorphism of $L(K - D)$ onto the space of ω' such that $(\omega') \geq D$. Thus by Proposition 12.15, $\delta(D) = l(K - D)$. Combining this with Eq. (12.4) gives Eq. (12.5). ∎

Proposition 12.17. *The dimension of the space of $\omega \in \Omega_{F/k}$ such that $(\omega) \geq 0$ is g.*

Proof. By Proposition 12.6, we have $l(0) = 1$. Substituting 0 for D in Eq. (12.4), we obtain $\delta(0) = g$. The conclusion follows from Proposition 12.15. ∎

Proposition 12.18. *The degree of the canonical class is $2g - 2$.*

Proof. By Proposition 12.17, we have $l(K) = \delta(0) = g$. On the other hand, $\delta(K) = l(0) = 1$ by Proposition 12.6. Now substituting K for D in Eq. (12.5), we obtain $g = l(K) = \deg(K) + 1 - g + 1$. ∎

Proposition 12.19. *A curve has genus zero if and only if it is isomorphic to $\mathbf{P}^1$.*

Proof. Let $P \in X$, and consider $L(P)$. By the Riemann-Roch Theorem, $l(P) = 2 + l(K - P)$, and $\deg(K - P) = -3$, so by Proposition 12.6, $l(K - P) = 0$. Thus $L(P)$ is two-dimensional. One dimension is accounted for by the fact that $k \subseteq L(P)$. Let f be a nonconstant function in $L(P)$. Then clearly f has a pole of degree one at P, and no other pole, so $(f)_\infty = P$. Now by Proposition 12.4, we have $[F : k(f)] = \deg\big((f)_\infty\big) = 1$. Hence $F = k(f)$, and therefore $X \cong \mathbf{P}^1$. ∎

Let X and Y are complete nonsingular curves with function fields F and E respectively. By Theorem 8.7, we associate with each morphism $\phi : Y \to X$ an injection of fields $F \to E$. Let P be a point of X and Q a point of $\phi^{-1}(P)$, and let v and w be the corresponding valuations of F and E. Let $e = e(w|v)$ be the ramification index. We call Q *wildly ramified* if e is divisible by the residue characteristic, which is the characteristic of the constant field k. If Q is ramified but not wildly ramified we say it is *tamely ramified*. This is consistent with the terminology Theorem 9.22.

Theorem 12.20 (Hurwitz' genus formula). *Let X and Y be complete nonsingular curves of genus g and g' respectively, and let $\phi : Y \to X$ be a nonconstant separable morphism of degree n. Let $Q_1, \cdots, Q_r$ be the points of Y which are ramified in this covering, and let $e_i = e\big(Q_i|\phi(Q_i)\big)$ be the ramification indices. Assume that the Q_i are tamely ramified, so no e_i is divisible by the characteristic of the constant field k. Then*

$$2g' - 2 = n\,(2g - 2) + \sum(e_i - 1).$$

Proof. By Proposition 12.18 the degree of a canonical divisor in Y is $2e' - 2$. Since the extension E/F of function fields is separable, the mapping $\Omega_{F/k} \to \Omega_{E/k}$ is injective by Proposition 12.6, so we may take $\omega \in \Omega_{F/k}$, and compare the divisors of ω in Y and X. Let Q be any point in Y, and let $P = \phi(Q)$. Let w and v be the valuations of E and F at Q and P respectively. Let $t \in E$ and $u \in F$ be local parameters at Q and P respectively. By definition, the coefficient of Q in (ω) is $w(\omega/du) = w(\omega/dt) + w(dt/du) = e\,v(\omega/dt) + w(dt/du)$, and by Exercise 12.3, $w(dt/du) = e(Q|P) - 1$. Now summing over all points Q of Y, we see that $2g' - 2 = e(2g - 2) + \sum(e_i - 1)$, as required. ∎

Example 12.21. Consider the curve X defined by $y^2 = x^3 + ax + b$, where the roots of $x^3 + ax + b$ are distinct, and the characteristic is not equal to two or three. (See Exercise 12.2.) We consider the ramified cover $\phi : X \to \mathbf{P}^1$ induced by the inclusion $k(x) \to F$. The degree of this cover is two. There are four ramified points, which may be described as follows. If e_i are the roots of $x^3 + ax + b$, we regard the e_i as points of $\mathbf{A}^1 \subseteq \mathbf{P}^1 = \mathbf{A}^1 \cup \{\infty\}$. The four ramified points are then the $\phi^{-1}(e_i)$, and $\phi^{-1}(\infty)$. In each case the ramification index is two. The degree of ϕ is also two. The genus of $\mathbf{P}^1$ is 0. Thus if g is the genus of X, the genus formula asserts that $2g - 2 = 2(-2) + 4 = 0$, so $g = 1$. For another proof see Exercise 12.2.

Example 12.22. This example shows that the Hurwitz genus formula fails when there is wild ramification. Suppose that the characteristic of k is $p > 0$. Let $F = k(x, y)$, where x and y satisfy the irreducible polynomial $xy^p - yx^p - 1 = 0$. Let X be the nonsingular complete curve with function field F. The inclusion $k(x) \to k(x, y)$ induces a separable morphism $X \to \mathbf{P}^1$ of degree p. Let $z = x^{-1}$, $t = y/x$, $u = y^{-1}$, $v = x/y$. We may cover X with three affine sets, having coordinate rings $k[x, y]$, $k[t, z]$ and $k[u, v]$. The equations of these three nonsingular affine curves are

$$xy^p - yx^p = 1,$$

$$t^p - t = z^{p+1},$$

$$v^p - v = -u^{p+1}.$$

Each of these affine curves is nonsingular by the criterion of Proposition 9.14.

If a is any nonzero element of k, the polynomial $aY^p - a^p Y - 1$ in $k[Y]$ is separable, and the first affine model shows that the fiber over any point in $\mathbf{P}^1$ except 0 and (possibly) ∞ in $\mathbf{P}^1$ consists of p distinct points. The fiber over ∞ also consists of p distinct points, as may be seen from the second affine model, for in the coordinates t and z, the fiber over infinity corresponds to the solutions to $t^p - t = 0$, $z = 0$. The value of t may be any element of the prime field, whose elements are the p distinct solutions to $t^p - t = 0$. Finally, we can compute the fiber over 0 from the third model—it consists of the single point $(u, v) = (0, 0)$. To see that this point maps to $0 \in \mathbf{P}^1$, note that u is a uniformizing parameter at $(0, 0)$ by Proposition 6.14 (i). Since $v^p - v = -u^{p+1}$, the valuation w at this point satisfies $w(u) = 1$, $w(v) = p + 1$, and so $w(x) = w(v/u) = p$. This shows that x lies in the local ring at this point, and moreover that the ramification index at this point is p. Ramification is wild.

The Hurwitz genus formula is not satisfied by this curve. The genus g of $\mathbf{P}^1$ is zero, and if the genus of X is g', the genus formula would have

$$2g' - 2 = p(2g - 2) + (p - 1) = -2p + (p - 1) = -1 - p,$$

so the genus g' of X would be $(1 - p)/2 < 0$, which is impossible.

Exercises

Exercise 12.1 Let X be a complete nonsingular curve with function field F, and let $0 \neq f \in F$. Suppose that the divisor $(f) = 0$. Prove that $f \in k$.

Hint: f is in every valuation ring of F containing k, so use Exercise 2.5 (i).

Exercise 12.2. Let X be the curve whose equation is $y^2 = x^3 + ax + b$, and let F be its function field. It is assumed that the roots e_1, e_2 and e_3 of $x^3 + ax + b = 0$ are distinct. Assume that the characteristic of k is not 2 or 3.

(i) Let $\phi : X \to \mathbf{P}^1$ be the covering induced by the inclusion $k(x) \to F$. As usual, write $\mathbf{P}^1 = \mathbf{A}^1 \cup \{\infty\}$. Show that the points of X may be enumerated as follows. Firstly, there are the points in $\phi^{-1}(\mathbf{A}^1)$. These may be identified with the ordered pairs (ξ, η) in $k \times k$ which are solutions to $\eta^2 = \xi^3 + a\xi + b$. Show that if $\eta \neq 0$, then $x - \xi$ is a local parameter at P. On the other hand, if $\eta = 0$, so $\xi = e_1$, e_2 or e_3, then show that y is a local parameter at P. (Use Proposition 6.14 (i).) Secondly, there is a single point $O = \phi^{-1}(\infty)$. Show that an affine neighborhood of O consists of the affine open set whose coordinate ring is $k[z, w]$, where $z = x^{-1}$ and $w = xy^{-1}$. (Verify that this ring is integrally closed, as in Example 9.15.) Show that w is a local parameter at O.

(ii) Show that the divisor of $y^{-1}\,dx$ is 0. (To compute the coefficient in $(y^{-1}\,dx)$ at O, use the identity $y^{-1}\,dx = 2\,(w^2\,g'(z) - 1)^{-1}\,dw$, where $g(z) = bz^3 + az + 1$.)

(iii) Show that X is not isomorphic to $\mathbf{P}^1$. (The canonical divisors of the two curves do not have the same degree.)

(iv) Show that the genus of X is one. (Use Proposition 12.18.)

Exercise 12.3. Let F be a field complete with respect to a discrete valuation v, and let E/F be a finite separable extension. Assume that E/F is totally and tamely ramified. Let w be the extension of v to E. Suppose that F contains a copy of the residue field K, as in Proposition 10.8. Let $t \in F$, $u \in E$ such that $v(t) = w(u) = 1$. Let $f \in E$ such that $dt = f\,du$. Show that $w(f) = e - 1$, where $e = [E : F]$ is the ramification index.

Hint: Suppose that E/F and E'/E are two such extensions. Show that if the statement is true for E'/E and for E'/F, then it is true for E/F. Taking E' to be the normal closure of E/F, we may assume that E/F is Galois. As in the proof of Theorem 11.19, E/F is then solvable by Exercise 10.7. Now show that if the statement is true for E/F and for E'/E then it is true for E'/E. As in the proof of Theorem 11.19, we may thus assume that E/F is cyclic of prime degree. As in Theorem 11.19, the cases where the characteristic does or does not divide $[E : F]$ are handled separately, using Proposition 11.17 and Proposition 11.18.

Exercise 12.4. If X be a complete nonsingular curve over $\mathbb{C}$, then it is a compact Riemann surface. Give a topological proof that the genus satisfies the same genus formula of Hurwitz as the algebraic genus, and conclude that the genus g in the Riemann-Roch Theorem is equal to the topological genus.

Exercise 12.5. Let $f(x)$ be a polynomial in $k[x]$ with distinct roots. Assume that the characteristic is not two. Use the genus formula to show that if the degree of f is n, then the genus of the hyperelliptic curve $y^2 = f(x)$ is $(n - 2)/2$ if n is even, and $(n - 1)/2$ if n is odd.

Hint: Consider the covering of degree two $X \to \mathbf{P}^1$ induced by $k(x) \to F$. Show that ∞ ramifies if and only if n is odd. You can deduce this without any work from the genus formula, since the genus is always an integer! Hence obtain another solution to Exercise 8.2.

Exercise 12.6. Suppose that $f : X \to Y$ is a purely inseparable morphism. Show that X and Y have the same genus. Are they necessarily isomorphic?

Exercise 12.7. Generalize Proposition 12.3 to function fields in an arbitrary number of variables.

13. Elliptic Curves and Abelian Varieties

Elliptic curves are an important topic in number theory. A useful basic reference is Silverman [27]. An *elliptic curve* is a complete nonsingular curve of genus one. Throughout this chapter, E will be an elliptic curve, and F will be its function field.

Proposition 13.1. *There exists an $\omega \in \Omega_{F/K}$, unique up to constant multiple, such that $(\omega) = 0$.*

Proof. By Proposition 12.18, the degree of (ω) is zero, so $(\omega) = 0$ if and only if $(\omega) \geq 0$. However by Proposition 12.17, the space of ω such that $(\omega) \geq 0$ is one-dimensional. ■

Proposition 13.2. *(i) For any divisor D on E, $l(D) = \deg(D) + l(-D)$.*

(ii) If $\deg(D) > 0$ then $l(D) = \deg(D)$, and if $\deg(D) < 0$, then $l(D) = 0$.

(iii) If D is a divisor of degree 1, there exists a unique point $P \in E$ such that $D - P$ is the divisor of a function.

(iv) If P_1 and $P_2 \in E$, then $P_1 - P_2$ is the divisor of a function if and only if $P_1 = P_2$.

Proof. The Riemann-Roch Theorem asserts that $l(D) = \deg(D) + l(K - D)$, where K is a canonical divisor. According to Proposition 13.1, the zero divisor is in the canonical class, whence (i).

By Proposition 12.6, $l(D) = 0$ if $\deg(D) < 0$, whence (ii).

As for (iii), it is enough to show that if $\deg(D) = 1$, there is a unique P such that $P - D$ is the divisor of a function. We have $l(-D) = 0$ by (ii), and so by (i), $L(D)$ is one-dimensional. Let f be a nonzero element of $L(D)$. Then $(f) + D$ is an effective divisor of degree 1, i.e. $(f) + D = P$ for some point $P \in X$. This proves the existence part. As for uniqueness, if $Q - D$ is the divisor of a function (g), then $(g) + D = Q$ is an effective divisor, so $g \in L(D)$. Therefore g is a constant multiple of f, so $(g) = (f)$ and $Q = P$.

Applying (iii) with $D = P_1$, there is a unique P such that $P_1 - P$ is the divisor of a function; since 0 is the divisor of a function, this $P = P_1$. ■

The most important property of elliptic curves is that they are groups. Our first task is to construct the group law. This requires selecting one point $O \in E$ to be the origin. We will fix this element once and for all. We will define a binary operation $\oplus$ on E with respect to which E is a commutative group. (We will use the symbol $\oplus$ instead of $+$ since $P + Q$ will continue to denote the divisor which is the sum of P and Q.)

If $P, Q \in E$, then applying Proposition 13.2 (iii) with $D = P + Q - O$, there is a unique R such that $P + Q - O - R$ is the divisor of a function. We define $P \oplus Q$ to be this R. Applying Proposition 13.2 (iii) with $D = 2O - P$, there is a unique Q such that $2O - P - Q$ is the divisor of a function. We denote by $\ominus P$ this point Q, and by $P \ominus Q$ the difference $P \oplus (\ominus Q)$.

Proposition 13.3. *With the group law thus defined, E is an abelian group.*

Proof. It is clear that $P \oplus Q = Q \oplus P$ and that $P \oplus O = O \oplus P = P$. Let us show that this group law is associative. Let P, Q and $R \in E$. Applying Proposition 13.2 (iii) with $D = P + Q + R - 2O$, there is a unique point S such that $P + Q + R - 2O - S$ is the divisor of a function. Now $P + Q - O - (P \oplus Q)$ and $(P \oplus Q) + R - O - ((P \oplus Q) \oplus R)$ are both divisors of functions, and adding these, we see that $P + Q + R - 2O - ((P \oplus Q) \oplus R)$ is the divisor of a function, so $S = ((P \oplus Q) \oplus R)$. Similarly so $S = (P \oplus (Q \oplus R))$. This proves that the group law is associative.

Since $P + (\ominus P) - 2O \sim O$, we have $P \oplus (\ominus P) = O$, so $\ominus P$ is an inverse of P with respect to this multiplication. Consequently E is a group. ■

Let X be a curve, and let $\mathcal{D}$ be its group of divisors. Let G be an abelian group. Then any mapping $i : X \to G$ may be extended to a map $i^* : \mathcal{D} \to G$ in a unique way so as to be a group homomorphism, namely $i^*(\sum P_i) = \sum i(P_i)$. In particular, let E be an elliptic curve. Then the identity map $E \to E$ may be extended to a map $c : \mathcal{D} \to E$ on divisors which is a group homomorphism.

Proposition 13.4. *A divisor D on E is the divisor of a function if and only if $\deg(D) = 0$ and $c(D) = O$.*

Proof. First we prove that if D is an effective divisor of degree $d > 0$, and if $P \in E$, then $D - (d - 1).O - P$ is the divisor of a function if and only if $P = c(D)$. By Proposition 13.2 (iii), there is a unique point P such that $D - (d-1)O - P \sim 0$. Hence it suffices to show that $D - (d-1)O - c(D) \sim 0$. If $d = 1$ then $D = Q$ for some Q, and since $c(Q) = Q$, this is true. Assume that $d > 1$ and that the assertion is true for divisors of lower degree. Write $D = D' + Q$ where D' is effective, and by induction the assertion is true for D', so denoting $c(D') = R$, we have $D' - (d - 2)O - R \sim 0$. By definition of the group law $\oplus$ we have $Q + R - O - (Q \oplus R) \sim 0$, and adding these two divisors gives $D' + Q + -(d-1)O - (Q \oplus R) \sim 0$, i.e. $D - (d-1)O - (Q \oplus R) \sim 0$. Now $Q \oplus R = Q + c(D') = c(Q + D') = c(D)$, which completes the induction.

Now we may prove the Proposition. By Proposition 12.1, the divisor of a function has degree zero, so we may assume that D has degree zero. Write $D = D_1 - D_2$ where D_1 and D_2 are effective divisors, and $\deg(D_1) = \deg(D_2) = d$. Let $P_1 = c(D_1)$ and $P_2 = c(D_2)$, so $c(D) = P_1 \ominus P_2$. We have already shown that $D_1 - (d-1).O - P_1$ and $D_2 - (d-1).O - P_2$ are the divisors of functions. Hence $D = D_1 - D_2 \sim P_1 - P_2$, and so D is the divisor of a function if and only if $P_1 - P_2$ is the divisor of a function. By Proposition 13.2 (iv), this is equivalent to $P_1 = P_2$, or $c(D) = O$. ∎

Proposition 13.5. *Let X and Y be complete nonsingular curves, and let $\phi : Y \to X$ be a separable morphism of degree two. Then the map $\tau : Y \to Y$ which interchanges the two points in the fiber $\phi^{-1}(P)$ for $P \in X$ (or fixes the unique point in the fiber if P is a point of ramification) is a morphism.*

Proof. Let F and K be the respective function fields of X and Y. Then ϕ induces an embedding of F into K, and K/F is a separable extension of degree two, hence Galois. Let $t : K \to K$ be the nontrivial element of the Galois group, and let τ be the morphism $Y \to Y$ which corresponds to this field map by Theorem 8.7. It follows from Proposition 9.16 that $\mathrm{Gal}(K/F)$ is transitive on the fiber $\phi^{-1}(P)$, which consists of two points unless P is a point of ramification. It follows that τ interchanges these two points. ∎

Proposition 13.6. *The maps $P \to \ominus P$ and (for fixed $Q \in E$) $P \to P \oplus Q$ and $P \to Q \ominus P$ are morphisms $E \to E$. If f is a nonconstant function in $L(O+Q)$, then the rational function field $k(f)$ is of index two in F, and is the field of functions invariant under the automorphism $P \to Q \ominus P$ of E.*

We call the map $P \to P \oplus Q$, the *translation map* with respect to Q, which is a morphism by this proposition.

Proof. It is sufficient to show that the map $P \to Q \ominus P$ is a morphism, because $P \to \ominus P$ is a special case of this map, and $P \to P \oplus Q$ is the composition of $P \to \ominus P$ and $P \to Q \ominus P$. The determination of the subfield of functions invariant by $P \to Q \ominus P$ will be a byproduct of the proof. We will denote the map $P \to Q \ominus P$ by τ.

Observe that by Proposition 13.2 (ii), $l(O+Q) = 2$. Let f be a nonconstant element of $L(O + Q)$. Since $L(O + Q)$ contains both k and $k.f$, and since its dimension is two, $L(O + Q) = k + k.f$, and so $k(f)$ contains $L(O + Q)$. By Proposition 12.4, $[F : k(f)] = 2$. Since we are assuming that the characteristic is not two, $F/k(f)$ is a separable extension. Let $\phi : E \to \mathbf{P}^1$ be the corresponding covering map. We will show that for every $P \in E$, $\{P, \tau(P)\} = \phi^{-1}(R)$ for some $R \in \mathbf{P}^1$. This will prove that τ is the map described in Proposition 13.5, which will prove that τ is a morphism.

Note that by Proposition 13.2 (i), any nonconstant element of $L(O + Q)$ must have at least two poles, so any nonconstant element of $L(O + Q)$ must have poles at both O and Q. By Proposition 13.4, $P + \tau(P) - O - Q$ is the divisor of a function g, which is therefore in $k + k.f$. There are now two cases.

In the first case, either $P = O$ or $P = Q$, in which case we may take $g = 1$. In this case, we consider the fiber $\phi^{-1}(\infty)$. Note that $v_P(f) = -1$ if either $P = O$ or $P = Q$, and so O and Q are both poles of f. This means that $\phi^{-1}(\infty) = \{O, Q\}$, as required.

In the second case, $P \neq O$ and $P \neq Q$. In this case we may take $g = f - a$ for some constant a. Now consider the fiber $\phi^{-1}(a)$ with $a \in \mathbf{A}^1$. Note that $v_P(f - a) = v_{\tau(P)}(f - a) = 1$. Thus $\phi^{-1}(a) = \{P, \tau(P)\}$, as required. ∎

Proposition 13.7. *The function field F of an elliptic curve E is generated by two elements x and y subject to the relation*

$$(13.1) \qquad y^2 + a_1 xy + a_3 y = x^3 + a_2 x^2 + a_4 x + a_6,$$

where a_1, a_2, a_3, a_4 and $a_6 \in k$. We may take the origin O to be the unique point where x and y have poles (of order two and three, respectively). In this case we may identify $E - O$ with the affine variety whose coordinate ring is $k[x, y]$. The affine curve in $\mathbf{A}^2$ whose equation is Eq. (13.1) will be nonsingular, and may be identified with $E - O$.

Equation (13.1) is called the *Weierstrass form* of the elliptic curve. There is no a_5—this numbering of $a_1, \cdots, a_6$ is standard. Note that we are *not* asserting that the affine curve Eq. (13.1) is nonsingular for an arbitrary choice of the a_i. The proof will show that if the a_i are such that Eq. (13.1) is singular, then complete nonsingular curve whose function field is $k(x, y)$ has genus zero.

Proof. By Proposition 13.2 (ii), $l(3O) = 3$, while $l(2O) = 2$. Let us pick nonconstant functions x and y such that $x \in L(2O)$, and $y \in L(3O) - L(2O)$. We have by Proposition 12.4 that $[F : k(x)] = 2$, and $y \notin k(x)$ since $[F : k(y)] = 3$, so $F = k(x, y)$.

Note that $1, x, x^2, x^3, y, xy$ and y^2 are all in $L(6O)$. Since $l(6O) = 6$, they are linearly dependent. There exist constants a_i, c and d therefore, not all zero, such that

$$(13.2) \qquad cy^2 + a_1 xy + a_3 y - dx^3 - a_2 x^2 - a_4 x - a_6 = 0.$$

On the other hand, $1, x, x^2, y$ and xy are linearly independent, because if a linear relation pertains among them, then we may express y in terms of x, so that $F = k(x)$, which is a contradiction since E is not rational. Therefore either c or d is nonzero. However if either c or d is nonzero both must be, since cx^3 and dy^2 are the unique terms in Eq. (13.2) having poles of order six at O, which must cancel. Multiplying x and y by constants, we may arrange things so that $c = 1$ and $d = -1$. Hence Eq. (13.1) is valid.

Now let us show that the affine curve Eq. (13.1) has no singularities. Making a change of variables $x \to x - a$, $y \to y - b$, we may move the singularity to the origin. By Proposition 9.14, denoting

$$f(X, Y) = Y^2 + a_1 XY + a_3 Y - X^3 - a_2 X^2 - a_4 X - a_6,$$

the three polynomials f, $\partial f / \partial X$ and $\partial f / \partial Y$ must all vanish at $(0, 0)$, so $a_6 = a_4 = a_3 = 0$. We will show that $k(x, y) = k(t)$ where $t = y/x$. Indeed, dividing Eq. (13.1) by x^2 gives $t^2 + a_1 t = x + a_2$, so $x \in k(t)$ and consequently $y = tx \in k(t)$. Since the rational function field $k(t)$ has genus zero, this contradicts the assumption that $k(x, y)$ is the function field of a curve of genus one.

Lastly, we must justify the claim that the points of the affine curve Eq. (13.1) correspond to the points of $E - O$. Since the affine curve U in $\mathbf{A}^2$ given by Eq. (13.1) is nonsingular, its coordinate ring $k[x, y]$ is integrally closed by Proposition 6.21. As in Chapter 8, the ring $k[x, y]$ is the intersection of all local rings of E which contain it, and it may be identified with an affine open set in E, and what we must show is that its complement consists of the single point O. In order to prove this, we note that corresponding to the inclusion $k(x) \to k(x, y)$ we have a morphism $\phi : E \to \mathbf{P}^1$ by Theorem 8.7. We first show that $\phi^{-1}(\mathbf{A}^1) = U$. It is clear that ϕ induces finite dominant morphism $U \to \mathbf{A}^1$ corresponding to the inclusion of coordinate rings $k[x] \to k[x, y]$. What we must show is that there are no points outside U in $\phi^{-1}(\mathbf{A}^1)$. Suppose that $P \in E$ such that $\phi(P) \in \mathbf{A}^1$. Let w be the valuation of $k(x, y)$ corresponding to the discrete valuation ring $\mathcal{O}_P \subset k(x, y)$. Since $\phi(P) \subset \mathbf{A}^1$, the function $x \in \mathcal{O}_P$. By Eq. (13.1) it follows that $y^2 + a_1 xy + a_3 y \in \mathcal{O}_P$. We will show that $y \in \mathcal{O}_P$ also. Indeed, if $y \notin \mathcal{O}_P$, then $w(y) < 0$. Now $w(y^2) < w(a_1 xy + a_3 y)$ because $w(a_1 xy + a_3 y) = w(a_1 x + a_3) + w(y) \geq w(y)$ and $w(y^2) < w(y)$. By Exercise 5.11 this implies that $w(y^2 + a_1 xy + a_3 y) < 0$, contradicting our observation that $y^2 + a_1 xy + a_3 y \in \mathcal{O}_P$. Now $x, y \in \mathcal{O}_P$, so $\mathcal{O}_P$ contains $k[x, y]$. This means that $P \in U$.

Since $\phi(U) = \mathbf{A}^1$, we see that $E - U = \phi^{-1}(\infty)$. We will show that ∞ is a ramified point for the morphism ϕ. Since the degree of ϕ is two, it will follow from Theorem 9.23 that the fiber over ∞ consists of a single point. Let $e = e(P|\infty)$ be the ramification index of P over ∞. Let w be the valuation of $k(x, y)$ corresponding to P. Since $\phi(P) = \infty$, by Proposition 9.11 we have $w(x) = -e < 0$. By Eq. (13.1) and Exercise 5.11, we have

$$w(y^2 + a_1 xy + a_3 y) = w(x^3 + a_2 x^2 + a_4 x + a_6) = -3e.$$

This leads to a contradiction of Exercise 5.11 unless $e = 2$ and $w(y) = 3$, proving that $e = 2$, so $P = O$ is the unique point in the fiber of ϕ above ∞. ■

Proposition 13.7 illustrates the discussion in Chapter 12 on the connection between linear systems and projective embeddings. The three functions $1, x$ and y form a basis of $L(D)$ where $D = 3O$. We get an embedding $E \to \mathbf{P}^2$ which may be described as follows. If $P \in E$, there exists $\lambda \in k(x, y)$ such that $\lambda, \lambda x, \lambda y \in \mathcal{O}_P$, and not all are in the maximal ideal (Proposition 4.16) $\mathfrak{m}_P$ of $\mathcal{O}_P$. Since the composition $k \to \mathcal{O}_P \to \mathcal{O}_P/\mathfrak{m}_P$ is an isomorphism, let $\phi_P : \mathcal{O}_P \to k$ be composition $\mathcal{O}_P \to \mathcal{O}_P/\mathfrak{m}_P \cong k$. We may then consider the point $\psi(P)$ in $\mathbf{P}^2$ with homogeneous coordinates $\phi_P(\lambda) : \phi_P(\lambda x) : \phi_P(\lambda y)$. Note that $\psi(P)$ does not depend on the choice of λ.

Recalling that the points of $\mathbf{P}^2$ correspond to lines through the origin in $\mathbf{A}^3$, a *line* in $\mathbf{P}^2$ is the set of points corresponding to lines through the origin lying in some fixed vector subspace of $\mathbf{A}^3$ of codimension one. A set of points is called *collinear* if they lie on a line.

Proposition 13.8. *The map* $\psi : E \to \mathbf{P}^2$ *is a morphism. The curve E is isomorphic to its image $\psi(E) \subset \mathbf{P}^2$. Identifying $E \subset \mathbf{P}^2$ with its image, three distinct points $P, Q, R \in E$ are collinear in $\mathbf{P}^2$ if and only if $P + Q + R = 0$.*

This shows that $L(3O)$ is indeed a linear system cut out by hyperplanes as discussed in Chapter 12. We will leave the proof to the reader. It may be skipped because we will not use this result in the sequel, though it is closely related to the proof of Proposition 13.12.

Proof. We leave this to the reader (Exercise 13.2). ■

Proposition 13.9. *With notations as in* Proposition 13.7, *the points P and Q of $E - O$ have equal x-coordinates if and only if $Q = P$ or $\ominus P$.*

Proof. If a is the x-coordinate of P, consider the divisor of the function $x - a$. This function has a pole of order 2 at O (since x does) and so it has two zeros. If these are P and Q, then $(x - a) = P + Q - 2O$. By definition $Q = \ominus P$. Hence P and $\ominus P$ are the only points with x-coordinate a. ■

Proposition 13.10. *If the characteristic of k is not equal to 2 or 3, then the Weierstrass equation of an elliptic curve E may be chosen in the form $y^2 = x^3 + ax + b$. A necessary and sufficient condition for the function field $k(x, y)$, where x and y satisfy $y^2 = x^3 + ax + b$ to have genus one is that the three roots of the cubic polynomial $x^3 + ax + b$ are distinct.*

Proof. Starting with the Weierstrass equation Eq. (13.1), replacing y by $y - \frac{1}{2}(a_1 x + a_3)$ gives us a new Weierstrass equation with a_1 and a_3 zero. Now replacing x by $x - \frac{1}{3}a_2$ gets rid of a_2, and the equation has the required form.

We next show that the roots of $x^3 + ax + b$ must be distinct. If two of them are equal, we may make a linear change of variables in x to move the multiple root to $x = 0$ and the other one to either $x = 0$ if the root at $x = 0$ has multiplicity three, or to $x = -1$ if the root at $x = 0$ has multiplicity two. After this, the Weierstrass equation then has the form $y^2 = x^3$ or $y^2 = x^2(x + 1)$. However in both these cases, as in the proof of Proposition 13.7, $k(x, y) = k(t)$ where $t = y/x$, so the function field cannot have genus one.

On the other hand, if the roots of $x^3 + ax + b$ are distinct, we proved in Example 12.21 that the function field $k(x, y)$ has genus one. (See also Exercise 12.1.) ∎

Lemma 13.11. *Let G be an infinite abelian group. Then there exist three distinct, nonzero elements x, y, $z \in \Gamma$ such that $x + y + z = 0$.*

Proof. Suppose not. Then for any two distinct, nonzero elements x and y one of $x + y$, $2x + y$ or $x + 2y$ is zero, since otherwise we could take $z = -(x + y)$.

Let $H = \{t \in G | 2t = 0\}$. Then H is a proper subgroup of G. Indeed, if $2t = 0$ for all $x \in G$, then for any two distinct nonzero elements x and y one of $x + y$, y or x is zero. Since G is infinite, we may choose x to be any nonzero element and y to be any element distinct from 0, x and $-x$ to obtain a contradiction.

If $2x \neq 0$ for some x, let $S = G - \{x, -x, -2x\}$. Then for any $y \in S$, both $x + y$ and $2x + y \neq 0$, so $x + 2y = 0$. Hence if $y_1, y_2 \in S$ then $y_1 - y_2 \in H$, so S is contained in a single coset of H, which is impossible since the complement of S is finite and H is a proper subgroup. ∎

Proposition 13.12. *If E is an elliptic curve, then the map $m : E \times E \to E$ given by $m(P, Q) = P \oplus Q$ is a morphism.*

Proof. It is sufficient to show that $(P, Q) \to \ominus P \ominus Q$ is a morphism $E \times E \to E$, since $P \to \ominus P$ is a morphism by Proposition 13.6.

Let

$$\Omega_1 = \{(P, Q) \in E \times E | P, Q, O \text{ are all distinct and } x(P) \neq x(Q)\}.$$

Clearly Ω_1 is open nonempty, since it is the intersection of two nonempty open sets, $\{(P, Q) | P, Q, O \text{ distinct}\}$ and $\{(P, Q) | x(P) \neq x(Q)\}$. We will define a morphism $\sigma : \Omega_1 \to E - O$ which will eventually turn out to satisfy $\sigma(P, Q) = \ominus P \ominus Q$. Let $(P, Q) \in \Omega_1$, and identify P and Q with the corresponding solutions $P = (x_P, y_P)$ and $Q = (x_Q, y_Q)$ to the Weierstrass equation

Eq. (13.1), consider the function

$$f = (y - y_Q) - (x - x_Q)\left(\frac{y_P - y_Q}{x_P - x_Q}\right).$$

Since y has a triple pole at O and x has a double pole, this has a triple pole at O. Moreover, it also vanishes at P and Q by construction. It follows from Proposition 13.4 that there is a unique point $R \in E$ such that $P + Q + R - 3O$ is the divisor of a function, namely $R = \ominus P \ominus Q$. Hence $(f) = P + Q + R - 3O$. We see that the line $f = 0$ in $\mathbf{A}^2$ must intersect the curve Eq. (13.1) in three points P, Q and $\ominus P \ominus Q$.

We define a morphism $\sigma : \Omega_1 \to E - O$ as follows. Substituting

$$y = y_Q + (x - x_Q)\left(\frac{y_P - y_Q}{x_P - x_Q}\right)$$

into Eq. (13.1) gives a cubic equation in x:

(13.3) $$x^3 + Ax^2 + Bx + C = 0$$

two of whose roots, x_P and x_Q are known. The coefficients A, B and C are polynomials in x_P, x_Q, y_P, y_Q. The third root $A - x_P - x_Q$ is therefore the x-coordinate of $\ominus P \ominus Q$, and the y-coordinate is then found by substituting the x-coordinate back into the linear equation $f(x, y) = 0$. This gives the coordinates of a point of $E - O$ as a rational functions in the coordinates of P and Q—we call this point $\sigma(P, Q)$. It is clear that $\sigma : \Omega_1 \to E - O$ is a morphism.

Suppose now that $(P, Q) \in \Omega_1$ and that P, Q and $R = \ominus P \ominus Q$ are distinct. In this case we will show that $\sigma(P, Q) = R$. Indeed, we have seen that the divisor of f is $P + Q + R - 3O$, so f vanishes at P, Q and R. It is clear that the x-coordinates of each point P, Q and R must satisfy Eq. (13.3), since the coordinates of each point satisfies both the Weierstrass equation Eq. (13.1) and the linear equation $f = 0$. The x-coordinates of P, Q and R must all be distinct by Proposition 13.9. For example, P and R cannot have equal x-coordinates because $P \neq R$ (by assumption) and $P \neq \ominus R$ since $Q \neq O$, while P and Q do not have equal x-coordinates because $(P, Q) \in \Omega_1$. Since Eq. (13.1) has exactly three roots, the third distinct root besides x_P and x_Q must be x_R, and it is now clear that $\sigma(P, Q) = R$.

Now let Ω be the subset of $(P, Q) \in \Omega_1$ such that $\sigma(P, Q) \neq P, Q$. Clearly Ω is open, and it is nonempty since by Lemma 13.11 there exist distinct nonzero P, Q and $R \in E$ such that $P \oplus Q \oplus R = O$; then $(P, Q) \in \Omega$. On the nonempty open set Ω, the morphism σ agrees with the map $(P, Q) \mapsto \ominus P \ominus Q$.

This shows that the restriction of $(P, Q) \mapsto \ominus P \ominus Q$ to a neighborhood of a point (P_0, Q_0) is a morphism of affine varieties provided $(P_0, Q_0) \in \Omega_1$. Since the group translation maps of E acts transitively (see Proposition 13.6), we may move an arbitrary point of $E \times E$ into Ω_1 by a pair of translations: choose (U, V) so that $(P_0 \oplus U, Q_0 \ominus V) \in \Omega_1$. Then $(P, Q) \mapsto (P \oplus U, Q \oplus V)$ is a morphism $E \times E \to E \times E$ by Proposition 13.6, so $(P, Q) \mapsto \ominus P \ominus U \ominus Q \ominus V$ is a morphism $E \times E \to E$ in a neighborhood of (P_0, Q_0). Composing this with another translation shows that $(P, Q) \mapsto \ominus P \ominus Q$ is locally a morphism of affine varieties in a neighborhood of any point of $E \times E$. By Proposition 7.8, it follows that $(P, Q) \mapsto \ominus P \ominus Q$ is a morphism. ■

An *algebraic group* is an algebraic variety G which is also a group, such that the maps $(x, y) \mapsto xy$ from $G \times G \to G$ and $x \mapsto x^{-1}$ from $G \to G$ are morphisms.

Proposition 13.13. *An algebraic group is a smooth variety.*

Proof. If G is an algebraic group, define a transitive action of G on itself, in which the group element $x \in G$ acts by the morphism $\lambda_x : G \to G$ defined by $\lambda_x(y) = x^{-1}y$. Since G admits a transitive group of automorphisms, all its local rings are isomorphic. Hence if one point is smooth, every point is smooth. But by Theorem 6.15, there exists at least one smooth point on every variety, so G is smooth. ■

We have just verified that an elliptic curve is an algebraic group. As another example, consider $GL(n)$, the group of $n \times n$ invertible matrices with coefficients in k. This is a principal open set in $\mathbf{A}^{n^2}$, being the complement of the locus of vanishing of the determinant, so this is an affine variety by Exercise 3.1. The group law and the inverse map are obviously morphisms in this case. There is one important difference between $GL(n)$ and the elliptic curve, however: $GL(n)$ is affine, while the elliptic curve is complete. An *abelian variety* is a complete algebraic group. Thus elliptic curves are abelian varieties. Many of the important features of elliptic curves generalize to abelian varieties.

Proposition 13.14. *Let X be an abelian variety. Then the group X is commutative.*

Proof. Define a morphism $f : X \times X \to X$ by $f(x, y) \to xyx^{-1}y^{-1}$. Then if $y = O$ is the origin in X, the image of f consists of a single point. Hence by Proposition 7.20 there is a morphism $g : X \to X$ such that $f(x, y) = g(y)$ for all $(x, y) \in X \times Y$. We have $g(y) = f(O, y) = O$, so $f(x, y) = g(y) = O$ for all x, y. This proves that X is commutative. ■

Having established the commutativity, we will denote the group law in X additively, using the symbol $\oplus$ as before for elliptic curves. If X is an abelian variety, let $O \in X$ denote the identity element.

Proposition 13.15. *Let X and Y be abelian varieties, and let $\phi : X \to Y$ be any morphism such that $\phi(O) = O$. Then ϕ is a group homomorphism.*

Proof. Define $f : X \times X \to Y$ by $f(x, y) = \phi(x \oplus y) \ominus \phi(x) \ominus \phi(y)$. Then if $y = O$ is the origin in X, the image of f consists of a single point. Hence by Proposition 7.20 there is a morphism $g : X \to X$ such that $f(x, y) = g(y)$ for all $(x, y) \in X \times X$. We have $g(y) = f(O, y) = O$, so $f(x, y) = g(y) = O$ for all x, y. This proves that ϕ is a group homomorphism. ■

If X and Y are abelian varieties, a *homomorphism* is by definition a morphism $X \to Y$ which maps O to O. By Proposition 13.15, it is a group homomorphism. We will denote the set of homomorphisms $X \to Y$ as $\mathrm{Hom}(X, Y)$, or $\mathrm{End}(X)$ if $X = Y$. This set has the natural structure of a group, namely, if $\phi_1, \phi_2 \in \mathrm{Hom}(X, Y)$, define $\phi_1 \oplus \phi_2$ by $(\phi_1 \oplus \phi_2)(P) = \phi_1(P) \oplus \phi_2(P)$. To see that this is morphism, observe that it is the composition of the map $X \to Y \times Y$ given by $P \to (\phi_1(P), \phi_2(P))$ (which is a morphism by Theorem 7.11) with the multiplication map $Y \times Y \to Y$, which is a morphism by the definition of an algebraic group.

The additive group $\mathrm{End}(X)$ has the further structure of a ring, in which the multiplication law is the composition of endomorphisms. As with any ring, there is a unique (unitary) ring homomorphism $\mathbb{Z} \to \mathrm{End}(X)$. If $m \in \mathbb{Z}$, we will denote by $[m]$ the image of m in $\mathrm{End}(X)$. Thus if $m > 0$, $[m]P = P \oplus \ldots \oplus P$ (m times). If there exist endomorphisms of X not in the image of $\mathbb{Z}$, we say the abelian variety has *complex multiplications*. This may or may not be the case.

Proposition 13.16. *Let X and Y be abelian varieties, and let $\phi : X \to Y$ be a morphism. Then the connected component X_0 of the identity in the kernel of ϕ, and image Y_0 of ϕ are abelian varieties, and $\dim(X) = \dim(X_0) + \dim(Y_0)$.*

Proof. The image Y_0 is a closed subgroup of Y since Y_0 is complete by Proposition 7.16 (iii), so it is an algebraic group. It is also complete by Proposition 7.16 (iii), and so it is an abelian variety. The kernel is closed since it is the preimage of the closed set $\{O\}$ under a morphism, and the connected component X_0 of the identity is complete by Proposition 7.16 (ii), so it too is an abelian variety.

To prove the dimension assertion, we may replace Y by Y_0, hence assume that ϕ is surjective, and that $Y_0 = Y$. Now by Exercise 7.9, there exist fibers of ϕ with dimension exactly $\dim(X) - \dim(Y_0)$. However, the fibers of ϕ are just the cosets of the kernel, and each has the same dimension, $\dim(X_0)$. Hence $\dim(X) = \dim(X_0) + \dim(Y_0)$. ∎

If X and Y are abelian varieties, a homomorphism $\phi : X \to Y$ is called an *isogeny* if the dimensions of X and Y are the same, and ϕ is surjective. It follows from Proposition 13.16 that this is equivalent to $\dim(X) = \dim(Y)$, and the kernel of ϕ being finite. It is clear that a morphism of elliptic curves is either zero, or is an isogeny.

Proposition 13.17. *(i) Let E and E' be elliptic curves, and let F, F' be their respective function fields. Let $\phi : E \to E'$ be a separable isogeny. Then ϕ induces an injection of F' into F. Identifying F' with its image in F, the extension F/F' is Galois, and the Galois group is isomorphic to the kernel of ϕ, which is finite.*

(ii) Suppose that E'' is a second elliptic curve, and $\psi : E'' \to E$ an isogeny such that $\ker(\phi) \subseteq \ker(\psi)$. There is induced a homomorphism of elliptic curves $\rho : E' \to E''$ such that $\psi = \rho \circ \phi$.

This is true also for abelian varieties.

Proof. To prove (i), recall that by Theorem 9.23, for any separable morphism $f : Y \to X$ of curves, the degree of the morphism is equal to the cardinality of all but a finite number of fibers $f^{-1}(P)$ ($P \in X$), the exceptions being the fibers above points of ramification. In the case of the morphism $\phi : E \to E'$ of elliptic curves, every fiber has the same cardinality, since the fibers are just the cosets of the kernel. Hence the cardinality of the kernel is equal to the degree of the cover. Note that if $P \in \ker(\phi)$, then the morphism $\lambda_P : Q \mapsto Q \ominus P$ of E to itself satisfies $\phi = \phi \circ \lambda_P$. Therefore the induced automorphism λ_P^* of the function field F of E leaves invariant the function field F' of E', injected by ϕ^*. This means that $P \mapsto \lambda_P^*$ is an injective homomorphism from $\ker(\phi)$ into the group of field automorphisms of F over F'. Since the cardinality of $\ker(\phi)$ equals the degree of the field extension, this shows that $\{\lambda_P^* | P \in \ker \phi\} = \mathrm{Gal}(F/F')$, and that the extension is Galois.

As for (ii), first assume that ψ is separable. Let F'' be the function field of E'', which we regard as a subfield of F. Our hypothesis shows that the Galois group of F/F' is a subgroup of the Galois group of F/F'', and hence there is induced a field homomorphism $F'' \to F'$. The corresponding morphism $\rho : E' \to E''$ of elliptic curves then satisfies $\psi = \rho \circ \phi$. Thus (ii) is valid in the special case where ψ is separable.

In the general case, we regard F'' as a subfield of F. Let F_1 be the maximal separable extension of F in F''. If $q = [F : F_1]$, then by Proposition 12.3, $F_1 = F^q$. The map $\mathcal{F}_q : F \to F_1$ is an isomorphism. It does not induce a morphism of curves, since it is not the identity map on the subfield k of constants. Nevertheless, the fact that the fields F and F_1 are isomorphic implies that the genus of the curve E_1 corresponding to F_1 is one, so E_1 is an elliptic curve. Corresponding to the inclusions $F'' \to F_1 \to F$, we have a factorization of the morphism ψ as $\psi = \psi_2 \circ \psi_1$, where $\psi_1 : E \to E_1$ is purely inseparable, and $\psi_2 : E_1 \to E''$ is separable. Let $F_1' = F'\,^q$. Clearly $\phi^*(F_1') \subseteq F_1$, and so we have a commutative diagram

$$
\begin{array}{ccc}
F & \xleftarrow{\ \phi^*\ } & F' \\
\uparrow & & \uparrow \\
F_1 & \xleftarrow{\ \phi_1^*\ } & F_1'
\end{array}
$$

where ϕ_1^* is the restriction of ϕ^* to F_1'. This gives rise to a map of elliptic curves:

$$
\begin{array}{ccc}
E & \xrightarrow{\ \phi\ } & E' \\
\downarrow & & \downarrow \\
E_1 & \xrightarrow{\ \phi_1\ } & E_1'
\end{array}
$$

The vertical morphisms are bijections (though not isomorphisms since the inverse maps are not morphisms). Since $\ker(\phi) \subseteq \ker(\psi)$, it follows that $\ker(\phi_1) \subseteq \ker(\psi_2)$. Moreover, ϕ_1 and ψ_2 are both separable, and so we may invoke the special case just discussed of ψ separable, and conclude that there exists a morphism $\rho' : E_1' \to E''$ such that $\psi_2 = \rho' \circ \phi_1$. Now we take $\rho = \rho' \circ \psi_1'$. ■

Next we would like to prove that if E is an elliptic curve, then the differential form ω of Proposition 13.1, which has no zeros or poles, is translation invariant. More generally, we will prove that there exist translation invariant differential forms on any algebraic group. This is the key step in constructing the Lie algebra of an algebraic group.

If X is a variety, $P \in X$, we define the tangent space $T_P(X)$ just as for affine varieties. Thus $T_P(X)$ is the space of linear maps $D : \mathcal{O}_P(X) \to k$, which are *derivations* in the sense that $D(fg) = f(P)\,D(g) + g(Q)\,D(f)$.

We will say that a derivation $D : F \to F$ of the function field F of any variety X is *regular* if whenever $f \in F$ is regular on an open set U, Df is regular on the same open set.

Let G be an algebraic group with function field F. We do not assume that G is commutative, so we will write its group law multiplicatively. The identity element of G will be denoted by e. Let $m : G \times G \to G$ be the multiplication map, so $m(x, y) = xy$. If $x \in G$, let $\lambda_x : G \to G$ be the morphism $\lambda_x(y) = x^{-1}y$. Let $L = T_e(G)$ be the tangent space at the identity.

If $D \in L$, we will associate with D a left invariant derivation $\widehat{D} : F \to F$. This means that $D(fg) = f\,Dg + g\,Df$, and that if $x \in G$, and if $\lambda_x^* : F \to F$ is the induced map of the function field, then $D \circ \lambda_x^* = \lambda_x^* \circ D$.

Proposition 13.18. *With notation as above, for $f \in F$, let U be an open subset of G on which f is regular. If $x \in U$, then $\lambda_{x^{-1}}f \in \mathcal{O}_e$, and the function $\widehat{D}f$, defined by $(\widehat{D}f)(x) = D(\lambda_{x^{-1}}f)$ is regular on U. Thus $\widehat{D}f \in F$. The map $\widehat{D} : F \to F$ is a regular left invariant derivation of F, and $D \to \widehat{D}$ defines an isomorphism between L and the k-vector space of left invariant regular derivations of F. The dimension of the space of left invariant derivations is equal to the dimension of G.*

With this result, we have constructed the *Lie algebra* of G, for the space of left invariant regular derivations of F is closed under the operation $[D_1, D_2] = D_1 \circ D_2 - D_2 \circ D_1$, and with this bracket operation, this space becomes a Lie algebra. The content of the Proposition is that as a vector space, this Lie algebra may be identified with the tangent space at the identity.

Proof. It is clear that if $x \in U$, then $f \in \mathcal{O}_x(G)$, so $\lambda_{x^{-1}}(f) \in \mathcal{O}_e(G)$, and so $(\widehat{D}f)(x)$ is defined. What requires justification is that $\widehat{D}f$ is a regular function on U. Since $\mathcal{O}(U) = \bigcap_{x \in U} \mathcal{O}_x(G)$, it is sufficient to show that $\widehat{D}f$ is regular at x for each $x \in U$. In other words, for each $x \in U$ we will show that there exists a function in $\mathcal{O}_x(G)$ which agrees with $\widehat{D}f$ in a neighborhood of x, and this implies that $\widehat{D}f \in \mathcal{O}(U)$.

We may assume that the open set U is affine. Let A be the coordinate ring of U. Let V be an affine neighborhood of the identity. Let W be a principal open set in $U \times V$ such that $m(W) \subseteq V$, and such that $(x, e) \in W$. Let A and B be the coordinate rings of U and V, respectively. Then the coordinate ring of $U \times V$ is $A \otimes B$, and the coordinate ring of W is a subring of the field of fractions of $A \otimes B$. Since $m^*(f)$ is regular on W, $m^*(f)$ is a ratio of elements of $A \otimes B$. We may thus write $m^*(f) = (\sum f_i \otimes g_i)/(\sum h_j \otimes k_j)$, where $f_i \otimes g_j$ and $h_j \otimes k_j$ are regular on W, and $\sum_j h_j \otimes k_j$ does not vanish at (x, e). We will show that $\widehat{D}f$ agrees, as a function, with the element

(13.4)
$$\frac{\sum_{i,j} \left(D(g_i)k_j(e) - g_i(e)D(k_j) \right) f_i h_j}{\left(\sum_j k_j(e)h_j \right)^2}$$

of $\mathcal{O}_x$ on a neighborhood of x.

For $(\xi, y) \in W$, we have

$$(\lambda_{\xi^{-1}}f)(y) = f(\xi y) = (m^*f)(\xi, y) = \frac{\sum_i f_i(\xi)g_i(y)}{\sum_j h_j(\xi)k_j(y)},$$

provided the denominator does not vanish. Therefore

$$\lambda_{\xi^{-1}} f = \left(\sum f_i(\xi) g_i \right) / \left(\sum h_j(\xi) k_j \right).$$

For ξ in a neighborhood of x the denominator is a unit in $\mathcal{O}_e$, since $\sum h_j \otimes k_j$ does not vanish at (x, e). Thus applying D, we get

$$(\widehat{D} f)(\xi) = D(\lambda_{\xi^{-1}} f) = D \left(\frac{\sum_i f_i(\xi) g_i}{\sum_j h_j(\xi) k_j} \right).$$

By the quotient rule this equals

$$\frac{\sum_{i,j} \left(D(g_i) \, k_j(e) - g_i(e) \, D(k_j) \right) f_i(\xi) \, h_j(\xi)}{\left(\sum_j k_j(e) h_j(\xi) \right)^2}.$$

This proves that $\widehat{D} f$ agrees with Eq. (13.4) in a neighborhood of x. Hence $\widehat{D} f$ is regular at each $x \in U$, hence $\widehat{D} f \in \mathcal{O}(U)$.

We have shown that $\widehat{D} f$ is a regular function on an open set in G, indeed on the same open set where f is regular, and therefore $\widehat{D} f$ may be interpreted as an element of F.

The fact that $\widehat{D}$ is a derivation of F follows from the fact that the tangent vector D is a derivation $\mathcal{O}_e \to k$. It is straightforward to check that $\widehat{D}$ is translation invariant. Hence $\widehat{D}$ is a regular left invariant derivation of F.

On the other hand, if $\widehat{D}$ is a given regular left invariant derivation of F, we may associate with $\widehat{D}$ a tangent vector $D \in T_e(G)$ as follows. The regularity of $\widehat{D}$ implies that $\widehat{D}$ maps $\mathcal{O}_e(G)$ to itself. Therefore we may define $D : \mathcal{O}_e \to k$ by $Df = (\widehat{D} f)(e)$. It is easily checked that the two correspondences $D \to \widehat{D}$ and $\widehat{D} \to D$ which we have defined are inverses to each other, and therefore $T_e(G)$ is isomorphic as vector space with the space of regular left invariant derivations of F. Since the dimension of $T_e(G)$ is equal to the dimension of G by Proposition 13.13, it follows that the dimension of the space of regular left invariant derivations of F also has the same dimension. ∎

Proposition 13.19. *Let E be an elliptic curve with function field F, and let $\omega \in \Omega_{F/k}$ be the differential of Proposition 13.1, which has neither zeros nor poles. Then ω is invariant under translations $P \to P \ominus Q$. Let $D : F \to F$ be a nonzero invariant regular derivation, whose existence is asserted by Proposition 13.18, and let $\rho : \Omega_{F/k} \to F$ be the induced map such that $D(f) = \rho(df)$, whose existence follows from the universal property of $\Omega_{F/k}$. Then $\rho(\omega) \in k$.*

Proof. Let $\lambda_Q : E \to E$ be the morphism $\lambda_Q(P) = P \ominus Q$. This morphism acts on the function field F, and therefore on $\Omega_{F/k}$. Since the divisor of $\lambda_Q^*(\omega)$ is a translate of the divisor of ω, which is zero, it follows from Proposition 13.1 that $\lambda_Q^*(\omega) = c\,\omega$ for some constant c. What is asserted here is that $c = 1$.

It is sufficient to show that there exists a nonzero translation invariant differential ω', since then the divisor of ω' is translation invariant, and the only translation invariant divisor is 0. Therefore the divisor of ω' is zero and, by Proposition 13.1, ω' is a constant multiple of ω.

The existence of a translation invariant differential follows from Proposition 13.18. For let $D : F \to F$ be a nonzero translation invariant derivation. By the universal property of $\Omega_{F/k}$, there is induced an F-module homomorphism $\rho : \Omega_{F/k} \to F$ such that $D(f) = \rho(df)$. Since by Proposition 11.10, $\Omega_{F/k}$ is one-dimensional as a vector space over F, there exists a differential ω' such that $\rho(\sigma)\,\omega' = \sigma$ for all $\sigma \in \Omega_{F/k}$. It is easily checked that the translation invariance of D implies the translation invariance of ω'.

Now observe that ω is a constant multiple of ω', and $\rho(\omega') = 1$. Therefore $\rho(\omega) \in k$. $\blacksquare$

We recall that if V_1 and V_2 are modules over a ring, and if $V = V_1 \oplus V_2$, then there are maps $i : V_1 \to V$, $j : V_2 \to V$, $p : V \to V_1$ and $q : V \to V_2$ such that

$$(13.5) \quad p \circ i = 1, \qquad q \circ j = 1, \qquad p \circ j = 0, \qquad q \circ i = 0, \qquad i \circ p + j \circ q = 1.$$

The maps are defined by $i(v_1) = (v_1, 0)$, $j(v_2) = (0, v_2)$, $p(v_1, v_2) = v_1$, and $q(v_1, v_2) = v_2$. Conversely, given modules V_1, V_2 and V, if we are supplied with maps i, j, p and q satisfying Eq. (13.5), then there is a unique isomorphism of V with $V_1 \oplus V_2$ realizing the maps i, j, p and q as injection and projection maps. In this setting, we say that V, together with the maps i, j, p and q, is an *external direct sum* of V_1 and V_2.

Proposition 13.20. *Let X and Y be varieties, $x_0 \in X$, $y_0 \in Y$. Let $p : X \times Y \to X$ and $q : X \times Y \to Y$ be the projection maps, and let $i : X \to X \times Y$ and $j : Y \to X \times Y$ be defined by $i(x) = (x, y_0)$, and $j(y) = (x_0, y)$. The tangent space $T_{(x_0, y_0)}(X \times Y)$ is naturally isomorphic to $T_{x_0}(X) \oplus T_{y_0}(Y)$, and in fact the maps $i_* : T_{x_0}(X) \to T_{(x_0, y_0)}(X \times Y)$, $j_* : T_{y_0}(Y) \to T_{(x_0, y_0)}(X \times Y)$, $p_* : T_{(x_0, y_0)}(X \times Y) \to T_{x_0}(X)$ and $q_* : T_{(x_0, y_0)}(X \times Y) \to T_{y_0}(Y)$ induce on $T_{(x_0, y_0)}(X \times Y)$ the structure of an external direct sum.*

Proof. The properties Eq. (13.5) are all clear except the last, that $i_* \circ p_* + j_* \circ q_*$ is the identity map on $T_{(x_0, y_0)}(X \times Y)$. To prove this, let D be an element of this tangent space. Then D is a derivation of the local ring on $X \times Y$ at (x_0, y_0). We will calculate its effect on an element f of this local ring. Observe that

$(D - (i_* \circ p_* + j_* \circ q_*)D)(f) = Df'$, where $f'(x,y) = f(x,y) - f(x,y_0) - f(x_0,y)$. We must show $Df' = 0$. We may add the constant $f(x_0,y_0)$ to f' without changing Df', and we see that what must be shown is that $Df = 0$ if $f(x_0,y) = f(x,y_0) = 0$ for $x \in U$, $y \in V$, where U and V are affine neighborhoods of x_0 and y_0, respectively. Let A and B be the coordinate rings of U and V, and let $\mathfrak{p}$ and $\mathfrak{q}$ be the maximal ideals corresponding to the points x_0 and y_0, respectively. Then if $f(x_0,y) = f(x,y_0) = 0$, we have, by Proposition 4.20 (ii), that $f \in \mathfrak{p} \otimes \mathfrak{q}$. Now let $\mathfrak{m}$ be the maximal ideal of (x_0,y_0) in $A \otimes B$. By Proposition 4.20 (i), we have $\mathfrak{m} = \mathfrak{p} \otimes B + A \otimes \mathfrak{q}$. Observe that $\mathfrak{p} \otimes \mathfrak{q} = (\mathfrak{p} \otimes B)(A \otimes \mathfrak{q}) \subseteq \mathfrak{m}^2$. Since by Proposition 12.1, D vanishes on $\mathfrak{m}^2$, we see that $Df = 0$, as required. ∎

Proposition 13.21. *Let G be an algebraic group with identity element e, X a variety, $x \in X$, and let $\phi, \phi' : X \to G$ be two morphisms such that $\phi(x) = \phi'(x) = e$. Define another morphism $\phi'' : X \to G$ by $\phi''(x) = \phi(x)\phi'(x)$. Let ϕ_*, ϕ'_* and $\phi''_* : T_x(X) \to T_e(G)$ be the induced maps on the tangent spaces. Then $\phi''_* = \phi_* + \phi'_*$.*

Proof. Let $r : X \to G \times G$ be the map $r(\xi) = (\phi(\xi), \phi'(\xi))$. We will show that if $T_{(e,e)}(G,G)$ is identified with $T_e(G) \oplus T_e(G)$ by Proposition 13.20, then $r_* : T_x(X) \to T_{(e,e)}(G,G)$ is the map $D \to (\phi_* D, \phi'_* D)$. Indeed, it is sufficient to check (in the notation of that Proposition) that $p_* \circ r_* = \phi_* : T_x(X) \to T_e(G)$, and that $q_* \circ r_* = \phi'_*$. These relations are clear, since $p \circ r = \phi$, and $q \circ r = \phi'$.

Next we show that if $T_{(e,e)}(G,G)$ is identified with $T_e(G) \oplus T_e(G)$ by Proposition 13.20, and if $m : G \times G \to G$ is the multiplication map $m(x,y) = xy$, then $m^* : T_{(e,e)}(G,G) \to T_e(G)$ is the map $(D_1, D_2) \mapsto D_1 + D_2$. For this, it is sufficient to verify that $m_* \circ i_*$ and $m_* \circ j_*$ are each the identity map on $T_e(G)$, and this follows since $m \circ i$ and $m \circ j$ are both the identity map on G.

The result now follows since ϕ'' is the composition $m \circ r$. ∎

Proposition 13.22. *Let E_1 and E_2 be elliptic curves with function fields F_1 and F_2. Let $\phi : E_1 \to E_2$ a nonzero homomorphism. Let $\omega_1 \in \Omega_{F_1/k}$ and $\omega_2 \in \Omega_{F_2/k}$ be invariant differential forms. Let $\phi^* : \Omega_{F_2/k} \to \Omega_{F_1/k}$ be the induced map. Then $\phi^*(\omega_2) = c\omega_1$ for some $c \in F$. If $\phi' : E_1 \to E_2$ is another homomorphism, and $\phi'^*(\omega_2) = c'\omega_1$, then $(\phi \oplus \phi')(\omega_2) = (c + c')\omega_1$.*

Proof. If $Q \in E_1$, let $\lambda_Q : E_1 \to E_1$ be the map $P \to P \ominus Q$, and if $R \in E_2$ let $\mu_R : E_2 \to E_2$ be the map $P \to P \ominus R$. We have $\phi \circ \lambda_Q = \mu_{\phi(Q)} \circ \phi$. Since ω_2 is $\mu_{\phi(Q)}$ invariant, we have $\lambda_Q^* \circ \phi^*(\omega_2) = \phi^* \mu_{\phi(Q)}^*(\omega_2) = \phi^*(\omega_2)$, and therefore $\phi^*(\omega_2)$ is translation invariant. It follows that $\phi^*(\omega_2)$ is a constant multiple of ω_1. Let $\phi^*(\omega_2) = c\omega_1$.

Let D_1 be a nonzero element of $T_O(E_1)$, and let $D_2 = \phi_* D_1 \in T_O(E_2)$, where $\phi_* : T_O(E_1) \to T_O(E_2)$ is the induced map on tangent vectors, so that if $f \in \mathcal{O}_O(E_2)$, then $D_2(f) = D_1(f \circ \phi)$. Let $\widehat{D}_1 \in \operatorname{Der}_k(F_1)$ and $\widehat{D}_2 \in \operatorname{Der}_k(F_2)$ be the associated invariant regular derivations, as described in Proposition 13.18. Unraveling the definitions involved shows that

$$(13.6) \qquad \qquad \phi^* \circ \widehat{D}_1 = \widehat{D}_2 \circ \phi^*,$$

where $\phi^* : F_2 \to F_1$ is the induced map of function fields. For $i = 1, 2$, the derivation $\widehat{D}_i$ factors as $\rho_i \circ d$, where $d : F_i \to \Omega_{F_i/k}$ is the canonical derivation, and $\rho_i : \Omega_{F_i/k} \to F_i$ is a homomorphism. By Proposition 13.19, $\rho_1(\omega_1) \in k$. After multiplying D_1 by a suitable constant, we may assume that $\rho_1(\omega_1) = 1$. Now Eq. (13.6) implies that the diagram

$$
\begin{array}{ccc}
\Omega_{F_2/k} & \xrightarrow{\rho_2} & F_2 \\
{\scriptstyle \phi^*}\downarrow & & \downarrow{\scriptstyle \phi^*} \\
\Omega_{F_2/k} & \xrightarrow{\rho_1} & F_1
\end{array}
$$

is commutative, which we may show by composing with the map $d : F_2 \to \Omega_{F_2/k}$. Indeed, we have $\rho_1 \circ \phi^* \circ d = \rho_1 \circ d \circ \phi^* = \widehat{D}_1 \circ \phi^*$, and using Eq. (13.6), the latter equals $\phi^* \circ \widehat{D}_2 = \phi^* \circ \rho_2 \circ d$, and since the image of $d : F_2 \to \Omega_{F_2/k}$ generates $\Omega_{F_2/k}$ as a vector space, it follows that $\rho_1 \circ \phi^* = \phi^* \circ \rho_2$. Now we have $c = \rho_1(c\omega_1) = \rho_1 \phi^*(\omega_2) = \phi^* \rho_2(\omega_2)$. Note that this is in k, and since $\phi^* : F_2 \to F_1$ is the identity on k, we have therefore simply

$$(13.7) \qquad \qquad c = \rho_2(\omega_2).$$

Now let $\phi' : E_1 \to E_2$ be another homomorphism, and let $\phi'' = \phi \oplus \phi'$. Let $D_2' = \phi'(D_1)$ and $D_2'' = \phi''(D_1) \in T_O(E_2)$. By Proposition 13.21, we have $D_2'' = D_2 + D_2'$. Hence if we define $\widehat{D}_2'$ and $\widehat{D}_2''$ by $\widehat{D}_2(f) = D_1(f \circ \phi')$ and $\widehat{D}_2(f) = D_1(f \circ \phi'')$, then we have $\widehat{D}_2'' = \widehat{D}_2 + \widehat{D}_2'$, and if we define ρ_2' and ρ_2'' so that $\widehat{D}_2' = \rho_2' \circ d$ and $\widehat{D}_2'' = \rho_2'' \circ d$, then $\rho_2'' = \rho_2 + \rho_2'$. Now if $\phi'^*(\omega_2) = c'\omega_1$ and $\phi''(\omega_2) = c''\omega_1$, it follows that $c'' = c + c'$, as required. ∎

Proposition 13.23 (Differential Criterion for Separability). *Let E_1 and E_2 be two elliptic curves with function fields F_1 and F_2, and let $\phi : E_1 \to E_2$ be a homomorphism. There is induced an injection $F_2 \to F_1$, and a map $\phi^* : \Omega_{F_2/k} \to \Omega_{F_1/k}$. Then ϕ^* is either injective or the zero map, and it is injective if and only if ϕ is separable.*

Proof. Since $\Omega_{F_2/k}$ is a one-dimensional vector space over F_2, and $\phi^* : \Omega_{F_2/k} \to \Omega_{F_1/k}$ is F_2-linear, it is clear that this map is either injective or zero. To test which it is, let $f \in F_2$ such that $F_2/k(f)$ is separable, so that $df \neq 0$ in $\Omega_{F_2/k}$ by Proposition 11.10. Then F_1 is separable over F_2 if and only if it is separable over $k(f)$, and by Proposition 11.10, a necessary and sufficient condition for this is that $df \neq 0$ in $\Omega_{F_2/k}$, that is, if and only if ϕ^* is injective and not zero. ∎

Proposition 13.24. *Let E_1 and E_2 be two elliptic curves, and let ϕ, $\phi' : E_1 \to E_2$ be two morphisms. Suppose that ϕ is separable, and ϕ' is inseparable. Then $\phi \oplus \phi'$ is separable.*

Proof. It follows from Proposition 13.23 that ϕ' induces the zero map on differential forms, while ϕ'' induces an injection. By Proposition 13.22, ϕ'' has the same effect as ϕ on a nonzero invariant differential form as $\phi \oplus \phi'$, and so $\phi \oplus \phi'$ induces a nonzero map on differential forms. By Proposition 13.23, it follows that $\phi \oplus \phi'$ is separable. ∎

Proposition 13.25. *Let E_1 and E_2 be elliptic curves, and let $\phi : E_1 \to E_2$ be a morphism of degree m. Then there exists a unique morphism $^T\phi : E_2 \to E_1$, also of degree m, such that $^T\phi \circ \phi = [m]$ in $\mathrm{End}(E_1)$, and $\phi \circ {}^T\phi = [m]$ in $\mathrm{End}(E_2)$.*

The endomorphism $^T\phi$ is called the *transpose* of ϕ.

Proof. Observe that if there exists a morphism $^T\phi : E_2 \to E_1$ such that $^T\phi \circ \phi = [m]$, then $^T\phi$ is unique, because ϕ is surjective. Suppose that the existence of such a ϕ is established. Then $\phi \circ {}^T\phi \circ \phi = \phi \circ [m] = [m] \circ \phi$ because ϕ is a homomorphism, therefore commutes with multiplication by m. Because ϕ is surjective, this implies that $\phi \circ {}^T\phi = [m]$ also. Therefore it is sufficient to show just the existence of $^T\phi$ such that $^T\phi \circ \phi = [m]$.

We may factor $\phi = \phi_1 \circ \phi_2$, where ϕ_1 is separable, and ϕ_2 is purely inseparable. It is sufficient to prove this for ϕ_1 and ϕ_2 separately, and then to define $^T\phi = {}^T\phi_2 \circ {}^T\phi_1$. We may therefore assume that either ϕ is separable, or purely inseparable.

First, let us assume that ϕ is separable. The order of the kernel of ϕ equals the degree m by Proposition 13.17 (i), and the order of any element of a finite group divides the order of the group. The kernel of $[m]$ is by definition the group of elements of order dividing m, and so the kernel of ϕ is contained in the kernel of $[m]$. Now by Proposition 13.17 (ii), there exists $^T\phi$ with the required property.

Now suppose that ϕ is purely inseparable. We may break ϕ up into morphisms each having degree equal to p, the characteristic, and prove the existence of the transpose for each of these. Therefore we may assume that $m = \deg(\phi) = p$. Let F_1 and F_2 be the function fields of E_1 and E_2. It follows from Proposition 12.3 that we may identify F_2 with F_1^p. Let us argue that $[p] : E_1 \to E_1$ is not separable. Indeed, it follows from Proposition 13.22 that $[m]$ induces multiplication by m on the invariant differential form on E_1. Thus by Proposition 13.23, $[m]$ is separable if and only if $[m]$ is prime to the characteristic. In particular, $[p]$ is not separable. We may therefore factor $[p] = \psi_2 \circ \psi_1$, where ψ_2 is purely inseparable, and ψ_1 is separable. Then

$[p]^* : F_1 \to F_1$ factors into $\phi_2^* : F_1 \to F_1'$ and $\phi_1^* : F_1' \to F_1$, where ϕ_1^* is purely inseparable of degree q, and by Proposition 12.3, we may identify F_1' with F_1^q in such a way that ϕ_1^* is just the inclusion of F_1^q in F_1. Now observe that $F_2 = F_1^p \supseteq F_1' = F_1^q$. Let $\phi_1'^* : F_1' \to F_2$ be the inclusion, and consider the composition $\phi_1'^* \circ \phi_2^* : F_1 \to F_2$. It is clear that if $^T\phi : E_2 \to E_1$ is the corresponding map of elliptic curves, we have $^T\phi \circ \phi = [p]$. ∎

Exercises

Exercise 13.1. In the proof of Proposition 13.8, it was shown how a formula could be obtained for the coordinates of $P \oplus Q$ as polynomials in terms of the coordinates of the points P and Q on an elliptic curve in Weierstrass form $y^2 = x^3 + ax + b$. Find this formula.

Exercise 13.2. Prove Proposition 13.8.

Hint: The proof of Proposition 13.12 should help.

Exercise 13.3. (i) Let E_1 and E_2 be elliptic curves, and let ϕ, $\phi' : E_1 \to E_2$ be morphisms. Show that $^T(\phi \oplus \phi') = {}^T\phi \oplus {}^T\phi'$.

(ii) Let E be an elliptic curve, and let $[m] : E \to E$ be multiplication by m. Show that $^T[m] = [m]$.

Hint: Use induction on m.

(iii) Show that $\deg [m] = m^2$.

(iv) Let $E(m)$ be the kernel of $[m] : E \to E$, consisting of the elements of E of order dividing m. Show that if m is not divisible by the characteristic, then the cardinality of $E(m)$ is m^2.

(v) Show that if m is not divisible by the characteristic, then $E(m)$ is isomorphic to $(\mathbb{Z}/m\mathbb{Z})^2$.

Exercise 13.4. (Weil's Skew-Symmetric Pairing). Let E be an elliptic curve. Let m be a positive integer not divisible by the characteristic. Let μ_m be the group of m-th roots of unity in k. Then the cardinality of μ_m is m. Let $E(m)$ be the subgroup of elements of E of order dividing m. Thus $E(m)$ is the kernel of $[m] : E \to E$. We will define a map $E(m) \times E(m) \to \mu_m$ as follows. If $P \in E(m)$, then it follows from Proposition 13.4 that $m P - m O$ is the divisor of a function f. Since $[m] : E \to E$ is surjective, there exists an $R \in E$ such that $[m]R = P$. Also by Proposition 13.4, there exists a function g whose divisor is

$$\sum_{S \in E[m]} ((R \oplus S) - R).$$

Note that this sum does not depend on the choice of R. Check that $f \circ [m]$ and g^m have the same divisor. After multiplying f by a constant, we have therefore $f \circ [m] = g^m$. Now let $Q \in E(m)$. Let $\lambda_Q : E \to E$ be the translation map $P \mapsto P \ominus Q$. Observe that $\lambda_Q^*(g^m) = g^m$, and conclude that $\lambda_Q(g) = e_m(P, Q)$ for some $e_m(P, Q) \in \mu_m$. Confirm the following properties of e_m:

(i) $e_m(P \oplus P', Q) = e_m(P, Q) e_m(P', Q)$;

(ii) $e_m(P, Q \oplus Q') = e_m(P, Q) e_m(P, Q')$;

(iii) $e_m(P, Q) = e_m(Q, P)^{-1}$;

(iv) If $e_m(P, Q) = 1$ for all $Q \in E(m)$, then $P = O$.

Exercise 13.5. Prove that an abelian variety of dimension one is an elliptic curve. That is, show that the genus is one.

Hint: The Hurwitz genus formula should be of some help to you, and the morphisms $[m] : E \to E$.

14. The Zeta Function of a Curve

Let us recall that the Riemann zeta function is defined by $\sum_{n=1}^{\infty} n^{-s}$. This series is convergent if $\mathrm{re}(s) > 1$. It has meromorphic continuation to all s. It has a simple pole at $s = 1$, and is analytic elsewhere. It has the *Euler product* representation

$$\zeta(s) = \prod_p (1 - p^{-s})^{-1},$$

also valid if $\mathrm{re}(s) > 1$. Finally, there is the functional equation. Let

$$\xi(s) = \pi^{-s/2} \, \Gamma\left(\tfrac{s}{2}\right) \, \zeta(s).$$

Then $\xi(s)$ is meromorphic for all s, with simple poles at just $s = 0$ and $s = 1$, and $\xi(s) = \xi(1 - s)$.

Riemann conjectured in 1859 that all the zeros of $\xi(s)$ are on the line $\mathrm{re}(s) = 1/2$. This important conjecture, the *Riemann hypothesis* remains unproved.

In his 1923 dissertation [2], Emil Artin defined an analog of the Riemann zeta function which is associated with a curve defined over a finite field, proved the functional equation in the special case of a hyperelliptic curve, and postulated that the analog of the Riemann hypothesis should be satisfied for these zeta functions. F. K. Schmidt, Hasse and Weil worked on the problem of generalizing Artin's work. For zeta functions of elliptic curves, the Riemann hypothesis was proved by Hasse [13], [14] in 1933. In order to prove the general Riemann Hypothesis for curves of arbitrary genus, Weil was led to recast the foundations of algebraic geometry. The proof of the functional equation and Riemann hypothesis are contained in his book [31]. We will follow Weil in deducing the functional equations from the Riemann-Roch theorem. We will prove the Riemann hypothesis only in the special case of an elliptic curve, the case originally due to Hasse.

Let R be a Dedekind domain with the property that for every nonzero ideal $\mathfrak{a}$ of R, the cardinality $R/\mathfrak{a}$ is finite. For example, $\mathbb{Z}$ has this property, or more generally, the integral closure of $\mathbb{Z}$ in any finite extension of $\mathbb{Q}$. We will denote the cardinality of $R/\mathfrak{a}$ by $\mathbb{N}\mathfrak{a}$. Consistent with the usage in algebraic number theory, we will refer to this cardinality as the *norm* of $\mathfrak{a}$. We define the *zeta function* of R to be

$$\zeta_R(s) = \sum_{\mathfrak{a}} N\mathfrak{a}^{-s}.$$

We assume that this Dirichlet series converges if $\mathrm{re}(s) > 1$. (For the particular class of examples R which are of concern to us—the zeta functions of algebraic curves—we will prove this later.) We will show that

$$(14.1) \qquad\qquad \zeta_R(s) = \prod_{\mathfrak{p}} \left(1 - N\mathfrak{p}^{-s}\right)^{-1},$$

where the product is over all prime ideals of R. Indeed, if the infinite product is expanded in a geometric series, this becomes

$$\prod_{\mathfrak{p}} \left(1 + N\mathfrak{p}^{-s} + N\mathfrak{p}^{-2s} + \dots\right)^{-1}.$$

By the unique factorization of ideals in R (Exercise 9.1), if this product is (formally) expanded, one obtains $N\mathfrak{a}^{-s}$ exactly once for every ideal, and hence the infinite product represents the zeta function. Thus if $R = \mathbb{Z}$, this gives us the Riemann zeta function.

To consider another example, this one illustrating the geometric theory to be discussed below, let $k_0 = \mathbf{F}_q$ be the finite field with q elements, where q is a power of the prime p. Let k be the algebraic closure of k_0. Let R be the polynomial ring $k_0[x]$. Then R is a Principal Ideal Domain, hence it is a Dedekind domain. It shares with $\mathbb{Z}$ the property of having finite residue fields. To see this, let us determine the ideals of R. Since R is a principal ideal domain, any ideal $\mathfrak{a}$ has the form (f), where f is a polynomial. f is uniquely determined if we ask that it be monic. $\mathfrak{a}$ is prime if and only if f is irreducible. A basis for $R/\mathfrak{a}$ as a vector space over k_0 consists of $1, x, \cdots, x^{d-1}$. Hence the cardinality $N\mathfrak{a}$ of $R/\mathfrak{a}$ is q^d, where $d = \deg(f)$. Moreover, it is clear from this description that the number of ideals with norm q^d is equal to the number of monic irreducible polynomials of degree d, which is q^d. Therefore

$$\zeta_R(s) = \sum q^d q^{-ds} = (1 - q^{1-s})^{-1}.$$

Now let $U = \mathbf{A}^1$ denote the affine line over k. We may identify U with k, and as such, we get an action of the (infinite) Galois group $G = \mathrm{Gal}(k/k_0)$ on U, and consequently on the divisors on U. Let D be an effective divisor. We may associate with D the monic polynomial whose roots are the coordinates of the points of D, with multiplicity equal to the multiplicity with which the point appears in D. If D is invariant under the action of G, then this polynomial has coefficients in k_0. Hence the ideals of R are in one-to-one correspondence with the Galois invariant effective divisors on $\mathbf{A}^1$. We will call such a divisor *prime* if the corresponding ideal is prime. It is clear that the prime divisors are a basis for the group of Galois invariant divisors. If D is an effective Galois invariant divisor on $\mathbf{A}^1$, we define $ND = q^{\deg D}$. This is equal to $N\mathfrak{a}$, where $\mathfrak{a}$ is the ideal of R associated with D. Therefore we may write

$$(14.2) \qquad \zeta_R(s) = \zeta_{\mathbf{A}^1}(s) = \sum_D ND^{-s} = \prod_{\mathfrak{P}} \left(1 - N\mathfrak{P}^{-s}\right)^{-1},$$

where now the sum is over effective Galois invariant divisors D, and the product is over effective *prime divisors* $\mathfrak{P}$.

Actually we should consider instead of the affine line the complete curve $\mathbf{P}^1$. The group of divisors for this curve is the same as for $\mathbf{A}^1$ except now there is one more prime divisor, namely, the point at infinity, having norm q. It follows from the Euler product representation Eq. (14.2) that the zeta function of $\mathbf{P}^1$ should be the zeta function of $\mathbf{A}^1$ times one more Euler factor, equal to $(1 - q^{-s})^{-1}$, corresponding to the point at infinity. Thus we should have

$$\zeta_{\mathbf{P}^1}(s) = (1 - q^{-s})^{-1} (1 - q^{1-s})^{-1}.$$

Note that this function shares with the Riemann zeta function the functional equation $\zeta_{\mathbf{P}^1}(s) = \zeta_{\mathbf{P}^1}(1-s)$. It satisfies the Riemann hypothesis for the trivial reason that it never vanishes. Our purpose is to generalize these facts.

Proposition 14.1. *Let K_0 be a perfect field, and let K/K_0 be an algebraic extension. Let F_0/K_0 be another field extension such that K_0 is algebraically closed in F_0. Let $F = F_0 \otimes_{K_0} K$. Then F is a field. If F_0 and K are embedded in F by $x \mapsto x \otimes 1$ and $y \mapsto 1 \otimes y$, then F_0 and K are linearly disjoint over K_0.*

Proof. Since F is the direct limit of the rings $F_0 \otimes K'$, where K' runs through the finite extensions of K_0 contained in K, we may assume that K/K_0 is a finite extension. Let $R = F_0 \otimes K$, and let $\mathfrak{m}$ be a maximal ideal in R. Let $F = R/\mathfrak{m}$. Note that in the statement of the theorem we defined F to equal R. We have defined F differently now, but since we will eventually show that $\mathfrak{m} = 0$, the two definitions will eventually be seen to be consistent.

We embed F_0 and K in F by composing the maps $x \mapsto x \otimes 1$ and $y \mapsto 1 \otimes y$ with the projection $R \to F$. We will show that $[F : F_0] = [K : K_0]$. Since K_0 is perfect, the extension K/K_0 is separable. By the Theorem of the Primitive Element (Lang [18], Theorem V.4.6 on p. 243), $K = K_0(\xi)$, where the minimal polynomial of ξ over K_0 has degree $n = [K : K_0]$. We have $F = F_0(\xi)$, and if $[F : F_0] < [K : K_0]$, then ξ must satisfy a polynomial g over F_0 of strictly smaller degree than n. The coefficients in $g(x)$ are symmetric functions of the roots, which are conjugates of ξ, hence these coefficients are algebraic over K_0. Since K_0 is algebraically closed in F_0, the coefficients of $g(x)$ are thus in K_0, which is a contradiction, since f is the minimal polynomial of ξ over K_0. Thus $[F : F_0] = [K : K_0]$.

This implies that $\mathfrak{m} = 0$, for we have $\dim_{F_0} R = \dim_{F_0} F_0 \otimes_{K_0} K = \dim_{K_0} K = \dim_{F_0} F$, and therefore the map $R \to F$ is injective, i.e. $\mathfrak{m} = 0$. Hence $F_0 \otimes K$ is a field. The linear disjointness of F_0 and K over K_0 is now clear, since in the ring R it is clear that a set of elements in K which are linearly independent over K_0 are linearly independent over F_0, and this is the definition of linear disjointness. ∎

Let k_0 be a perfect field (later to be assumed finite), and let k be its algebraic closure. Let F_0 be a finitely generated extension of k_0 of transcendence degree one. We assume that k_0 is algebraically closed in F_0. Let $F = F_0 \otimes_{k_0} k$, which is a field by Proposition 14.1. We will identify F_0 and k with their images in F. These data will be fixed throughout the discussion.

It follows from linear disjointness (Proposition 14.1) that F/F_0 is an (infinite) Galois extension, and that if $\mathcal{G}$ is the (infinite) Galois group of F/F_0, then by restriction to k, there is induced an isomorphism of $\mathcal{G}$ onto the Galois group of k/k_0. We will identify $\mathcal{G}$ with the Galois group of k/k_0.

It is clear that F is a finitely generated extension of k of transcendence degree one. We let X be the associated nonsingular complete curve. We know that the points of X are in one-to-one correspondence with the valuation rings of F fixing k. We will refer to X as a complete nonsingular curve *defined over* k_0. Since $\mathcal{G}$ acts on F, it also acts on these valuation rings. Therefore $\sigma \in \mathcal{G}$ induces a mapping $\sigma : X \to X$ such that ${}^\sigma f({}^\sigma P) = f(P)$. This mapping σ of X is not a morphism. (An important exception: we will see later that if k_0 is a finite field with q elements, then the action of the particular element $\sigma_q \in \mathrm{Gal}(k/k_0)$ defined by $\sigma_q(f) = f^q$ is a morphism $X \to X$.) The Galois group therefore acts on divisors.

Let us point out that if k_1/k_0 is any finite extension (contained in k), then we may also regard X as a curve defined over k_1. Indeed, let F_1 be the field generated by F_0 and k_1. Then k_1 and F_1 satisfy the same axioms as k_0 and F_0.

Now suppose that k_0 is finite, and that q is its cardinality. If D is an effective Galois invariant divisor, we define $\mathbb{N}D$ to be q^d, where $d = \deg(D)$, and we consider

$$\zeta_X(s) = \sum \mathbb{N}D^{-s},$$

summing over effective Galois invariant divisors. We will prove later that this sum is convergent if $\mathrm{re}(s) > 1$.

We may partition X up into orbits under $\mathcal{G}$. Each orbit is finite, and we call a divisor $\mathfrak{P}$ *prime* if it is the sum of the points in a single $\mathcal{G}$ orbit. If $\mathfrak{P}$ is a prime divisor of degree one, then it corresponds to a point of X which is invariant under $\mathcal{G}$. We call such a point of X a k_0-*rational point*. We will say that such a point is *rational* or *defined* over k_0. We will also call a Galois invariant divisor k_0-*rational*. While k_0 is fixed, we may omit it from the terminology, and simply refer to a point or divisor as *rational*. It is clear that the prime divisors are a $\mathbb{Z}$-basis for the group of rational divisors. We therefore have (pending discussion of the convergence issue) that

$$\zeta_X(s) = \prod_{\mathfrak{P}} (1 - \mathbb{N}\mathfrak{P}^{-s})^{-1}.$$

If k_1 is a finite extension of k_0 contained in X, then we may consider the points of X which are rational over k_1. If $P \in X$, let $\mathcal{G}_P$ be its stabilizer in $\mathcal{G}$, and let k_P be the fixed field of $\mathcal{G}_P$. The orbit of P under the Galois group is finite, hence the index of $\mathcal{G}_P$ in $\mathcal{G}$ is finite. Therefore k_P is a finite extension of k_0. We see that P is defined over k_1 if and only if $k_1 \supseteq k_P$. Hence k_P is the *minimal field of definition* of P. Similarly, a divisor on X has a minimal field of definition.

Since all the coefficients in the Dirichlet series for $\zeta_X(s)$ are powers of q, we may as well work with the power series

$$(14.3) \qquad Z_X(T) = \sum T^{\deg D} = \prod_{\mathfrak{P}} (1 - T^{\deg \mathfrak{P}})^{-1},$$

so that $\zeta_X(s) = Z_X(q^{-s})$.

We would like to use the Riemann-Roch theorem to count the number of effective rational divisors of a given degree. We must therefore show that if D is rational, then $l(D)$ has a reinterpretation as the dimension of a suitable vector space over k_0.

Proposition 14.2. *Let L/K be a Galois extension of fields, and let $\sigma : \mathrm{Gal}(L/K) \to GL(n, L)$ be a mapping satisfying the cocycle relation*

$$(14.4) \qquad B(\sigma).^{\sigma}B(\tau) = B(\sigma\tau)$$

for $\sigma, \tau \in \mathrm{Gal}(L/K)$. Then there exists a matrix $A \in GL(n, L)$ such that $B(\sigma) = A^{-1}.^{\sigma}A$.

This result is a basic one in Galois cohomology. It says that the cohomology set $H^1\big(\mathrm{Gal}(L/K), GL(n, L)\big)$ is trivial. This is a generalization of Hilbert's Theorem 90.

Proof. Suppose that we can find $C \in GL(n, L)$ such that $\sum_\sigma B(\sigma)^\sigma C$ is nonsingular. It is easily checked that if A is the inverse of this matrix, then $AB(\sigma) = {}^\sigma A$, as required. We will construct C to be diagonal, with diagonal entries $c_i = c_{ii}$.

If $1 \leq i \leq n$, let m_i be the minor

$$\begin{vmatrix} \sum b_{11}(\sigma)\,^\sigma c_1 & \sum b_{12}(\sigma)\,^\sigma c_2 & \cdots & \sum b_{1i}(\sigma)\,^\sigma c_i \\ \sum b_{21}(\sigma)\,^\sigma c_1 & \sum b_{22}(\sigma)\,^\sigma c_2 & \cdots & \sum b_{2i}(\sigma)\,^\sigma c_i \\ \vdots & & \ddots & \vdots \\ \sum b_{i1}(\sigma)\,^\sigma c_1 & \sum b_{i2}(\sigma)\,^\sigma c_2 & \cdots & \sum b_{ii}(\sigma)\,^\sigma c_i \end{vmatrix},$$

where each summation is over σ. We will construct c_i recursively so that $c_i \neq 0$. Suppose that c_{i-1} is constructed. To get c_i, observe that expanding the above determinant in the last column gives us

$$(14.5) \qquad m_i = \sum_{j=1}^{i} M_j \left(\sum_\sigma b_{ji}(\sigma)\,^\sigma c_i \right)$$

$$= \sum_\sigma \left(\sum_{j=1}^{i} M_j \, b_{ji}(\sigma) \right) {}^\sigma c_i,$$

where the M_j are minors formed with the first $i - 1$ columns. Since we have constructed $c_1, \cdots, c_{i-1}$, these are given, and in particular, $M_i = m_{i-1} \neq 0$ by induction hypothesis. Substituting $\sigma = \tau = 1$ into Eq. (14.4), we see that $B(1)$ is just the identity matrix. Therefore the coefficient of $c_i = {}^1 c_i$ in Eq. (14.5) is just M_i. Hence the coefficients in Eq. (14.5) are not all zero. It follows from Artin's theorem on the linear independence of characters (cf. Lang [18], Theorem VI.4.1 on p. 283) that we may choose c_i such that $m_i \neq 0$.

Now $m_n = \det \sum_\sigma B(\sigma)^\sigma C$. We have shown that we may construct C so that this is nonzero, as required. ∎

If D is a divisor, let $L_0(D) = \{f \in F_0 | (f) \geq -D\}$. This is a vector space over k_0.

Proposition 14.3. *If D is a $\mathcal{G}$-invariant divisor, then $L(D)$ has a k-basis consisting of elements of $L_0(D)$, and therefore $l(D) = \dim_{k_0} L_0(D)$.*

This result will allow us to apply the Riemann-Roch theorem to answer questions about F_0, instead of F.

Proof. Let $n = l(D)$, and let v be a column vector whose entries are a basis for $L(D)$. These functions all lie in some finite Galois extension L of F. Evidently if $f \geq -D$, then ${}^\sigma f \geq -{}^\sigma D = -D$, so $L(D)$ is invariant under the Galois action. Therefore if $\sigma \in \mathrm{Gal}(L/F)$, we have $v = B(\sigma)\,{}^\sigma v$ for some $B(\sigma) \in GL(n, L)$. It is clear that $B(\sigma)$ satisfies the cocycle relation Eq. (14.4), and therefore there exists a nonsingular matrix A such that $B(\sigma) = A^{-1}.{}^\sigma A$. Now let $u = A\,v$. Then u is $\mathrm{Gal}(L/K)$ invariant, and therefore the components of u are in F_0. This gives us a k-basis of $L(D)$ which is contained in $L_0(D)$. The components of u are also a k_0-basis of $L_0(D)$, and so $l(D) = \dim_{k_0} L_0(D)$. ∎

For the remainder of this chapter, we will assume that k_0 is finite, of cardinality q. We will denote $Z_X(T)$ as $Z(T)$. Let g be the genus of X.

Proposition 14.4. *(i) Let k_1 be a finite extension of k_0 which is contained in k. Then the set of points of X which are rational over k_1 is finite.*

(ii) The set of effective rational divisors of given degree d is finite. In particular, the set of rational points is finite.

(iii) The group of divisors of degree zero rational over k_0 modulo the subgroup of divisors of functions in F_0 is a finite group.

The finite group in (iii) is called the *divisor class group*. It may be identified with the group of rational points in the Jacobian of X, which however we have not defined.

Proof. Let us prove (i). Since F_0 is finitely generated over k_0, let $f_1, \cdots, f_n$ be a set of generators. Let A be the integral closure of $k[f_1, \cdots, f_n]$ in F. Then by Proposition 6.18, A is an affine algebra. Let U be the corresponding affine open set of X. Since $X - U$ is finite, it is sufficient to show that only a finite number of $P \in U$ are rational over k_1. It is clear from the definition of the action on $\mathcal{G}$ on X that $f_i(P)$ must be in k_1. Moreover, the point P is determined by the values of $f_1(P), \cdots, f_n(P)$. Since k_1 is finite, there are only a finite number of possible such values, hence only a finite number of rational points.

Let us prove (ii). If D is an effective divisor of degree d over k_0, then there are at most d points occurring with nonzero multiplicity in D. Let P be such a point. Then the $\mathcal{G}$-orbit of X is contained in the support of D, hence has cardinality $\leq d$. This means that P is defined over an extension of k_0 of degree less than or equal to d. There are only a finite number of such fields, and by (i), there are only a finite number of points defined over each. Hence there are only a finite number of divisors of degree d.

Now we may prove (iii). Let m be a positive number greater than $2g - 2$ such that there exists a rational divisor of degree m. Let $D_1, \cdots, D_n$ be the distinct effective divisors of degree m which are rational over k_0. There are only a finite number of such divisors by (ii). We will prove that any divisor D of degree zero, rational over k_0, is linearly equivalent to $D_i - D_1$ for some i.

According to the Riemann-Roch Theorem,

$$l(D + D_1) = m + 1 - g + l(K - D - D_1).$$

The degree of the canonical class is $2g - 2$, so $\deg(K - D - D_1) < 0$, and therefore $l(K - D - D_1) = 0$. We see that $l(D + D_1) > 0$. Let $f \in L_0(D + D_1)$. Then $(f) + D + D_1$ is an effective divisor of degree m, invariant under $\mathcal{G}$, and hence is one of the D_i. This shows that $D \sim D_i - D_1$. This shows that the group of rational divisors of degree zero, modulo linear equivalence, is a finite group. ∎

Proposition 14.5 (the Functional Equation). *The power series $Z(T)$ is absolutely convergent if $|T| < 1/q$, and represents a rational function in T. It has simple poles at $T = 1$ and $T = 1/q$, and no other poles. It satisfies the functional equation:*

(14.6) $$Z\left(\frac{1}{qT}\right) = q^{1-g} T^{2-2g} Z(T).$$

If X has a point rational over k_0, then there exist complex numbers $\alpha_1, \cdots, \alpha_{2g}$ such that

(14.7) $$Z(T) = \frac{\prod_{i=1}^{2g}(1 - \alpha_i T)}{(1 - T)(1 - qT)}.$$

For $1 \le i \le 2g$, the quantity q/α_i is equal to α_j for some j.

The assumption that X has a rational point over k_0 is actually automatic, though we will not prove this.

Proof. Since the degree map from the group of rational divisors to $\mathbb{Z}$ is a homomorphism, there exists a positive integer δ such that there exists a rational divisor of degree n if and only if $n|\delta$. (It may be deduced from the Riemann hypothesis that $\delta = 1$, but this is not needed to prove the functional equation.) Moreover, the canonical divisor is rational, since if $f \in F_0$, (df) is a Galois invariant element of the canonical class. Thus the degree $2g - 2$ of the canonical class is a multiple of δ, and we may write $2g - 2 = \rho\delta$.

Let C be a fixed rational divisor of degree δ. By Proposition 14.4 (iii) the group of rational divisors of degree zero modulo linear equivalence is finite. $D_1, \cdots, D_h$ be a set of representatives for these divisor classes. Then for every rational divisor D, there exist unique integers n and i, $1 \le i \le h$, such that $D \sim n.C + D_i$. (Indeed, $n = \deg(D)/\delta$.)

If D is a given rational divisor, how many effective rational divisors linearly equivalent to D are there? If $0 \ne f \in F_0$, then $f + D$ is such a divisor if and only if $f \in L_0(D)$. A different element f' of $L(D)$ will determine the same divisor if and only if $f = c.f'$ for $c \in k_0^\times$. The number of nonzero elements of $L_0(D)$ is $q^{l(D)} - 1$. The number of nonzero elements of k_0 is $q - 1$. Hence the number of effective rational divisors linearly equivalent to D is

$$\frac{q^{l(D)} - 1}{q - 1}.$$

It follows that the number of rational effective divisors of degree $n\delta$ is

$$\sum_{i=1}^{h} \frac{q^{l(n.C+D_i)} - 1}{q - 1},$$

and therefore

(14.8) $$(q - 1) Z(T) = \sum_{n=0}^{\infty} \left(\sum_{i=1}^{h} q^{l(n.C+D_i)} - 1 \right) T^{n\delta}.$$

Now observe that if $n > \rho$, then $\deg(K - n.C - D_i) < 0$, and therefore by the Riemann-Roch Theorem,

$$l(n.C + D_i) = n\delta + 1 - g.$$

Following Weil, we may thus split Eq. (14.8) into two sums: $(q - 1) Z(T) = F(T) + h.R(T)$, where

$$F(T) = \sum_{n=0}^{\rho} \sum_{i=1}^{h} q^{l(n.C+D_i)} T^{n\delta},$$

and

$$R(T) = - \left(\sum_{n=0}^{\rho} T^{n\delta} \right) + \sum_{n=\rho+1}^{\infty} (q^{n\delta+1-g} - 1) T^{n\delta}.$$

From this expression, it is clear that $Z(T)$ is absolutely convergent if $|T| < 1/q$, since $F(T)$ is a finite sum, and $R(T)$ is the sum of two geometric series, one with radius of convergence 1, and the other with radius of convergence $1/q$. $Z(T)$ is evidently entire, and if X has a rational point over k_0, then $\delta = 1$, and in this case the only poles of $R(T)$ are at $T = 1$ and $1/q$. To prove the functional equation Eq. (14.6), we will show that both $F(T)$ and $R(T)$ satisfy Eq. (14.6).

As for $R(T)$, summing the geometric series, we have

$$R(T) = q^{1-g} \frac{(qT)^{(\rho+1)\delta}}{1 - (qT)^\delta} - \frac{1}{1 - T^\delta}.$$

Now since $\rho\delta = 2g - 2$, we verify directly that

$$R\left(\frac{1}{qT}\right) = q^{1-g} T^{2-2g} R(T).$$

Now as for $F(T)$, we use the Riemann-Roch Theorem:

$$l(nC + D_i) = n\delta + 1 - g + l(K - nC - D_i).$$

Note that for fixed $1 \leq n \leq \rho$, as $nC + D_i$ runs through the distinct linear equivalence classes of rational divisors of degree $n\delta$, $K - nC - D_i$ runs through the distinct linear equivalence classes of rational divisors of degree $(\rho - n)\delta$. Therefore we may write

$$l(nC + D_i) = n\delta + 1 - g + l((\rho - n)C + D_i'),$$

where the D_i' are a permutation of the D_i. Therefore,

$$F(T) = \sum_{n=0}^{\rho} \sum_{i=1}^{h} q^{n\delta+1-g+l((\rho-n).C+D_i)} T^{n\delta},$$

and substituting $\rho - n$ for n, this equals

$$\sum_{n=0}^{\rho} \sum_{i=1}^{h} q^{g-1-n\delta+l(n.C+D_i)} T^{2g-2-n\delta} = T^{2g-2} q^{g-1} F(1/qT).$$

This proves that $F(T)$ as well as $R(T)$ also satisfies Eq. (14.6).

Assuming now that X has a rational point over k_0, we have noted that $\delta = 1$ and in this case $Z(T)$ has simple poles at 1 and $1/q$, and no other poles. Thus $P(T) = (1 - T)(1 - qT) Z(T)$ is a polynomial in T. The functional equation for $Z(T)$ implies that

$$(14.9) \qquad\qquad P(T) = q^g T^{2g} P\left(\frac{1}{qT}\right).$$

It is easy to see that a polynomial satisfying such a relation has degree $2g$. Moreover, $Z(0) = 1$ by Eq. (14.3), and so $P(T)$ has the form

$$\prod_{i=1}^{2g} (1 - \alpha_i T).$$

Now the functional equation Eq. (14.9) implies that

$$P(T) = \prod_{i=1}^{2g} \left(1 - \frac{q}{\alpha_i T}\right).$$

Hence $\alpha_i \mapsto q/\alpha_i$ is a permutation of the α_i. ∎

For the remainder of the chapter, *we will assume that X has a rational point over k_0*. As we have already noted, this can actually be proved, though we shall not do so. There is no serious loss of generality in making this assumption, since if X did not have a rational point over k_0, it would still have a rational point over a finite extension k_1, and we could just as easily work over the latter field.

Now let us show that the zeta function gives information about the number of rational points on the curve. Let d be any integer. We know from the theory of finite fields that there exists a unique extension $\mathbf{F}_{q^d}$ of degree d over $k_0 = \mathbf{F}_q$ contained in k. Let N_d be the number of points on the curve which are rational over $\mathbf{F}_{q^d}$.

Proposition 14.6. *We have*

$$(14.10) \qquad T\frac{d}{dT}\log Z(T) = \sum_{k=1}^{\infty} N_k\, T^k,$$

and

$$(14.11) \qquad N_k = 1 - \left(\sum_{i=1}^{2g}\alpha_i^k\right) + q^k.$$

Proof. We will make use of the formal identity

$$(14.12) \qquad T\frac{d}{dT}\log(1-T^d)^{-1} = d\sum_{n=1}^{\infty} T^{nd},$$

which follows since the left-hand side equals $dT^d/(1-T^d)$. By Eq. (14.3) and Eq. (14.12), the left-hand side in Eq. (14.10) equals

$$\sum_{\mathfrak{P}}\sum_{n=1}^{\infty}\deg(\mathfrak{P})\, T^{n\,\deg(\mathfrak{P})} = \sum_{d=k}^{\infty}\left[\sum_{\deg\,\mathfrak{P}|k}\deg\mathfrak{P}\right] T^k.$$

To prove Eq. (14.10), we must show that the term in brackets equals N_k. To see this, consider that N_k is the union over all $m|k$ of the set of points of X whose minimal field of definition is $\mathbf{F}_{q^m}$, since these are the subfields of $\mathbf{F}_{q^k}$. Now, to count the number of points whose minimal field of definition is $\mathbf{F}_{q^m}$, note that these are the points whose Galois orbit is a prime divisor of degree m, and each such orbit contains m points. Hence the number of points is

$$\sum_{\deg\,\mathfrak{P}=m}\deg\mathfrak{P},$$

whence Eq. (14.10).

Now to prove Eq. (14.11), apply Eq. (14.12) to Eq. (14.7). We get

$$T \frac{d}{dT} \log Z(T) = \sum_{k=1}^{\infty} T^k + q^k T^k - \sum_{i=1}^{2g} \alpha_i^k T^k.$$

Equating coefficients between this and Eq. (14.10) gives Eq. (14.11). ∎

Proposition 14.7. *Let E be an elliptic curve defined over the finite field k_0. Then E has a rational point over k_0.*

Proof. Suppose not, so $N_1 = 0$. By Eq. (14.11), we have $1 - \alpha_1 - \alpha_2 + q = 0$. Since $\alpha_2 = q/\alpha_1$, this implies that either $\alpha_1 = 1$ and $\alpha_2 = q$ or vice versa. In either case, Eq. (14.11) implies that $N_d = 0$ for all d, which is clearly impossible. ∎

The Riemann hypothesis was first established by Weil using the ring of correspondences of the curve, a substitute for the endomorphism ring of the Jacobian. To present Weil's proof, we would first have to develop intersection theory on the algebraic surface $X \times X$, which is the foundation of the theory of correspondences. However in the case of an elliptic curve, the curve is its own Jacobian, so we may prove the Riemann hypothesis with no more machinery than we already have.

Let A be an abelian group. By a *quadratic form* we mean a real valued function Q on A such that the symmetric function $B(x, y) = Q(x + y) - Q(x) - Q(y)$ is additive in x and y, that is, $B(x + x', y) = B(x, y) + B(x', y)$, and $B(x, x) = 2Q(x)$. We say that $B(x, y)$ is the *polarization* of Q. We say that the form Q is *positive definite* if $Q(x) \geq 0$, with equality if and only if $x = 0$.

Proposition 14.8. *Let A be an abelian group, and let Q be a positive definite quadratic form on A with polarization B. Then if $x, y \in A$, we have $|B(x, y)| \leq \frac{1}{2}\sqrt{Q(x)\, Q(y)}$.*

This is a version of the familiar Cauchy-Schwartz inequality.

Proof. We have

$$2m^2 Q(x) + 2mn B(x, y) + 2n^2 Q(y) = B(mx + ny, mx + ny) \geq 0$$

for all integers m and n, so

$$\lambda^2 Q(x) + \lambda B(x, y) + Q(y) \geq 0$$

when λ is rational, and so by continuity, when λ is real. Thus the discriminant of this quadratic polynomial in λ must be negative, so $B(x, y)^2 - 4Q(x)Q(y) < 0$, which implies that $|B(x, y)| \leq 2\sqrt{Q(x)\, Q(y)}$, as required. ∎

If E and E' are elliptic curves, and $\phi : E \to E'$ is a homomorphism, we want to define the degree of ϕ. Recall that if $\phi : E \to E$ is an isogeny, the degree of ϕ has already been defined to be the field degree of $[F : \phi^*(F')]$. If ϕ is the zero map, we then define $\deg(\phi) = 0$. Since every homomorphism of elliptic curves is either zero or an isogeny, we have defined the degree map in every case.

Proposition 14.9. *Let E be an elliptic curve. The degree map on* $\mathrm{End}(E)$ *is a positive definite quadratic form.*

Proof. We recall from Proposition 13.25 that $\phi \circ {}^\top\phi = [\deg \phi]$ in $\mathrm{End}(E)$. This formula remains true if we define the transpose of the zero map to be zero. Therefore if

$$B(\phi, \phi') = \deg(\phi \oplus \phi') - \deg(\phi) - \deg(\phi')$$

is the polarization, we have

$$[B(\phi, \phi')] = \phi \circ {}^\top\phi' \oplus \phi' \circ {}^\top\phi,$$

which shows that B is bilinear. Moreover,

$$[B(\phi, \phi)] = \phi \circ {}^\top\phi \oplus \phi \circ {}^\top\phi,$$

which shows that $B(\phi, \phi) = 2 \deg(\phi)$. Thus the polarization is bilinear, and hence the degree map is a quadratic form. Positive definiteness is clear. ∎

By Eq. (14.10), the number of points on a curve is related to the zeta function, which in fact may be regarded as a generating function for N_d. The proof of the Riemann hypothesis depends on being able to count the points on a curve. This is accomplished by means of the *Frobenius endomorphism*, as we shall now explain. Let us start with an example. Let E be an elliptic curve, defined over k_0. By Proposition 14.7, E has a rational point over k_0, which we may take to be the origin. Then we may go through the proof of Proposition 13.7, but using Proposition 14.3 to choose x and y in $L_0(3O)$. Then the parameters a and b in the Weierstrass equation $y^2 = x^3 + ax + b$ will lie in k_0. This implies that $a^q = a$, and that $b^q = b$. Therefore if $\xi = x^q$ and $\eta = y^q$, we also have $\eta^2 = \xi^3 + a\xi + b$. There is therefore an automorphism $\mathcal{F}^*$ of $F = k(x, y)$ which is trivial on k, but which satisfies $\mathcal{F}^*(x) = \xi$ and $\mathcal{F}^*(y) = \eta$. Note that it is *not* the case that $\mathcal{F}^*(\alpha) = \alpha^q$ for any $\alpha \in F$, since we have asked that $\mathcal{F}^*$ be the identity on k. It *is* the case that $\mathcal{F}^*(\alpha) = \alpha^q$ for any $\alpha \in F_0$, since for $\alpha \in k_0$, we have $\alpha = \alpha^q$, and $F_0 = k_0(x, y)$.

To generalize this, let X be an arbitrary curve defined over k_0. Let $\mathcal{F}^* : F_0 \to F_0$ be the Frobenius endomorphism, $\mathcal{F}^*(x) = x^q$. Note that this is the identity on k_0, hence by linear disjointness (Proposition 14.1) can be extended to an endomorphism $\mathcal{F}^*$ of F which is trivial on k. Again, it is important to note that it is not true arbitrary for $x \in F$ that $x = x^q$. Let $\mathcal{F} : X \to X$ be the corresponding morphism.

Proposition 14.10. $\mathcal{F}$ *is a purely inseparable morphism of degree* q. *If* $P \in X$, *then* $\mathcal{F}(P) = P$ *if and only if* P *is rational over* k_0.

Proof. To show that $\mathcal{F}$ is a purely inseparable morphism of degree q, we must show that $F/\mathcal{F}^*(F)$ is a purely inseparable field extension of degree q. Indeed, since $\mathcal{F}^*$ is $\alpha \mapsto \alpha^q$ on F_0, and the identity on k, $\mathcal{F}^*(F)$ is the field generated by k and F_0^q. Since $k = k^q$, this is the same as the field generated by k^q and F_0^q, which is the image under $\alpha \mapsto \alpha^q$ of the field generated by k and F_0, that is, $\mathcal{F}^*(F) = F^q$. It is now clear that $F/\mathcal{F}^*(F)$ is purely inseparable. That its degree is equal to q follows from Proposition 12.3.

It remains to be shown that if $P \in X$, then $\mathcal{F}(P) = P$ if and only if P is rational over k_0. Recall that if P is a point of X, we defined the action of $\mathcal{G} = \mathrm{Gal}(k/k_0)$ on X in such a way that $^{\sigma}f(^{\sigma}P) = f(P)$ for $f \in F$ and $P \in X$. In particular, let $\sigma_q \in \mathrm{Gal}(k/k_0)$ be the particular endomorphism $x \to x^q$. We will show that the effect of σ_q on X is the same as the action of $\mathcal{F}$, although in general the action of an element of $\mathrm{Gal}(k/k_0)$ will not be a morphism. To see this, note that σ_q is the identity on F_0, and the q-th power map on k, while $\mathcal{F}^*$ is the identity on k, and the q-th power map on F_0. It follows from linear disjointness (Proposition 14.1) that $\sigma_q \circ \mathcal{F}^*(f) = f^q$ for any $f \in F$. Now $f(\mathcal{F}P) = 0$ if and only if $(\mathcal{F}^*f)(P) = 0$, and by the definition of the action of $\sigma = \sigma_q$, this is equivalent to the vanishing of $(\sigma \circ \mathcal{F}^*f)(^{\sigma}P) = f(^{\sigma}P)^q$, or of $f(^{\sigma}P)$. Since this is true for all $f \in F$, we see that $\mathcal{F}P = {}^{\sigma}P$.

We see that P is Galois invariant, that is, defined over k_0, if and only if it is fixed by $\mathcal{F}$. ∎

Theorem 14.11 (the Riemann Hypothesis for elliptic curves). *Suppose that* X *is an elliptic curve defined over a field* k_0 *with* q *elements. If* N_1 *is the number of* k_0-*rational points of* X, *then*

$$(14.13) \qquad\qquad |N_1 - 1 - q| \le 2\sqrt{q}.$$

The parameters α_1 *and* α_2 *in Eq. (14.7) both have absolute value* $\sqrt{q}$, *and are complex conjugates. If* $\zeta_X(s) = 0$, *then* $\mathrm{re}(s) = 1/2$.

Proof. Let us show first that every statement of this theorem follows from the first: that $N_1 \le 2\sqrt{q}$. Indeed, suppose that we have proved this. We may also regard X as defined over the unique extension of degree d of k_0, and therefore we have $N_d - 1 - q^d \le 2q^{d/2}$ for all d. Since $\alpha_1 \alpha_2 = q$, either $|\alpha_1| = |\alpha_2| = \sqrt{q}$, or else either $|\alpha_1| > \sqrt{q}$ or $|\alpha_2| > \sqrt{q}$. Assume that $|\alpha_1| > \sqrt{q}$. Then we have by Eq. (14.7) that

$$|N_d - 1 - q^d| \ge |\alpha_1|^d,$$

and if d is sufficiently large, this will exceed $q^{d/2}$, which will be a contradiction.

Once we know that $|\alpha_1| = |\alpha_2| = \sqrt{q}$, it is clear that $Z(t) = 0$ implies that $|t| = q^{-1/2}$, and if $\zeta_X(s) = Z(q^{-s}) = 0$, then $\mathrm{re}(s) = 1/2$.

Thus if we can only show that $|N_1 - 1 - q| \leq 2\sqrt{q}$, we will be done. Now we see by Proposition 14.10 that the group of rational points of E is the kernel of the morphism $1 \ominus \mathcal{F}$, where $\mathcal{F}$ is the Frobenius map. Since 1 is separable and $\mathcal{F}$ is inseparable, $1 \ominus \mathcal{F}$ is separable by Proposition 13.24. By Proposition 13.17, it follows that the cardinality N_1 of the kernel of $1 \ominus \mathcal{F}$ is equal to its degree. Now let $B(\phi, \psi)$ be the polarization of the degree map, which is a quadratic form on $\mathrm{End}(E)$ by Proposition 14.9. Thus

$$B(1, \mathcal{F}) = -B(1, \ominus\mathcal{F}) = \deg(1 \ominus \mathcal{F}) - \deg(1) - \deg(\mathcal{F}) = N_1 - 1 - q$$

by Proposition 14.10. By Proposition 14.8, we therefore have

$$|N_1 - 1 - q| \leq 2\sqrt{\deg(1)\,\deg(\mathcal{F})} = 2\sqrt{q},$$

as required. ∎

Exercises

Exercise 14.1. Let notations be as in this chapter. Let $\mathfrak{P}$ be a prime divisor on X. Let $[\mathfrak{P}]$ be the set of points of X which have nonzero coefficient in $\mathfrak{P}$, so that $[\mathfrak{P}]$ is a single orbit in the action of $\mathcal{G}$ on X, and $\mathfrak{P}$ is the sum of the points in $[\mathfrak{P}]$, each with multiplicity one. Let $\mathcal{O}_{\mathfrak{P}}$ be the set of $f \in F_0$, such that f is in $\mathcal{O}_P(X)$ for each $P \in \mathfrak{P}$.

(i) Show that $\mathcal{O}_{\mathfrak{P}}$ is a discrete valuation ring of F_0 containing k_0.

(ii) Now assume that k_0 is finite of cardinality q. Let $\mathfrak{M}_{\mathfrak{P}}$ be the maximal ideal of $\mathcal{O}_{\mathfrak{P}}$. Show that $\mathcal{O}_{\mathfrak{P}}/\mathfrak{M}_{\mathfrak{P}}$ is a finite field with cardinality $q^{\deg \mathfrak{P}}$.

Show that $\mathfrak{P} \mapsto \mathcal{O}_{\mathfrak{P}}$ is a bijection between the prime divisors of X and the discrete valuation rings of F_0 which contain k_0.

Exercise 14.2. Let X be an elliptic curve defined over k_0, let $P_1, \cdots, P_h$ be the points of X which are defined over k_0. Show that if the origin O of X is one of the P_i, say $O = P_1$, then the set of rational points is closed under the group law, hence forms a finite subgroup. Show that every Galois invariant divisor of degree zero of X is linearly equivalent to exactly one of the divisors $P_i - O$. Hence the cardinality of the group of k_0-rational points is equal to the cardinality of the divisor class group. Show that in fact $P_i \mapsto P_i - O$ is an isomorphism between these groups.

Exercise 14.3. Let E be the elliptic curve $y^2 = x(x^2 - 1)$ over the field $k_0 = \mathbb{Z}/p\mathbb{Z}$, where p is a prime not equal to 2. Let $\omega \in k$ be a solution to $\omega^2 + 1 = 0$.

(i) Verify that $c : (x, y) \mapsto (-x, \omega y)$ is an endomorphism of E. It is a *complex multiplication*. Show that $c \circ \sigma = \sigma \circ c$ for $\sigma \in \mathrm{Gal}(k/k_0)$, acting on E as explained in the text, if and only if $p \equiv 1 \bmod 4$.

(ii) Show that if $p \equiv 3 \bmod 4$, then the number of rational points on E is $q + 1$.

Hint: Show that if $0 \neq a \in k_0$, and if $a = x(x^2 - 1)$ has a solution x, then there are exactly two solutions (x, y) to $y^2 = x(x^2 - 1)$ with $x(x^2 - 1) = \pm a$.

(iii) Show that if $p \equiv 1 \bmod 4$, then the number of rational points on E is divisible by eight.

Hint: Using $\omega \in k_0$, explicitly construct a subgroup of order eight.

(iv) (More difficult) If $p \equiv 1 \bmod 4$, show that α_1 and α_2 in (14.7) are Gaussian integers.

References

[1] S. Abhyankar, *Resolution of singularities of embedded algebraic surfaces*, Second edition, Springer-Verlag, (1998).

[2] E. Artin, Quadratische Körper im Gebiete der höheren Kongruenzen, I, II, *Math. Zeitschrift* **19** (1924), 153–206, 207–246.

[3] M. Atiyah and I. Macdonald, *Introduction to Commutative Algebra*, Addison-Wesley (1969).

[4] M. Auslander and D. Buchsbaum, Unique factorization in regular local rings, *Proc. Nat. Acad. Sci. USA.* **45** (1959) 733–734.

[5] S. Balcerzyk and T. Józefiak, *Commutative Rings: Dimension, Multiplicity and Homological Methods*, Ellis Horwood (1989).

[6] A. Borel, *Linear Algebraic Groups*, Second edition, Springer-Verlag (1991).

[7] P. Cartier, Une nouvelle opération sur les formes différentielles, *C. R. Acad. Sci. Paris*, **244** (1957), 426-428.

[8] D. Eisenbud, *Commutative Algebra, With a View Toward Algebraic Geometry*, Springer-Verlag (1995).

[9] W. Fulton, *Introduction to Intersection Theory in Algebraic Geometry*, CBMS Regional Conference Series in Mathematics **54** (1983).

[10] W. Fulton, *Intersection Theory*, Springer-Verlag (1984).

[11] E. Kunz, *Introduction to Commutative Algebra and Algebraic Geometry*, Birkhäuser (1985).

[12] R. Hartshorne, *Algebraic Geometry*, Springer-Verlag (1977).

[13] H. Hasse, Beweis des Analogons der Riemannschen Vermutung für die Artinschen und F. K. Schmidtschen Kongruenzzetafunktionen in gewissen elliptischen Fällen, *Nachr. Ges. d. Wiss. Math. Phys. Kl.*, Göttingen (1933), 253–262

[14] H. Hasse, Zür Theorie der abstrakten elliptischen Funktionenkörper, I, II, III, Abhandl. Math. Sem. Hamburg **10** (1936), 55–62, 69–88 and 193–208.

[15] H. Hironaka, Resolution of singularities of an algebraic variety over a field of characteristic zero, I and II, *Ann. of Math.* **79** (1964), 109–203 and 205–326.

[16] S. Lang, *Introduction to Algebraic Geometry*, Interscience (1958).

[17] S. Lang, *Introduction to Algebraic and Abelian Functions*, Second edition, Springer-Verlag (1982).

[18] S. Lang, *Algebra*, Third edition, Addison Wesley (1993).

[19] H. Laufer, *Normal Two-Dimensional Singularities*, Ann. of Math. Studies **71**, Princeton University Press (1971).

[20] J. Lipman, Introduction to resolution of singularities, in *Algebraic Geometry*, AMS Proc. Sympos. Pure Math., **29** (1975), pp. 187–230.

[21] D. Mumford, *The Red Book of Varieties and Schemes* Lecture Notes in Mathematics **1358**, Springer-Verlag (1988).

[22] P. Samuel, La notion de multiplicité en algèbre et en géométrie algébrique I and II, *J. Math. Pures Appl.* **30** (1951), 159–278.

[23] J.-P. Serre, Faisceaux *algébriques cohérents*, Ann. of Math. **61**, (1955), 197–278.

[24] J.-P. Serre, *Algèbre Locale · Multiplicités*, Lecture Notes in Mathematics **11**, Springer-Verlag (1965).

[25] J.-P. Serre, *Local Fields*, Springer-Verlag (1979).

[26] J.-P. Serre, *Algebraic Groups and Class Fields*, Springer-Verlag (1988).

[27] J. Silverman, *The Arithmetic of Elliptic Curves*, Springer-Verlag (1986).

[28] G. van der Geer, *Hilbert Modular Surfaces*, Springer-Verlag (1988).

[29] A. Weil, Zur algebraischen Theorie der algebraischen Funktionen, *Crelle's J.* **179** (1938), 129–133.

[30] A. Weil, *Foundations of Algebraic Geometry*, The American Mathematical Society (1946).

[31] A. Weil, *Sur les Courbes Algebriques et les Varietes qui s'en Deduisent*, Hermann (1948).

[32] A. Weil, *Varietes Abeliennes et Courbes Algebriques*, Hermann (1948).

[33] A. Weil, *Basic Number Theory*, Third edition, Springer-Verlag (1974).

[34] G. Whaples, Local theory of residues, *Duke Math. J.* **18** (1951), 683–688.

[35] O. Zariski, *Algebraic Surfaces,* Second edition, with appendices by S Abhyankar, J. Lipman, and D. Mumford, Springer-Verlag (1971).

[36] O. Zariski and P. Samuel, *Commutative Algebra* v. 1 and 2, Corrected reprinting of the 1958 edition, Springer-Verlag (1975).

Index